SEMICONDUCTOR JUNCTIONS AND DEVICES

WILLIAM B. BURFORD III
McDonnell Aircraft Corporation, St. Louis, Mo.

H. GREY VERNER
Mallinckrodt Chemical Works, St. Louis, Mo.

SEMICONDUCTOR JUNCTIONS AND DEVICES

Theory to Practice

McGRAW-HILL BOOK COMPANY

New York   San Francisco   Toronto   London   Sydney

SEMICONDUCTOR JUNCTIONS AND DEVICES

Copyright © 1965 by McGraw-Hill, Inc. All Rights Reserved. Printed in the United States of America. This book, or parts thereof, may not be reproduced in any form without permission of the publishers. *Library of Congress Catalog Card Number* 65-18745

08950

1234567890MP7210698765

Preface

This book was undertaken for a specific purpose: to demonstrate in simple terms the link between the basic concepts of solid-state theory and the practical aspects of contemporary semiconductor electronics. It is intended to serve as a reading guide and outline for scientists and engineers of all ages and technical backgrounds whose contact with some facet of the new technology has stimulated their interest and who seek a broader understanding. This group is large, and its needs are acknowledged by many employers, research supervisors, and production engineers; yet the available remedies are, as of this writing, inadequate. It is not presumed that this brief effort will close so large a gap, but it will be counted a success if it provides a measure of assistance to its readers and suggests a path along which further efforts may be directed.

The present situation in semiconductor technology represents an inevitable result when a major new field of knowledge becomes established over a short period of time. Such a development cannot, in most cases, be related to any one of the existing specialties, but embraces several. In the present case, a new branch of science made its appearance in 1948 with the discovery of the transistor. The next few years of spontaneous and irresistible growth were characterized by effective interdisciplinary cooperation between leading physical scientists and engineers who "grew up" with the new concepts and their application. In the mid-1950s this had resulted in a curious mixing of talents and backgrounds: electrical engineers had become practicing metallurgists; chemists were performing basic measurements of semiconductor properties; and production engineers were turning to physicists for help with recalcitrant manufacturing processes. By the end of the first decade, however, the impact of solid-state electronics had reached all segments of the industry, and the need for additional qualified personnel had become pressing. It had also become evident that there was no adequate source of suitably trained individuals.

The immediate answer was, as it has always been under such circumstances, to enlist workers from the related specialties and to provide some form of on-the-job training. This approach was successful in overcoming the most obvious bottlenecks, but the training was limited to learning through participation in a very narrow section of the overall field. Such methods were of no lasting value, because there was no provision for the technical growth of the individual and therefore no basis for future contributions. If these long-range objectives are to be reached, an entirely different kind of educational process is required. The individual must acquire a basic theoretical and practical familiarity with the entire subject so that he can evaluate his own previous training in terms of the larger problem and focus his effort accordingly. The real issue, then, is one of learning not by daily routine exercise, but by effective study. For most individuals, this requires a teacher or guide to point out the most rewarding sources and to provide continuity. Unfortunately, it is in this area that the would-be student faces a major problem.

The universities offer no quick or direct solution. Formal courses are still taught within the framework of the traditional departments, and a course in transistor theory in the physics department often has little resemblance to one with the same title offered in the electrical engineering school. Neither has very much practical orientation, and each is subject to the particular viewpoint of the professor in charge. Furthermore, an electrical engineer is not normally found in a solid-state theory course, any more than a chemistry major enrolls in an advanced course in electronics. Each would find his previous training somewhat inadequate, and yet each has a need to understand the principles involved if he is to be effective in the semiconductor field.

Another training possibility is the literature, which has become rather formidable and diversified. There is no lack of material or viewpoints, but the periodical material is difficult to use. Much of it is written by experts for experts, and the student finds it hard to comprehend or to evaluate in terms of his needs. Among published books and monographs there is a good degree of continuity and selectivity, yet the classics in the field are, in general, those of a rather theoretical nature and usually assume that the reader is already familiar with much of the very material he is seeking. Finally, there are survey volumes at various levels of technical difficulty. Many of these are excellent for orientation purposes but, because of the large mass of material covered and the noncritical presentation, they are usually little better than outlines or bibliographies.

In an effort to provide a coherent outline which could be used by professionals of widely divergent backgrounds, the present work was begun in 1960. The principal emphasis was on clarity and continuity, starting from basic ideas which should be familiar to all physical scientists and engineers. By concentrating on planar junctions and current phenomena therein, it was hoped that the progression from theory to practice would be found logical and straightforward. Physicists or chemists will be at home in the initial chapters, and they may move quickly to the equivalent-circuit analysis in Chaps. 10 and 11. Conversely, readers with an electronics background may find it desirable to move slowly at the outset to assimilate the physical theory which will later be applied to familiar electronic performance. Both groups will find the subject matter limited and the level of presentation rather elementary; additional material is suggested by the reading list and references. Finally, although not intended as a teaching text, the method of presentation has been followed with good results in informal seminars.

The outline of this book and about 80 per cent of the text were completed in 1960–1961 by direct collaboration between the authors while both were employed by the Mallinckrodt Chemical Works. The decision to complete and edit the manuscript was arrived at during the winter of 1962–1963, owing to interest within the Research Division of McDonnell Aircraft Corporation, with whom one of the authors was then affiliated.

The authors wish to thank Profs. Ralph Bray and Robert Buschert, both of the Physics Department of Purdue University, who gave unstintingly of their time and patience in the development of the basic outline for the original manuscript. More recently, the burden of scientific consultation was assumed by Prof. W. A. Barker of the Physics Department of St. Louis University. His help and counsel during the final editing were invaluable.

It is a privilege to express our gratitude and to acknowledge the contribution of John R. Ruhoff, Vice President and Technical Director, Mallinckrodt Chemical Works. Without his support, encouragement, and technical advice the evolution of the first draft from its initial concept could not have been accomplished.

The helpful criticism and material aid given by Albert E. Lombard, Jr., Director of Research of McDonnell Aircraft Corporation, have been a major factor in the completion and final revision of the text. His assistance is gratefully acknowledged.

The authors thank McDonnell Aircraft Corporation and Mallinckrodt Chemical Works for permission to include in this book original research and data resulting from their employment by these companies.

A particular word of appreciation is due to John J. Tentor, whose creative advice and editorial assistance are keenly appreciated. In addition, the authors express a genuine note of thanks to the many colleagues, both at Mallinckrodt and McDonnell, who provided constructive comment and tangible support in the mechanics of preparation. Notable in this group is Miss Janice Eades, who worked willingly and endlessly through the many revisions. Finally, to our long-suffering wives go thanks for encouragement and unselfish cooperation.

William B. Burford III
H. Grey Verner

Contents

Preface v
List of Symbols and Abbreviations xiii

1. INTRODUCTION 1

1-1 Scope and Purpose 1
1-2 Historical 2
1-3 Definition of a Semiconductor 3
1-4 Reading List 5
REFERENCES 6

2. CONDUCTIVITY AND BAND THEORY 8

2-1 Conductivity 8
2-2 Band Theory—Historical 9
2-3 Energy Levels—Single Atom 11
2-4 Band Theory—Conductors 13
2-5 Band Theory—Insulators 14
2-6 Band Theory—Semiconductors 15
REFERENCES 17

3. CONDUCTIVITY IN SEMICONDUCTORS 18

3-1 Intrinsic Conductivity 18
3-2 Impurity Conductivity 21
3-3 Single Impurity, p-type 22
3-4 Single Impurity, n-type 23
3-5 Two or More Impurities—Compensation 24
3-6 Conductivity as a Function of Temperature 25
3-7 Constancy of the p-n Product 29
REFERENCES 32

4. FERMI STATISTICS AND FERMI LEVEL 33

4-1 The Fermi Probability Function 33
4-2 Dependence on Temperature 40
4-3 Constancy of the Fermi Level 42
4-4 Metals 42
4-5 Intrinsic Semiconductors 42

4-6	One-impurity Semiconductors	44
4-7	The Compensated Semiconductor	46
	REFERENCES	47

5. SYMMETRICAL $p\text{-}n$ JUNCTIONS ... 48

5-1	Introduction	48
5-2	Forming a $p\text{-}n$ Junction	48
5-3	Equilibrium Currents in a $p\text{-}n$ Junction	52
5-4	Energy, Field, and Charge Density	57
5-5	Junction Width	62
5-6	Junction or Barrier Capacitance	65
5-7	The Effect of Temperature on a $p\text{-}n$ Junction	70
5-8	Breakdown in $p\text{-}n$ Junctions	71
	REFERENCES	76

6. BIASED $p\text{-}n$ JUNCTIONS ... 77

6-1	Biased Symmetrical Junction (No Injection)	77
6-2	Symmetrical $p\text{-}n$ Junction under Reverse Bias	79
6-3	Symmetrical $p\text{-}n$ Junction under Forward Bias	81
6-4	The Rectifier Equation	82
6-5	Unsymmetrical Junctions with Injection	85
6-6	The Unbiased Case	86
6-7	Forward Bias	88
6-8	Reverse Bias	89
	REFERENCES	90

7. SINGLE-JUNCTION DEVICES ... 91

7-1	Introduction	91
7-2	Junction Fabrication	91
7-3	Rectifiers	97
7-4	Controlled Rectifiers	99
7-5	Zener Diodes	100
7-6	Tunnel or Esaki Diodes	103
7-7	Photocells	107
7-8	Solar Cells	107
	REFERENCES	109

8. SIMPLE TRANSISTORS OR TWO-JUNCTION DEVICES ... 111

8-1	Introduction	111
8-2	The $n\text{-}p\text{-}n$ Structure and Its Operation	111
8-3	Transistor Design Requirements	115
8-4	Current Relationships	117
	REFERENCES	128

9. TRANSISTOR AMPLIFIER CIRCUITS ... 129

9-1	Introduction	129
9-2	Connection Arrangements	129

9-3	General Comparisons	133
	REFERENCES	138

10. BLACK-BOX EQUIVALENT—GROUNDED BASE ... 139

10-1	Method	139
10-2	Results—Grounded Base	150
	REFERENCES	160

11. BLACK-BOX EQUIVALENT—GROUNDED EMITTER AND GROUNDED COLLECTOR 161

11-1	Results—Grounded Emitter	161
11-2	Results—Grounded Collector	167
	REFERENCES	170

12. TRANSISTOR PERFORMANCE AND FREQUENCY EFFECTS ... 171

12-1	Introduction	171
12-2	Current Patterns	172
12-3	Typical Low-frequency Performance	177
12-4	High-frequency Problems	178
12-5	Resistance Parameter Inadequacies	179
12-6	Hybrid Parameters	180
	REFERENCES	183

13. COMMERCIAL TRANSISTOR SPECIFICATIONS ... 184

13-1	Introduction	184
13-2	Selected Parameter Symbols	184
13-3	Specification Sheet—2N332 Transistor	187
13-4	Specification Sheet—2N2193 Transistor	193
13-5	Transistor Structure Terminology	196
	REFERENCES	198

14. SWITCHING APPLICATIONS OF TRANSISTORS ... 199

14-1	Introduction	199
14-2	The Transistor Amplifier as a Switch	200
14-3	Large-signal D-C Response	203
14-4	Switching for Values of Alpha Greater than Unity	208
14-5	Switch Characteristics, Load Lines, and Stability	212
	REFERENCES	216

15. MULTIJUNCTION FOUR-LAYER DEVICES ... 217

15-1	Introduction	217
15-2	The p-n Hook Transistor	218
15-3	The p-n-p-n Transistor Switch	220
15-4	The p-n-p-n Controlled Rectifier	225
15-5	The p-n-p-n Switching Diode	229
15-6	Thyristors	230
	REFERENCES	233

xii   *Contents*

16. OTHER SPECIAL DEVICES 234

    16-1   Introduction 234
    16-2   Avalanche Devices 234
    16-3   Intrinsic-region Devices 237
    16-4   The Double-base Diode or Unijunction Transistor 240
    16-5   The Junction Tetrode Transistor 242
    16-6   Field-effect Devices 243
    16-7   Tunnel or Majority Carrier Injection Devices 249
           REFERENCES 254

17. METAL-SEMICONDUCTOR CONTACTS 256

    17-1   Introduction 256
    17-2   Metal-Metal Contacts 260
    17-3   Ideal Metal-Semiconductor Contacts 263
    17-4   The Rectification Series 269
    17-5   Surface States and Inversion Layers 270
    17-6   Metal-Semiconductor Contacts 273
           REFERENCES 279

18. POINT-CONTACT DEVICES 281

    18-1   Introduction 281
    18-2   The Rectifier-Detector 281
    18-3   The Bonded Rectifier 283
    18-4   The Type A Transistor 285
    18-5   Commercial Transistors 287
           REFERENCES 290

19. DEVICE AND SYSTEM RELIABILITY 292

    19-1   Introduction 292
    19-2   Reliability and Production 293
    19-3   Device Testing and Failure Mechanisms 295
    19-4   Failure Distribution Data 297
    19-5   Screening Tests and Burn-in 305
    19-6   Device-System Interaction 307
    19-7   System Reliability Improvement 309
           REFERENCES 313

Appendix: CONDUCTION IN FILLED BANDS 315

Author Index 317
Subject Index 319

List of Symbols and Abbreviations

$\mathcal{A}$	Helmholtz free energy
A	area
Å	angstrom unit
a-c	alternating-current
amp	ampere
apb	atoms per billion (atoms)
B	base (generally used with C and E for collector and emitter)
BV, BV_R	breakdown voltage
BV_{XYZ}	breakdown voltage, conditions given by XYZ
C	capacitance per unit area
C	collector (generally used with B and E for base and emitter)
C_c	collector junction capacitance
C_{cb}, C_{ob}	output capacitance, collector to base
C_{eb}	emitter-to-base capacitance
C_{oe}	output capacitance, collector to emitter
C_T	total capacitance
cm	centimeter
cps	cycles per second
$\mathcal{D}$	disorder index (statistics)
D	diffusion constant, diffusivity
D_b	diffusivity of minority carriers in base region
D_e	diffusivity of minority carriers in emitter region
d	diameter
db	decibel
d-c	direct-current
$\mathcal{E}$	electric field strength
E	electron energy

E	emitter (generally used with B and C for base and collector)
E_a, E_d	energy level of acceptors, donors
E_c	energy level of lower edge of conduction band
E_F	Fermi energy level
E_F^A, E_F^B	Fermi energy levels for metals A and B
E_g	width of forbidden gap
E_∞	energy of electron at an infinite distance from nucleus
E_∞^A, E_∞^B	reference zero energies for metals A and B
E_L	lattice energy
E_L^A, E_L^B	lattice energies for metals A and B
E_L^M, E_L^S	lattice energies for metal and semiconductor, general
E_{POT}	potential energy
E_v	energy level of upper edge of valence band
e_G	voltage generator having zero resistance
emf	electromotive force, voltage
ev	electron volt
e^-	electron, electron flow
e^+	hole, hole flow
$\mathfrak{F}$	free energy, identical to μ, Gibbs function
$\bar{\mathfrak{F}}$	partial molar free energy
F	force
f	frequency
f_{hfb}, f_{ab}	alpha cutoff frequency
$f(I)$	current function for nonohmic device
$f_{\text{MAX}}, f_{\text{OSC}}$	maximum frequency of oscillation
$f\left(\dfrac{n(x)}{g}\right)$	concentration function for electrons in energy levels of density, g
$f(t)$	probability of failure to time t
$f(x)$	function of quantity x
G	number of energy states
G_A	anode gate
G_C	cathode gate
g	concentration of states
g_{yz}	conductance parameters
$\bar{h}$	Planck's constant divided by 2π
h_{FE}	static forward current transfer ratio
h_{yz}	hybrid parameters
I	total current

List of Symbols and Abbreviations

I_{AC}	anode-cathode current
I_B	base current (external)
I_{BO}	breakover current
I_C	collector current (external)
I_{CBO}, I_{CO}	collector current, emitter open
I_{CEO}	collector current, base open
I_{CER}	collector current, resistance R between base and emitter
I_{CES}	collector current, base shorted to emitter
I_D	drain current
I_E	emitter current (external)
I_{EBO}, I_{EO}	emitter current, collector open
I_{ECS}	emitter current, base shorted to collector
I_f	equilibrium junction diffusion current
I_g	equilibrium junction drift current
I_H	holding current
I_J	net current at point J (controlled rectifier)
I_{MAX}	emitter current at which α' is a maximum
I_P	pinch-off current
I_R	reverse current
I_s	reverse saturation current
I_{SAT}	leakage current (controlled rectifier)
i_c	collector current (internal)
i_e	emitter current (internal)
i_G	current from generator e_G
i_i	input current
i_o	output current
i_1, i_2	current (internal)
i'_1, i'_2	current (internal)
in.	inch
J	common electrode, general
j	current density
K, K'	arbitrary constants
k	Boltzmann's constant
kc	kilocycles per second
kg	kilogram, footnote page 60
k_g	generation rate constant
k_n, k_p, k'_n, k'_p	integration constants
k_r	recombination rate constant
L_b, L_e	diffusion length of minority carriers in base and emitter
M	collector junction multiplication factor
MADT	microalloy-diffused transistor

m	electron mass
m	empirical exponent, Eq. (8-9)
m	meter
ma	milliampere, 10^{-3} amp
μa	microampere, 10^{-6} amp
mμa	millimicroampere, 10^{-9} amp
$\mu\mu$f	micromicrofarad, 10^{-12} farad (see pf)
mc	megacycle, 10^{6} cps
mho	reciprocal ohm
μmho	micromho, 10^{-6} mho
mil	one thousandth of an inch
mm	millimeter
mv	millivolt, 10^{-3} volt
mw	milliwatt, 10^{-3} watt
$\mathfrak{N}$	number of electrons
N_A	concentration of ionized acceptors
N_D	concentration of ionized donors
n	electron density
n_i	equilibrium concentration of electrons in conduction band of an intrinsic semiconductor
n^i	n-type, near-intrinsic
n_n, n_p	electron concentration in n-type, p-type regions
n_0	concentration of combined electrons
nsec	nanosecond, 10^{-9} sec
n_1, n_2	n-type areas
n^+	n-type, heavily doped
$\mathcal{P}$	number of holes
P	pressure
PADT	post-alloy-diffused transistor
$P(E)$	probability (Fermi) that state of energy E is occupied
PG	power gain, usually grounded base
PG_c	power gain, grounded collector
PG_e	power gain, grounded emitter
PIV	peak inverse voltage
P_m	maximum output power
P_o	output power
P_T	total average continuous power dissipation
p	hole density
pf	picofarad, 10^{-12} farad
p_i	equilibrium concentration of holes in valence band of intrinsic semiconductor
p^i	p-type, near-intrinsic

p_n, p_p		hole concentration in n-type, p-type regions
p_0		concentration of combined holes
p_1, p_2		p-type areas
p^+		p-type, heavily doped
Q		total electronic charge
q		unit electronic charge
R		resistance, fixed or variable
R_A		bias resistor
R_B		base resistor
R_C		collector resistor
R_G		resistance of voltage source
R_g, R_r		rate of hole-electron pair generation, recombination
R_i		input resistance, grounded base
R_{ic}		input resistance, grounded collector
R_{ie}		input resistance, grounded emitter
$R_{kj,(\text{SAT})}, R_{ks}$		saturation resistance
R_L		load resistance
R_o		output resistance, grounded base
R_{oc}		output resistance, grounded collector
R_{oe}		output resistance, grounded emitter
r		spreading resistance
r_b		base resistance (internal)
r_c		collector resistance (internal)
r_e		emitter resistance (internal)
r-f		radio-frequency
r_{jk}		resistance parameter (general)
r_m		mutual resistance (internal)
$r_{11}, r_{12}, r_{21}, r_{22}$		resistance parameters
S		entropy
sec		second
T		absolute temperature, °K
T_A		ambient operating temperature
TFT		thin-film transistor
T_J		junction temperature
T_{STG}		storage temperature
t		time
t_f		pulse fall time
t_r		pulse rise time
t_s		pulse storage time
U		thermodynamic internal-energy function

xviii *List of Symbols and Abbreviations*

u	velocity
υ	volume
V	voltage, bias voltage
V_A	voltage drop across nonlinear device
V^{AB}	Volta potential between metals A and B
V_{AC}	anode-cathode voltage
V_B	battery voltage
V_{BB}	interbase voltage (double-base diode)
$V_{BE(\text{SAT})}$	base saturation voltage, grounded emitter
V_{BO}	breakover voltage
V_C	collector voltage
V_{CE}	collector voltage, grounded emitter
$V_{CE(\text{SAT})}$	collector saturation voltage, grounded emitter
V_D	drain voltage
V_{DS}	drain-source voltage (field-effect transistor)
V_E	emitter voltage (double-base diode)
V_{EB}	emitter-base bias (double-base diode)
V_F	voltage equivalent of Fermi level energy
VG	voltage gain, usually grounded base
VG_c	voltage gain, grounded collector
VG_e	voltage gain, grounded emitter
VG_m	maximum theoretical voltage gain
V_G	gate bias voltage (TFT)
V_{GS}	gate bias voltage (field-effect transistor)
V_{MAX}	maximum rated voltage
V^{MS}	Volta potential between metal and semiconductor (general)
V_n, V_p	electron potential in n-type region, p-type region
V_0	potential difference between n- and p-type regions in an unbiased p-n junction
V_P	pinch-off voltage
V_R	reverse bias voltage
V_S	sustaining voltage
V_T	total potential difference between n- and p-type regions in a biased p-n junction
v_1, v_2	small-signal equivalent-circuit input and output voltages
W	work, footnote page 60
W	work function
W^A, W^B	work functions for metals A and B
W^{AB}	work function difference, metals A and B
W_b	base width

W^M	work function of a metal, general
W^S_B	bulk work function of a semiconductor
W^S_S	surface work function of a semiconductor
X	unknown variable
x	distance
x_T	junction width
Y	alternative notation for g parameters
Y_0	barrier height
Z	alternative notation for r parameters
α	output-input current ratio (usually grounded base)
α'	output-input current ratio as a function of η only
α_{cb}	output-input current ratio, grounded emitter (β)
α_{ce}	output-input current ratio, grounded base
α_{eb}	output-input current ratio, grounded collector
β	output-input current ratio (usually grounded emitter)
β_{cb}	output-input current ratio, grounded emitter
β_{eb}, β_c	output-input current ratio, grounded collector
ϵ	permittivity
ϵ_0	permittivity of free space
ζ	base transport efficiency
η	emitter injecting efficiency
θ	Galvani potential, footnote page 261
κ	dielectric constant, ϵ/ϵ_0
λ	lagrangian undetermined multiplier
μ	micron
μ	mobility
μ_n, μ_p	mobility of electron, hole
μ	electrochemical potential or Gibbs function
$\bar{\mu}$	electrochemical potential or Gibbs function (general)
ρ	resistivity
ρ	charge density
σ	conductivity
σ_b	conductivity of base region
σ_e	conductivity of emitter region
τ	lifetime of minority carriers

τ_b	lifetime of minority carriers in base region
τ_e	lifetime of minority carriers in emitter region
ϕ_S^S, ϕ_B^S	inner potentials, surface and bulk of semiconductor
χ^A, χ^B	surface dipole potentials, metals A and B
χ^S, χ^M	surface dipole potentials, metal and semiconductor, general
Ω	ohm
Ω-cm	ohm-centimeter

1: Introduction

1-1. SCOPE AND PURPOSE

This book will present the basic principles of semiconductor physics in nonrigorous form, proceed to the development of conventional semiconductor-device parameters, and conclude with a description of typical units in simple circuit applications. The scope has been arbitrarily limited. For example, the theoretical discussion will be confined to a single representative semiconductor, silicon, and the common-junction configurations in which it is used. These choices are dictated by the increasing number of commercial devices which are made from silicon and the fact that both theory and application are best explained in terms of an elemental semiconductor in junction configurations.

To emphasize the relationship between semiconductor theory and the problems of device fabrication and application, an isolated ideal p-n junction will be explained by means of basic theoretical principles. An elementary band theory approach will be employed. The requirements for specific applications will then be discussed in terms of such ideal structures, and the limitations imposed by fabrication problems will be assessed. This approach will demonstrate the interdependence of fundamental solid-state theory and device technology and will lead directly to discussion of such topics as metal-semiconductor contacts, multijunction structures, and device reliability.

An additional limitation in the selection of material to be included in this book involves the omission of several interesting and important classifications of semiconductor devices. Photoelectric, radiation, and luminescent effects are largely omitted, as are thermoelectric, optical, and magnetic properties, since these topics may be studied elsewhere once the fundamentals are understood. Preparative and testing methods are not included, nor is any attempt made to embrace compounds such as the 3-5 semiconductors, ferrites, or organic materials. Polycrystalline devices, such as thermistors, and a variety of others variously used for temperature sensing, voltage limitation, rectification, and current control are also left to works of more encyclopedic ambitions. The once-numerous class known as *varistors*, where this term was used to embrace all devices, including rectifiers, having a nonlinear or unsymmetrical voltage-current characteristic (which will be seen to apply to almost any junction device as well as the original examples such as polycrystalline Cu_2O and SiC) is represented herein by a single current-limiter diode, where the nomenclature has persisted. Finally, some important and useful types, such as microwave diodes, are omitted as being extensions of other classifications which are similar in operation.

Before attempting to define a semiconductor as a preliminary to the theoretical discussion, a brief historical survey will be helpful as a means of orientation.

1-2. HISTORICAL*

The earliest recorded observations of what we would now recognize as semiconductor behavior are over a century old. Faraday (1833) found that Ag_2S exhibited a negative temperature coefficient of resistance [2]. Photoelectric effects involving selenium were reported by Smith as early as 1873 [3]. Rectification by metal-semiconductor contacts was first observed in 1874 by Braun [4], but it did not come into prominence until the early 1900s when crystal detectors for radio waves were widely used, only to be supplanted by vacuum tubes. Later, after 1920, rectifiers of selenium and copper oxide became commercially important and supplied the impetus for the development of a host of other polycrystalline devices such as thermistors, varistors, and photocells on an empirical basis, with little or no elucidation of the theory.

Finally, however, the ultimate stimulus which may be said to have truly started semiconductor physics upon the path to fundamental understanding was the desperate need for radar detection systems in the British Isles during the period of German air supremacy in the early days of World War II. The high range of frequency in radar, and the deficiencies of the then-available tubes, suddenly found the point-contact crystal diode back in business. With the ability to produce high-purity germanium (and later silicon) in single crystalline form, the basic theory began to be understood, although, until the substitution of the junction device for the point-contact configuration, progress was relatively slow.

The monumental achievement, in 1948, of Drs. Bardeen, Brattain, and Shockley and others at the Bell Telephone Laboratories in predicting

* Adapted and quoted (in part) with permission from J. N. Shive, courtesy of D. Van Nostrand Company, Inc. [1]. Numbers in brackets indicate references listed at the ends of chapters.

transistor action, demonstrating it with point-contact devices, and founding the theory upon which the science is based was the turning point [5-7]. In rapid succession came better materials, the junction device [8], and finally the publication of Shockley's book, "Electrons and Holes in Semiconductors," in 1950 [9]. The new discipline was now on firm theoretical grounds, and the subsequent developments have been in the nature of a mushrooming growth spurred by the tremendous commercial impact of transistors and a host of other devices replacing the vacuum tube, producing new circuits, and leading to present concepts of rugged reliability, subminiature size, and low cost.

In the years since the transistor saw the light of day in the Bell Laboratories, technological advances have been so rapid and diversified as to have rendered concepts and techniques obsolete in periods of months or a few years. Reflecting this situation, Dr. Shive, of Bell Telephone Laboratories, wrote in 1959 concerning a display of some fifteen semiconductor devices both old (copper-oxide rectifiers, photocells, etc.) and new (recently developed multijunction transistors, capacitors, etc.) that "fully half of the devices . . . have been developed since 1950; and the number of electrical engineers who are familiar with all of them is relatively small." Nor is the present situation any less fluid. Some of the devices of five years ago are already museum pieces, and new ones appear almost monthly either to assume a permanent place or to be briefly tested, evaluated, and discarded.

1-3. DEFINITION OF A SEMICONDUCTOR

The term "semiconductor" is defined as "a rather poor conductor where the conductivity may be changed radically by small changes in its physical condition" [10]. Such a material usually has a conductivity lying between that of good conductors such as metals and very poor conductors such as insulators. This definition does not really give us very much practical information about what semiconductors really are or what about them is important. Even the basic parameter, conductivity, does not supply us with a good general measuring stick by which to determine whether or not a substance is truly a semiconductor. For example, the useful range of conductivity, which is a characteristic of many semiconductors, extends roughly from 10^2 to 10^{-9} ohm^{-1} cm^{-1} at room temperature [11]. If the temperature is varied, the above range extends from 11 to about 16 orders of magnitude. By this extension, the conductivity overlaps that of very good metallic conductors as well as some recognized insulators.

Another property of semiconductors is that of a negative temperature coefficient of electrical resistance. This is somewhat better as a criterion, but there are numerous materials not generally classed as semiconductors, such as Nb_3Sn at superconducting temperatures, CuS, and a number of borides, nitrides, and carbides, which exhibit a negative temperature coefficient at some region on the temperature scale [12]. Semiconductors also display an extreme sensitivity in their electrical resistivity to some chemical impurities. In silicon, for example, one atom of boron per billion atoms of silicon, in an otherwise pure sample, changes the resistivity by a factor of about 1,000. These effects are so pronounced and the contamination levels are so low in terms of "chemical purity" by ordinary standards as to have demanded an entirely new definition of purity and

TABLE 1-1. TYPICAL SEMICONDUCTOR MATERIALS*

Elements	Germanium
	Silicon
	Selenium
	Boron
Oxides	Cuprous oxide
	Zinc oxide
	Manganous oxide
	Lead peroxide
Sulfides	Lead sulfide
	Cadmium sulfide
	Thallous sulfide
Mixed oxides	Zinc ferrite
	Magnesium titanate
	Ferrous ferrite
Miscellaneous compounds	Silver telluride
	Cuprous iodide
	Boron carbide
	Silicon carbide
	Zinc selenide
3-5 Compounds	Gallium arsenide
	Indium phosphide
	Indium antimonide

* After J. N. Shive, courtesy of D. Van Nostrand Company, Inc. [13].

new techniques, involving electrical behavior, to indirectly determine impurity levels. Although this degree of sensitivity falls rapidly with increasing impurity content, it is still a useful basis for distinction between semiconductors and other solids. Other conditions affecting the resistivity of semiconductors to a greater degree than that of metals or insulators include (1) previous high-temperature history of the sample, particularly in certain atmospheres, (2) state of crystallinity, that is, single crystalline, microcrystalline, or macropolycrystalline, and (3) surface conditions [12]. Unfortunately, none of the properties listed above are unique, except in degree, to any well-defined group of materials. It therefore becomes somewhat academic to devise a necessary and sufficient set of readily measurable characteristics which will provide a clear distinction between semiconductors and either metals or insulators.

The mechanism of conduction provides another means of identifying semiconductors. Conduction in a solid may be ionic, in which case charged or ionized atoms move through the lattice under the influence of an electric field and result in fundamental disturbances of the structure, owing to transport of matter, changes in stoichiometry, and polarization. Such a material is semiconducting AgCl [13]. In another mode of conduction, the current is carried by electrons, leaving the basic structure and properties of the material unchanged. This is so both in metals and electronic semiconductors such as germanium or silicon. However, in semiconductors, unlike metals, there is, in addition to forward electron motion, the reverse motion of electron vacancies (holes) which behave like positive electrons and make a significant, or controlling, contribution to the conduction of electricity. Detection of hole conduction, however, requires rather refined measurement techniques such as determination of the Hall coefficient or rectifying properties of the material.

In view of the above, it is not surprising that many and diverse types of materials are classed as semiconductors; examples are the widely different elements and compounds listed in Table 1-1.

1-4. READING LIST

In the belief that the average reader will find it useful from the outset, a list of selected sources in book form is included here. These range from qualitative introductions to standard university texts and outstanding monographs. Most of the material to be presented in this volume was adapted from these sources. Specific references from these and other sources are included at the end of each chapter.

Bridgers, H. E., J. H. Scaff, and J. N. Shive (eds., vol. 1); F. J. Biondi (ed., vols. 2 and 3): "Transistor Technology," D. Van Nostrand Company, Inc., Princeton, N.J., 1958.
Coblenz, A., and H. L. Owens: "Transistors: Theory and Applications," McGraw-Hill Book Company, New York, 1955.
Dummer, G. W. A., and J. W. Granville: "Miniature and Microminiature Electronics," John Wiley & Sons, Inc., New York, 1961.
Dunlap, W. C.: "An Introduction to Semiconductors," John Wiley & Sons, Inc., New York, 1957.
Evans, J.: "Fundamental Principles of Transistors," D. Van Nostrand Company, Inc., Princeton, N.J., 1958.
Gibson, A. F., et al. (eds.): "Progress in Semiconductors," John Wiley & Sons, Inc., New York, vol. 1, 1956; vol. 2, 1957; vol. 3, 1958; vol. 4, 1960; vol. 5, 1961; vol. 6, 1962.
Hannay, N. B. (ed.): "Semiconductors," Reinhold Publishing Corporation, New York, 1959.

Henisch, H. K.: "Rectifying Semiconductor Contacts," Oxford University Press, London, 1957.
Holm, R.: "Electric Contacts Handbook," Springer-Verlag OHG, Berlin, 1958.
Holmes, P. J. (ed.): "The Electrochemistry of Semiconductors," Academic Press Inc., New York, 1962.
Hunter, L. P. (ed.): "Handbook of Semiconductor Electronics," 2d ed., McGraw-Hill Book Company, New York, 1962.
Ioffe, A. F.: "Physics of Semiconductors," 2d ed., Academic Press Inc., New York, 1960.
Jonscher, A. K.: "Principles of Semiconductor Device Operation," G. Bell & Sons, Ltd., London, 1960.
Kingston, R. H. (ed.): "Semiconductor Surface Physics," University of Pennsylvania Press, Philadelphia, 1957.
Kittel, C.: "Introduction to Solid State Physics," 2d ed., John Wiley & Sons, Inc., New York, 1956.
Middlebrook, R. D.: "Introduction to Junction Transistor Theory," John Wiley & Sons, Inc., New York, 1957.
Nussbaum, A.: "Semiconductor Device Physics," Prentice-Hall, Inc., Englewood Cliffs, N.J., 1962.
Seitz, F.: "The Modern Theory of Solids," McGraw-Hill Book Company, New York, 1940.
Shive, J. N.: "Properties, Physics and Design of Semiconductor Devices," D. Van Nostrand Company, Inc., Princeton, N.J., 1959.
Shockley, W.: "Electrons and Holes in Semiconductors," D. Van Nostrand Company, Inc., Princeton, N.J., 1950.
Slater, J. C.: "Quantum Theory of Matter," McGraw-Hill Book Company, New York, 1951.
Spenke, E.: "Electronic Semiconductors," McGraw-Hill Book Company, New York, 1958.
Tillman, J. R., and F. F. Roberts: "An Introduction to the Theory and Practice of Transistors," John Wiley & Sons, Inc., New York, 1961.
Torrey, H. C., and C. A. Whitmer: "Crystal Rectifiers," McGraw-Hill Book Company, New York, 1948.
Turner, R. P.: "Transistors, Theory and Practice," Gernsback Library, Inc., New York, 1954.
Valdes, L. B.: "The Physical Theory of Transistors," McGraw-Hill Book Company, New York, 1961.
Warschauer, D. M.: "Semiconductors and Transistors," McGraw-Hill Book Company, New York, 1959.

REFERENCES

1. J. N. Shive: "Properties, Physics and Design of Semiconductor Devices," pp. 3–8, D. Van Nostrand Company, Inc., Princeton, N.J., 1959.
2. M. Faraday: "Experimental Researches in Electricity," vol. 1, p. 122, Bernard Quaritch, London, 1839.
3. W. Smith: The Action of Light on Selenium, *J. Soc. Telegraph Engr.*, **2:** 31 (1873).
4. F. Braun: Uber die Stromleitung durch Schwefelmetalle, *Ann. Physik u. Chem.*, **153:** 556 (1874).
5. J. Bardeen and W. H. Brattain: The Transistor: A Semiconductor Triode, *Phys. Rev.*, **74:** 230, 231 (1948).

6. W. H. Brattain and J. Bardeen: Nature of the Forward Current in Germanium Point Contacts, *Phys. Rev.*, **74**: 231, 232 (1948).
7. W. Shockley and G. L. Pearson: Modulation of Conductance of Thin Films of Semiconductors by Surface Charges, *Phys. Rev.*, **74**: 232, 233 (1948).
8. W. Shockley: The Theory of *p-n* Junctions in Semiconductors and *p-n* Junction Transistors, *Bell System Tech. J.*, **28**: 435–489 (1949).
9. W. Shockley: "Electrons and Holes in Semiconductors," D. Van Nostrand Company, Inc., Princeton, N.J., 1950.
10. D. E. Gray (ed.): "American Institute of Physics Handbook," 2d ed., p. 5-5, McGraw-Hill Book Company, New York, 1963.
11. C. Kittel: "Introduction to Solid State Physics," 2d ed., p. 347, John Wiley & Sons, Inc., New York, 1956.
12. E. Spenke: "Electronic Semiconductors," pp. 3, 4, McGraw-Hill Book Company, New York, 1958.
13. Ref. 1, p. 11.

2: Conductivity and Band Theory

In this chapter we shall compare the structures and electrical conductivities of conductors, semiconductors, and insulators [1].

2-1. CONDUCTIVITY

The flow of electricity through a solid actually consists of electrons (small negatively charged particles of low but known mass) moving from regions of negative to positive potential. (We shall see later that in some cases positively charged particles, known as "holes," share in the conduction, but that these are special cases and can also be explained in terms of electrons.)

All substances are to some degree conductors of electricity. If they are subjected to a voltage gradient, there will be a flow of electrons from the region of higher negative voltage to the region of lower voltage. Most substances obey Ohm's law, which states that the current density and the electric field strength are linearly related. The proportionality constant relating these two quantities is called the conductivity σ.

$$j = \sigma \mathcal{E} \qquad (2\text{-}1)$$

where j = current density, amp cm^{-2}
$\mathcal{E}$ = electric field strength, volts cm^{-1}

Hence, σ has units of (amperes/volt) cm^{-1} or ohm^{-1} cm^{-1}. (It will be recalled that current density may also be expressed as coulombs sec^{-1} cm^{-2}, from the definition of the ampere. Further, coulombs, or charge, may be written as $\mathfrak{N}q$, where $\mathfrak{N}$ is the number of electrons and q is the charge on an electron in coulombs.)

Because the conductivity of pure semiconductors is frequently an extremely small quantity, it is convenient to express this electrical parameter as the reciprocal of conductivity, that is, resistivity ρ:

$$\rho = \frac{1}{\sigma} \tag{2-2}$$

where ρ now has units of ohm-centimeters.

The above expression for conductivity [Eq. (2-1)] can also be written in terms of the electron concentration, the charge on the electron, and the velocity of the electron under the influence of a field equivalent to one volt per centimeter. The importance of electron (and hole) velocity will be pointed out later, in the discussion of semiconductors. For good conductors the concept of velocity is not important, as may be seen from the following simple analogy. Consider a pipe having a very large cross section relative to the quantity of liquid flowing through it. Large volumes of liquid may be transferred in spite of a relatively low linear velocity. This case is analogous to a good conductor with many free electrons in which large quantities of current may flow without dependence on a large electron velocity. As we shall see, a semiconductor has a limited number of charge carriers and corresponds to a pipe having a relatively small cross section so that the flow of liquid is dependent upon the velocity which can be achieved. Hence the flow of current in a semiconductor will be shown to be dependent upon the velocity of the current carriers per unit field. This velocity is known as *mobility*. The foregoing expressions for conductivity will be derived in terms of mobility in later chapters.

2-2. BAND THEORY—HISTORICAL*

The band theory approach to atomic structures was inspired, in part, by the Rutherford experiments in which gold foil was bombarded with alpha particles from natural radium. Since each alpha particle carries two positive charges, it was established, by measuring alpha-particle deflection, that the gold nucleus is exceedingly small but has a very high positive charge.

In an effort to interpret these and subsequent results, many physicists suggested atomic models which they hoped would explain the observed experimental facts. According to an early model, proposed by Niels Bohr, an atom was made up of a small, dense nucleus plus a cloud of electrons surrounding it which neutralized its positive charge and occupied most of the space generally accredited to the solid atom. Since each element was characterized by a nucleus with a different number of positive charges (atomic number), the corresponding number of electrons varied from element to element and, in addition to binding the atoms together, gave each element its unique characteristics.

* Adapted with permission from R. D. Middlebrook [2].

Unfortunately, the theoretical physicist soon recognized that this simple model suffered from certain weaknesses. It was argued that the electrons were not stationary, since under such conditions electrostatic attraction would draw the electrons into the nucleus. Therefore, it was postulated that the electrons must revolve around the nucleus with the electrostatic attraction toward the nucleus being exactly balanced by the centrifugal force of the motion of the electrons in their orbits.

However, the physicists reasoned that the electrons must radiate energy as a result of their motion through the electric field, and that this energy would be lost. Therefore, the electrons would gradually get closer and closer to the nucleus and finally fall into it. Since this is contrary to the experimental evidence, another model was adopted. It was postulated that there were certain orbits in which the electrons did not radiate, that is, certain discrete values of energy which the electrons could possess and yet remain in stable orbits. Various rules were devised by Bohr, Sommerfeld, Heisenberg, and others to determine these stable energy levels. These rules defined quantum numbers which described each stable state for an electron, but these numbers were difficult to use and to relate to the experimental evidence.

The attempt to replace these numerous and cumbersome rules by a much smaller number of more fundamental assumptions was made by Schrödinger and Heisenberg. Schrödinger postulated that the state of an electron could be described by a wave function. On this basis, the quantum numbers appeared as a necessary consequence of the theory of wave mechanics, and not as a separate assumption. The solution of Schrödinger's wave equation defined discrete energy levels which an electron may occupy without loss of energy due to radiation, and it thereby replaced Bohr's picture of a stable atom comparable to a miniature solar system.

These allowed energy levels for a single atom broaden into bands when many atoms form a periodic array. In this case, the solutions of the Schrödinger wave equation give allowed regions of energy (or bands) for the electrons as well as forbidden gaps. This quantitative picture of the structure of matter, known as *band theory*, has been widely used to explain experimental data. There are more accurate—and more complex—theories which deal with the same problem, but the simple band theory due to Bloch will serve our purpose.

A further important contribution was made by Pauli, who postulated that only two electrons of opposite spin may be described by the same wave function resulting from a solution of the Schrödinger equation. If this were not the case, all electrons would fall to the energy level next to the nucleus, and there would be no distinction between atomic species. This is the well-known *Pauli exclusion principle*.

It is beyond the scope of the present work to include the derivation and solutions of these various equations. The reader may pursue these matters in greater detail by reference to standard works.

2-3. ENERGY LEVELS—SINGLE ATOM

As a result of the foregoing argument, we are now ready to consider the electron energies in a single atom. We have learned that an atom is made up of a positively charged nucleus and negative electrons in discrete energy states. The number of electrons in a given level is governed by the states available (as determined by the solution of the Schrödinger equation) and the Pauli exclusion principle.

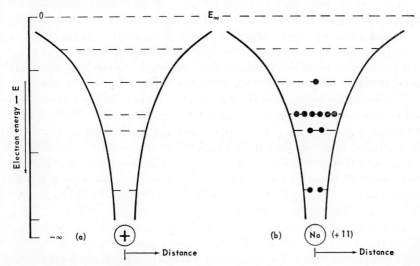

FIG. 2-1. Electron energy levels for single atoms. (*After J. N. Shive, courtesy of D. Van Nostrand Company, Inc.* [3].)

Figure 2-1a depicts the energy level scheme for an atom at room temperature. In this figure, E_∞ represents the zero of energy for an electron at an infinite distance from the nucleus. At this distance the electrostatic attraction is zero and the electron no longer must move in an orbit to avoid falling into the nucleus. The energy scale shows negative values for the electron energy; they increase numerically as the electron nears the nucleus. These energies are measured in electron volts.* The dotted

---

* By convention, the zero of electron energy is taken at a point very far away from the surface of the solid. By Coulomb's law, the binding energy is negative and becomes equal to minus infinity at the center of the positive nucleus. However, since energy differences are the only significant quantities in most cases, the absolute zero of energy is immaterial. Note, in Fig. 2-1a, that an increase in electron energy moves up on the energy scale (becomes less negative).

lines represent discrete stable configurations where the electrostatic attractions (forces) of the nucleus for an electron are balanced by centrifugal forces and where the electron energy corresponds to an allowed solution of the Schrödinger wave equation. The position of the electron, that is, the level which it occupies, is a strong function of the temperature. In this section we shall limit ourselves to a temperature range from about −70 to 150°C (200 to 420°K). Over this limited temperature range, deviations from room-temperature conditions are minor.*

In Fig. 2-1b we have indicated schematically the first five energy levels which are allowed for the sodium atom, which has an atomic number of 11 and therefore 11 electrons in the orbital structure. Here we see that an energy level may embrace two, six, or more states differing very slightly in energy (such as the electron spin mentioned in connection with the Pauli exclusion principle). We shall assume in the following discussion that the term "energy level" includes all allowed states resulting from wave equation solutions for that level. Note that the fourth level contains only one electron, while the fifth level contains none.

The electrons in the first three levels, being closer to the nucleus, have lower (more negative) potential energies as a result of their greater electrostatic attraction to the nucleus. The single electron in the fourth level, being much farther from the nucleus and therefore least tightly held, has the highest energy and is most readily removed by interaction with another atom to form the traditional sodium ion with a positive charge of unity. For this reason, this single electron in the upper level is known as the *valence electron*.

In the preceding paragraphs we have noted the existence of levels of discrete energy at varying distances from the nucleus. The spaces between these levels which never contain electrons are known as *forbidden gaps*. It is possible for an electron of lower energy to move up into an unoccupied state at a higher level, but this process requires the absorption of energy by the electron in a quantity equivalent to the difference in energy of the two levels, known as the *gap*. This energy absorption may take the form of thermal excitation, radiation or applied voltage.

The physicist has given a definite designation to each of the allowed

---

* In this work we shall consider the effect of a number of externally controlled variables upon the properties of semiconductors. These will include temperature, purity (presence or absence of various controlled and adventitious impurities), external stimuli (radiation, applied electric field, etc.), crystalline perfection, and the nominal chemical composition of the material.

To avoid repetition, it will be assumed, unless otherwise stated, that we are concerned with a pure (no impurities of any type) semiconductor, namely, silicon, at room temperature (or within its useful range as a semiconductor, given above as −70 to 150°C). It will be further assumed that the specimen is a perfect single crystal and that no external stimuli are present.

levels. In our discussion of the sodium atom (Fig. 2-1b) we called attention primarily to the upper level (valence level) and the single electron therein. In the following section we shall observe the change of this energy level system to a band system under the influence of the surrounding atoms in a real sample of metallic sodium.

2-4. BAND THEORY—CONDUCTORS

Consider now what happens when a large number of sodium atoms are combined to form a crystal of sodium metal. Figure 2-1b might then be expanded to give a potential-energy profile along a line of atoms somewhat like that shown in Fig. 2-2, where the levels have become bands.

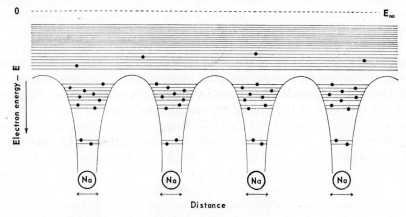

FIG. 2-2. The distribution of electrons in the allowed bands of sodium metal. (*After J. N. Shive, courtesy D. Van Nostrand Company, Inc.* [4].)

Because the atoms are close together, the electron energy barrier does not rise to E_∞ as it did for the single atom. When sodium atoms are close together, as they are in a piece of solid metallic sodium, the curves of electron energies as functions of distance from each sodium nucleus intersect at values well below E_∞ and, as a matter of fact, somewhat below the electron energy corresponding to the valence band. Consequently, a valence electron may easily leave its parent atom and wander freely through the crystal. In particular, if an electric field is imposed on the crystal, free electrons will be pulled in one direction by the field, resulting in an electric current. Thus we have used the modified Bohr model to provide a simple picture explaining the conduction of electricity in a typical metal like sodium.

Although the foregoing discussion, using sodium metal as an example, is fairly logical and presents the essential ideas needed for a basic understanding of electrical conduction in a metal, a number of oversimplications were made. These must be recognized and separated from the qualitative picture in order that future discussions will be consistent with this introductory treatment.

It should first be pointed out that the vertical scale in the figures is considerably foreshortened. In actuality, the low-lying bands near the nucleus are much farther down in the valley and the separation between bands becomes smaller as the energy of the allowed states or electrons increases. At the same time, the bands become much wider and embrace many more available states, particularly when the valence band is reached. Further, the energy gap (vertical distance) between the valence band and the E_∞ line is much greater than is shown, and there are many more bands of higher energy above the valence band. While these are of little concern to us now in this preliminary discussion, their existence will be significant in later, more quantitative considerations.

Confining ourselves, therefore, to the valence band, representing the highest one occupied by electrons in sodium metal under normal conditions at room temperature, we must next consider the distribution of electrons in this band. As drawn in Fig. 2-2, it will be noted that the valence electrons of the several sodium atoms do not all have the same energy. This is correct for the following reasons. First, the valence electrons no longer belong to any particular atom but must be considered as belonging to the crystal at large. Therefore, if the crystal itself, considered as a single system, were to contain all these electrons with the same energy, there would be a violation of the Pauli exclusion principle, since no more than two electrons may occupy the same energy state at any level. Second, when a multiplicity of atoms is arranged in a crystalline array, all the energy levels (and allowed states comprising them) associated with each atom are preserved, ensuring enough levels in the valence band to take care of all of the valence electrons in the crystal without violating the Pauli exclusion principle. This multiplying of levels occurs in all bands when atoms are grouped into crystals, resulting in a large number of allowed levels and broadening of the bands on the energy scale. As will become apparent later, the quantitative determination of electron energies and available energy levels becomes a very complicated problem in real solids, requiring a rather sophisticated quantum-statistical approach.

2-5. BAND THEORY—INSULATORS

The foregoing discussion has shown, in terms of simple band theory, the mechanism of electrical conduction in a metallic conductor, sodium. In this case it was found that the valence band lies in an energy region well above the potential hills of the atomic electron energy curves and that it is continuous throughout the crystal. In this special case, therefore, the valence electrons in the valence band can also be regarded as conduction electrons and the band might be also called the conduction band. In other words, the two bands are indistinguishable.

Consider, however, a situation wherein the valence band lies below the tops of the potential hills and the valence electrons cannot move freely through the crystal. Since this is the case for all other solids, that is, insulators and semiconductors, a more general picture is useful, and the case of metallic conduction by valence electrons from the valence band is found to be a special circumstance.

We consider, therefore, two bands near the top of the potential hills. The lower of these, known as the valence band, is generally below the top of the potential hills and contains all the valence electrons, which are now constrained and move only with difficulty under the influence of an external field at room temperature. Above the valence band we find a second group of allowed energy levels, known as the conduction band, which lies generally above the potential hills at a varying distance (energy gap) from the valence band. The energy gap, known as a forbidden gap, represents an energy hurdle which an electron must surmount to enter the conduction band and become free to move about the crystal under the influence of a voltage gradient. To make this transition, the electron must acquire energy from heat, radiation, or an electric field. If the forbidden gap is large, there is little chance of an electron acquiring the necessary energy except under extremes of temperature, radiation, or applied field, and we characterize the material as an insulator.

Any substance may, therefore, be classified according to the relative locations (on the energy scale) of the valence band, the tops of the interatomic potential barriers, the conduction band, and the forbidden gap between the valence and conduction bands. At one extreme are the insulators, described in this section, and at the other extreme are the metals, where the forbidden gap is nonexistent and the valence band is indistinguishable from the conduction band.

2-6. BAND THEORY—SEMICONDUCTORS

The transition from conductor to insulator is not clear-cut, and it is the materials in the transition region in which we are most interested. For obvious reasons these materials are known as semiconductors. The semiconductors include that class of materials which might normally be considered as insulators but which have a forbidden gap of such magnitude that it is possible to energize a certain portion of the valence electrons into the conduction band levels, thereby converting them temporarily into conductors. Because this process may be controlled, these materials have found application as semiconductor devices which have assumed widespread importance in the electronics industry.

The upper energy-position diagrams in Fig. 2-3 show the various energy situations discussed in the preceding paragraphs. Figure 2-3a depicts the energy bands in a conductor. Here we have a nonexistent forbidden gap or a situation in which the valence band overlaps the conduction band. That is, there are energy states from both bands at approximately the same energy level. However, note in Fig. 2-3b, which presents the

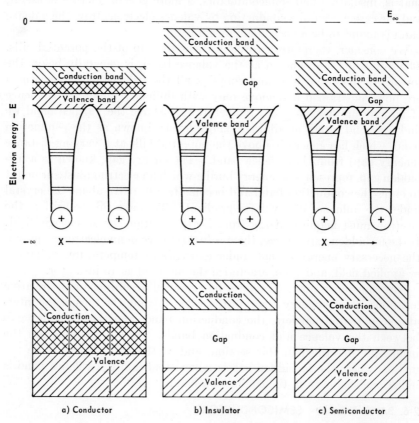

FIG. 2-3. Band energy diagrams for a conductor, insulator, and semiconductor.

energy diagram of an insulator, the very large amount of energy needed to raise an electron from the constrained valence band to the conduction band. Figure 2-3c pictures a semiconductor with a band gap between the conduction and valence energy levels whose width is intermediate between insulators and conductors.

The lower block diagrams of Fig. 2-3 show the band representation in a more schematic form, which will be found frequently in the literature. In these drawings, the horizontal scale is no longer of great significance, and it may be regarded as an arbitrary distance through the crystal in any direction. The energy bands are generalized, and they no longer show any influence due to discrete atomic nuclei.

The semiconductor achieves practical significance because electrons may be raised to the conduction band by either of two mechanisms: (1) the electrons may be easily excited by adding energy or (2) electrons may be put into the conduction band by "doping" the semiconductor with controlled amounts of suitable impurities.

Since the remainder of this work will be devoted to semiconductors, the terms "valence band" and "conduction band" will be reserved from this point forward to indicate energy levels below and above the forbidden gap in a semiconductor.

REFERENCES

1. Among the numerous general references on this subject, the following are recommended:

    Hannay, N. B. (ed.): "Semiconductors," Reinhold Publishing Corporation, New York, 1959.

    Kittel, C.: "Introduction to Solid State Physics," 2d ed., John Wiley & Sons, Inc., New York, 1956.

    Seitz, F.: "The Modern Theory of Solids," McGraw-Hill Book Company, New York, 1940.

    Shive, J. N.: "Properties, Physics and Design of Semiconductor Devices," D. Van Nostrand Company, Inc., Princeton, N.J., 1959.

    Shockley, W.: "Electrons and Holes in Semiconductors," D. Van Nostrand Company, Inc., Princeton, N.J., 1950.

    Spenke, E.: "Electronic Semiconductors," McGraw-Hill Book Company, New York, 1958.

2. R. D. Middlebrook: "Introduction to Junction Transistor Theory," pp. 40–42, John Wiley & Sons, Inc., New York, 1957.
3. J. N. Shive: "Properties, Physics and Design of Semiconductor Devices," pp. 265, 267, D. Van Nostrand Company, Inc., Princeton, N.J., 1959.
4. Ref. 3, p. 269.

3: Conductivity in Semiconductors

3-1. INTRINSIC CONDUCTIVITY

In Fig. 2-3c a certain amount of oversimplification was used in the interest of presenting the important ideas of band theory and conductivity. The diagram indicates that all of the electrons are in the valence band, where they are not free to conduct. However, for all temperatures greater than absolute zero, some electrons from the valence band are excited to the conduction band by thermal activation, their number being a strong function of temperature and relatively small compared to the number in a conductor.

This thermal generation of conducting electrons in semiconductors introduces another way in which semiconductors differ from conductors. In an ordinary metallic conductor such as copper all the valence electrons are in the conduction band at all temperatures and are free to move. Electrical neutrality results from the statistical equality between free conduction electrons and positively charged atomic nuclei. As previously mentioned, the conduction of electricity takes place as the result of the flow of electrons when an external field is applied. When the electric field is removed, the specimen contains the original number of electrons and is again electrically neutral.

In a semiconductor the excitation of an electron into the conduction band leaves behind a vacancy which behaves like a unit positive charge in the valence band. The vacancy created in the valence band (known as a hole) preserves electrical neutrality and acts as though it were a mobile, positively charged current carrier. Since in intrinsic (pure) semiconductors a hole is created for every electron excited into the conduction band, the holes make a significant contribution to the overall conductivity.

Conductivity, as discussed in the preceding chapter, is defined by an empirical relationship known as Ohm's law. This was obtained from overall observations of macroscopic samples. Now, based on the concepts of band theory and the mechanism of conduction by valence electrons and/or electron-hole pairs, we may reexamine this phenomenon from a microscopic point of view.

We now find that electrons and holes move through the semiconductor at different speeds under the influence of an applied field. (Electrons in conductors also have a characteristic speed for a given applied field.) Here the concept of velocity, or carrier mobility (velocity per unit field) becomes useful. Mobility is given by

$$\mu = \frac{u}{\mathcal{E}} \qquad (3\text{-}1)$$

where u = velocity parallel to the applied field, cm sec^{-1}
$\mathcal{E}$ = electric field strength, volts cm^{-1}
Therefore, the units for mobility are cm^2 volt^{-1} sec^{-1}.

Current density [Eq. (2-1)] may also be written in terms of carrier velocity

$$j = nqu \qquad (3\text{-}2)$$

where n = concentration of electrons, $\mathfrak{N}$ cm^{-3}
q = charge on the electron, coulombs
From Ohm's law it follows that

$$\sigma = nq\frac{u}{\mathcal{E}} = nq\mu \qquad (3\text{-}3)$$

The conductivity of a semiconductor is the sum of the contributions of the holes and the electrons. The current carried by each is the product of the carrier density (n for electrons and p for holes), their respective mobilities μ_n and μ_p, and the electrical charge associated with each (charge on a hole = charge on an electron = q). Because the charges are opposite in sign, electrons and holes move in opposite directions. However, their contribution to total current carried is additive, because the movement of a hole in one direction is equivalent to movement of an electron in the opposite direction.

From the above, we may write the following expression for conductivity:

$$\sigma = nq\mu_n + pq\mu_p \qquad (3\text{-}4)$$

If we ignore the right-hand term above (the contribution of holes to the conductivity), Eq. (3-4) is found to be consistent with the expression for conductivity as derived by using Ohm's law in Chap. 2.

It is of interest to compare the electron densities, hole densities, and mobilities of a good metallic conductor with those of an intrinsic semiconductor at room temperature. Table 3-1 gives the values of the foregoing parameters for copper and highly purified silicon. From this table it is

TABLE 3-1. CARRIER DENSITIES AND MOBILITIES FOR INTRINSIC SILICON AT ROOM TEMPERATURE AND FOR COMMERCIAL COPPER [1,2]

Parameter	Intrinsic silicon at room temperature	Commercial copper
Conductivity, $ohm^{-1}\, cm^{-1}$	4.4×10^{-6}	6×10^5
Number of electrons, $\mathfrak{N}\, cm^{-3}$	1.6×10^{10}	10^{23}
Number of holes, $\mathcal{P}\, cm^{-3}$	1.6×10^{10}	None
Electron mobility, $cm^2\, volt^{-1}\, sec^{-1}$	1,300	40
Hole mobility, $cm^2\, volt^{-1}\, sec^{-1}$	500	

apparent that the electrons in silicon move more than 30 times faster than the electrons in copper (1,300/40) and the holes over 10 times as rapidly as the electrons in copper; yet the conductivity of copper is 1.4×10^{11}

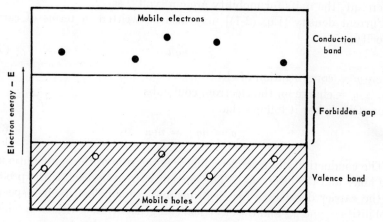

FIG. 3-1. Band energy diagram of an intrinsic semiconductor.

times that of silicon. The answer lies in the much larger number of electrons available for conduction in copper. Despite the fact that silicon has both holes and electrons which serve as current carriers, their sum is still below the electron concentration in copper by a factor of 10^{12}. (No contribution by holes is indicated for copper because the current-carrying electrons are already present in the conduction band and there is no generation of current-carrying holes.)

Figure 3-1 illustrates the mechanism of thermal activation in an intrinsic semiconductor at room temperature. Every time an electron is excited into the conduction band it creates a hole in the valence band. The case of the truly intrinsic semiconductor is exceedingly rare—never seen for silicon at ordinary temperatures. However, for practical purposes, as will be demonstrated, the intrinsic contribution of thermally generated charged particles is of great importance and must be adequately allowed for, despite the experimental impossibility of achieving a truly intrinsic sample.

In the next section it will be shown that silicon devices which function at room temperature are chiefly dependent upon electrically active impurities for their semiconductor action. Under such conditions the conductivity is many times what it would be if the silicon were intrinsic. However, we shall see that as the temperature increases the conductivity attributable to the impurities remains essentially constant, whereas the conductivity attributable to thermal activation increases and eventually becomes controlling. For silicon this occurs between 150 and 200°C. Once the intrinsic condition of silicon becomes controlling, the concentrations of electrons and holes are essentially equal (the impurity contribution becomes negligible).

3-2. IMPURITY CONDUCTIVITY

Certain materials when added to silicon exert a profound influence on its electrical conductivity. By carefully regulating the amount of certain desirable impurities in silicon and by utilizing the electrical effects produced by their presence, we can control the conductivity of a silicon single crystal over a wide range. It is this phenomenon which makes possible the fabrication of silicon semiconductor devices.

For the silicon devices being made today, the impurity concentration of the selected impurity varies from a few tenths of an atom per billion silicon atoms (apb) up to about 10,000 apb. An impurity level of 1 apb produces a carrier density (holes or electrons per cubic centimeter) of 5×10^{13}. Table 3-1 indicates electron and hole densities for intrinsic silicon at room temperature of 1.6×10^{10}, or about a thousandfold lower. In view of this large difference we are justified in ignoring the intrinsic contribution to the conductivity of silicon containing impurities in the range mentioned above.

There are, however, important areas where the intrinsic contribution can play a vital role. One such case, which was mentioned earlier, concerns the conductivity as a function of temperature. In addition, we shall learn that the behavior of minority carriers (opposite in electrical charge to the carriers produced by the impurity added, for example, holes in the electron-dominated impure silicon) is the controlling mechanism in the operation of almost all devices. In these instances it will be necessary to calculate the density of the majority carriers arising from both the intrinsic properties of the sample and the added impurities. From such a calculation we shall then be able to determine the minority carrier concentration. The details of this calculation will be explained in the section on the constancy of the p-n product.

The impurities that are used most frequently are those which are found in the periodic table of chemical elements under Groups III and V. The Group III elements (boron, aluminum, gallium, and indium) contain one less valence electron than silicon; the Group V impurities (phosphorus, arsenic, antimony, and bismuth) have five valence electrons, one more than silicon. We shall describe the effect of adding a typical Group III element, boron, to a silicon crystal and then present the parallel situation for the addition of a Group V element, phosphorus.

3-3. SINGLE IMPURITY, p-TYPE

When a boron atom with its deficiency of one valence electron enters the silicon lattice, there is one less electron than would be required to

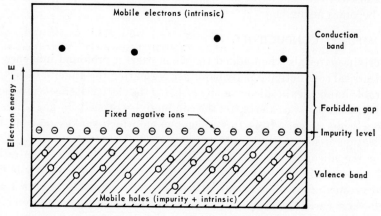

FIG. 3-2. Band energy diagram of a p-type impurity semiconductor.

form bonds with the four neighboring silicon atoms. This deficiency is overcome at the expense of an electron from the valence band of the silicon, so that four normal electron-pair bonds are formed. The vacancy thus created in the valence band behaves like a positively charged particle, which we call a hole, since it results from the removal of an electron by a new mechanism analogous to the excitation of an electron to the conduction band. This mechanism depends on the fact that boron atoms have a stronger affinity for electrons than silicon atoms have, with the result that when the boron atom acquires an electron to achieve completion of four bonds with the neighboring silicon atoms, the electron involved is immobilized. Since it contributes a negative charge to the previously neutral boron atom, this results in an immobile boron ion having a unit negative charge and a corresponding hole which is free to move under the influence of an electric field as shown in Fig. 3-2.

Finally, the electron forming the boron ion now has a higher energy than other electrons in the valence band. This fact has two important consequences: (1) the boron ion is more stable than the neutral silicon atom in the crystal and (2) this electron is now in the forbidden gap, although still immobilized by the boron atom. At ordinary temperatures the boron ion is stable, but at low temperatures the electron in the forbidden gap leaves the boron ion and returns to the valence band, thereby destroying a hole.

Because this type of impurity enters the crystal lattice in such a way as to produce deficiencies among the valence electrons, it is known as an acceptor. Silicon crystals in which the boron-type impurity predominates are classically known as p-type crystals, where the letter p might refer to the creation of "positive" holes.

3-4. SINGLE IMPURITY, n-TYPE

Phosphorus and the other elements from Group V of the periodic table have five valence electrons—one more than necessary to combine with

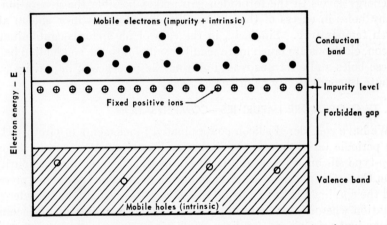

FIG. 3-3. Band energy diagram of an n-type impurity semiconductor.

four silicon atoms in the crystal lattice. The average energy of the phosphorus electrons corresponds to a level very near the bottom edge of the conduction band, that is, at the top of the forbidden gap. As a result, the excess phosphorus electron readily migrates into the conduction band, where it is free to move under the influence of an electric field and thereby conduct a current. The resultant phosphorus ions with a positive charge of unity remain in fixed positions in the lattice (Fig. 3-3).

The valence electrons associated with the phosphorus ion therefore have a higher energy than the silicon valence electrons in the valence band or indeed than the valence electrons associated with the boron ion, which is just barely in the forbidden gap. At the highest energy level of all are the excess electrons which have left the phosphorus ions and are well into the conduction band.

Since impurities like phosphorus have an excess electron when incorporated into the silicon lattice, they are known as donors; and the silicon is classically designated as n-type, where the letter n might refer to the addition of "negative" electrons.

It should now be obvious that for both boron and phosphorus the electron which created the ion lies at a higher energy level than before. In the boron case this has resulted in the generation of holes, while in the phosphorus case we find electrons in the conduction band. The effect of temperature is the same in both cases. At ordinary temperatures the stability of both ions will remain essentially unchanged. At low temperatures, however, the phosphorus ion will revert to a neutral condition when its fifth valence electron falls out of the conduction band, just as a boron ion becomes neutral and destroys a hole.

We have seen in the case of boron-contaminated silicon that the presence of energy levels in the forbidden gap makes possible the generation of many holes in excess of those normally present in intrinsic silicon at a given temperature. Similarly, in the case of phosphorus-contaminated silicon, electrons are much more readily excited into the conduction band. These holes and electrons resulting from the presence of impurities contribute to conductivity as discussed above.

3-5. TWO OR MORE IMPURITIES—COMPENSATION

We have considered silicon contaminated by elements of Group III of the periodic table (boron, aluminum, gallium, indium, etc.) and known as p-type silicon, as well as the situation when one or more impurities from Group V (phosphorus, arsenic, antimony, bismuth, etc.) are present and the specimen is known as n-type silicon. We shall now consider the situation when one or more impurities from both groups simultaneously contaminate the silicon. Silicon containing unequal amounts of impurities from both families (p-type and n-type) exhibits electrical behavior which would be expected in a sample containing an amount of impurity equal to the excess of p-type over n-type or vice versa. In other words, a silicon sample containing 10 apb of phosphorus and 7 apb of boron will behave for most practical purposes the same as silicon containing 3 apb of phosphorus (the net difference). This phenomenon, called *compensation*, is characterized by a neutralization of p-type impurities by n-type impurities on an atom-for-atom basis.

A simplified explanation of the mechanism of compensation states that the excess electron of the Group V element, instead of entering the conduction band, falls through the energy gradient across the forbidden gap and is captured by a Group III impurity, thereby satisfying the latter's electron deficiency and preventing it from capturing an electron from the lattice and forming a hole. This process produces two immobile ions without changing the current carrier concentration in the semiconductor. When this compensation process has neutralized all the atoms of the less concentrated impurity, the remaining impurity atoms react in the manner described above for a one-impurity semiconductor.*

Figure 3-4 depicts the situation in a compensated semiconductor, showing the immobile ions and the net impurity contribution to the carrier concentration. In the case shown, the silicon contains an excess of

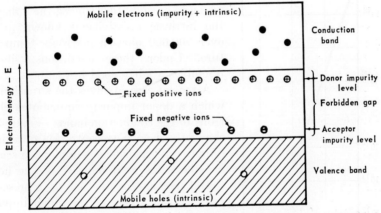

FIG. 3-4. Band energy diagram of a compensated n-type impurity semiconductor.

phosphorus and is therefore n-type after compensation is complete. The situation for a p-type compensated sample is analogous.

3-6. CONDUCTIVITY AS A FUNCTION OF TEMPERATURE

In the foregoing discussion of conductivity we limited ourselves to a relatively narrow temperature range within which useful semiconductor properties were preserved. We mentioned the fact that intrinsic silicon above 150°C begins to lose its normal resistivity owing to thermal generation of hole-electron pairs and is of no interest for device fabrication. Yet, at a lower temperature, say, 100°C, the intrinsic conductivity is so low as to be almost useless as a conductor. Obviously, conductivity must be a strong function of temperature, and its exact dependence on this external variable must be of considerable importance.

In Fig. 3-5 we show a semiquantitative plot giving the relationship

* There are important differences in samples of silicon having different amounts of compensation with the same net difference of impurity. These practical differences are, however, beyond the scope of this volume. Device makers favor the use of uncompensated silicon because it gives better reproducibility and yields.

between the carrier concentration in intrinsic silicon and the reciprocal of the absolute temperature. Over the range 300 to 1000°K for silicon (which is comparable to other semiconductors) it will be noticed that the carrier concentration increases rapidly as a function of temperature (8 orders of magnitude in the 700° interval). The linear portion of the original experimental curve has been extrapolated without change in slope to the region of room temperature. Although the room temperature resistivity calculated from this curve would be higher than the value derived from Table 3-1, owing to second-order effects, the value of the intrinsic resistivity is known to be over 200,000 ohm-cm at room temperature. Under these conditions, silicon resembles a fairly good insulator.

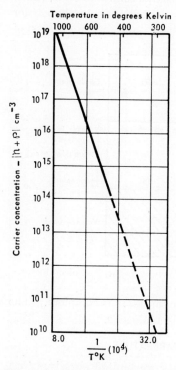

FIG. 3-5. Temperature dependence of carrier concentration in intrinsic silicon. (*After F. J. Morin and J. P. Maita, by permission* [3].)

A second case involves the situation in which a donor impurity contributes electrons to the conduction band. Figure 3-6 shows the carrier concentration of a one-impurity n-type silicon sample; the scales are the same as in Fig. 3-5. The horizontal portion of the curve represents complete ionization of the donor impurities, which, it will be recalled, is an experimental fact over a wide temperature range. The dotted portion of the curve, corresponding to temperatures in excess of about 100°C for silicon, has been included to show, by comparsion with Fig. 3-5, how rapidly the intrinsic process of hole-electron pair generation overcomes the carrier contribution due to donor impurities.

The right-hand portion of the curve in Fig. 3-6 corresponds to a decrease in concentration as a result of incomplete ionization of the donor impurity. As the temperature drops, the conduction electrons reassociate with the corresponding ions of the donor impurities and become immobilized, and hence they are unable to take part in the conduction process. At 0°K there are no ionized donor impurities and no conductivity from this source. However, the very low temperature range for many semiconductors, including silicon, is imperfectly understood because of second-order effects such as lattice imperfections and random impurities.

For p-type semiconductors containing only an acceptor impurity, a curve analogous to Fig. 3-6 results, with no significant change except a slight shifting of the carrier concentration scale. We shall henceforth represent both single-impurity types of silicon by the same curve, recognizing that the carrier concentration scale represents an average value.

In the case of a semiconductor containing both n- and p-type impurities at sufficiently low concentrations a curve resembling the single-impurity

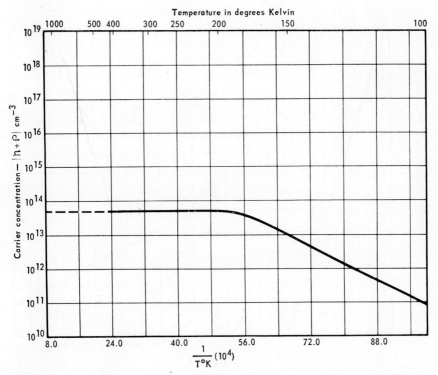

FIG. 3-6. Temperature dependence of carrier concentration in impurity-controlled silicon [4].

case will result. The low-temperature portion of the curve will deviate from the single-impurity case depending upon the chemical identity of the two impurities, the absolute level of concentration of each, and the net excess of one over the other.

We may now plot the carrier concentration as a function of the reciprocal of the absolute temperature from 100 to 1000°K. This curve (combining Figs. 3-5 and 3-6) plotted for silicon appears as Fig. 3-7, showing the extrapolated portions of the intrinsic and ionized impurity contributions to the carrier concentration. In discussing curves of this type, it has become customary to speak of three different regions corresponding to the processes which are controlling. Thus, the high-temperature region, known as the *intrinsic range*, is characterized by its exceedingly steep slope. The horizontal portion in the intermediate-temperature range, known as the *exhaustion* or *saturation region*, corresponds to complete ionization of the impurities. The low-temperature range is actually part of the impurity-dominated range, but it corresponds to a temperature

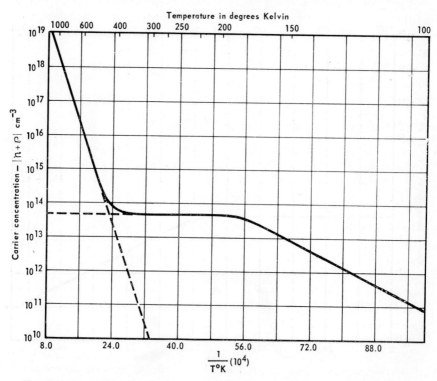

FIG. 3-7. Composite plot: temperature dependence of carrier concentration [4].

region where the degree of ionization of the impurity falls from 100 per cent to zero. The change in carrier concentration in this range takes place much more slowly as a function of temperature, resulting in less slope than that of the intrinsic range.

Since the conductivity is directly related to the carrier concentrations n and p by

$$\sigma = q(n\mu_n + p\mu_p) \tag{3-5}$$

where μ_n and μ_p are the mobilities and q is the charge on the electron, we see that we could construct curves in terms of conductivity as well as free carrier concentration. However, although q is a constant, μ_n and μ_p are themselves functions of temperature, and the resulting curves would depart slightly from the more general case shown in Fig. 3-7. Also, since $\mu_n \neq \mu_p$, the scale would be shifted slightly for n- versus p-type material. These errors are relatively small, and no loss of meaning has occurred as a result of plotting the absolute value of $n + p$ versus absolute temperature.

Attention is also directed to the continuous increase in conductivity as a function of increasing temperature in the intrinsic range. This behavior is opposite to that of metals, and semiconductors are said to have a negative coefficient of resistance with temperature. This behavior with respect to temperature (and, as will be seen later, applied voltage) is an important property of semiconductors and one which forms the basis of numerous applications of these materials in electronic devices.

3-7. CONSTANCY OF THE p-n PRODUCT

In considering the three types of impurity semiconductors, we have confined our attention to the dominant impurity. This is sufficient basis for a discussion of conduction, but it is not sufficient to explain the operation of a semiconductor device involving a p-n junction. In the study of junction devices it is necessary to consider the action of both p- and n-type carriers regardless of which is in excess. (The carrier in excess, whether p- or n-type, is known as the *majority carrier;* conversely, its complement is known as the *minority carrier.*) The presence of minority carriers in any impurity semiconductor results from the generation of hole-electron pairs exactly as described for the intrinsic semiconductor above. This process continues as a function of temperature regardless of the presence of an impurity.

For any given temperature there is an average rate of generation of hole-electron pairs, and in order that the concentration remain constant at this temperature there must be an equal rate of recombination which destroys hole-electron pairs. Let us represent the equilibrium concentration of electrons in the conduction band of an intrinsic sample as n_i and the concentration of holes in the valence band as p_i. Since electrons and holes are always generated in pairs, it follows that n_i equals p_i; and since their concentration is constant for a given temperature, n_i times p_i equals a constant. Therefore,

$$n_i^2 = p_i^2 = n_i p_i = K' \tag{3-6}$$

The p-n product $(n_i p_i)$ has a fixed value for intrinsic silicon at any given temperature.

In an impurity semiconductor at the same temperature one might expect (for n-type material) that the number of electrons would be equal to the sum of the electrons provided by the impurity atoms plus those that are thermally generated by the process mentioned above (n_i) and that the number of holes would remain the same (p_i). Experimentally, it was learned that this is not true and that the concentration of holes p_n multiplied by the concentration of electrons n_n is a constant at the temperature involved. Surprisingly, this was the same constant as for intrinsic silicon, $p_i n_i$. In a similar manner it was shown that for p-type silicon the product $p_p n_p$ was also equal to the same constant ($p_i n_i$) at a given temperature.

It follows, therefore, that the density of holes p_n in the valence band of an n-type semiconductor is less than p_i. The physical significance of the phenomenon is not immediately obvious. It is analogous to the constant solubility product familiar to chemists, with its common-ion effect, and the repressed ionization associated with buffered solutions. We may provide some additional physical insight into this phenomenon with the following semiquantitative discussion in terms of semiconductors.

In an intrinsic semiconductor at equilibrium for a given temperature there is simultaneous generation and recombination of hole-electron pairs. At equilibrium, these processes must occur at the same rate. If we now reason by analogy to the laws of mass action and solubility product, we may let the rate of generation R_g of hole-electron pairs be represented by

$$R_g = k_g n_0 p_0 \tag{3-7}$$

where k_g is a proportionality constant and $n_0 p_0$ is the product of the concentrations of electrons and holes in the combined state, that is, for our model, the concentration of electrons in the valence band. Since this concentration is very large (at normal temperatures) compared to the concentration of thermally generated free pairs, we may regard it as a constant and, by combining $p_0 n_0$ with k_g, obtain (for a given temperature)

$$R_g = K \tag{3-8}$$

where K is a constant.

Likewise, the rate of recombination R_r will be expressed by

$$R_r = k_r n_i p_i \tag{3-9}$$

where n_i and p_i are the concentrations of thermally generated electrons and holes in an intrinsic material. At equilibrium,

$$R_g = R_r = k_r n_i p_i = K \tag{3-10}$$

We now come to the case of an n-type impurity semiconductor in which the concentration of electrons is n_n, where

$$n_n > n_i \tag{3-11}$$

but where the value of n_n is small enough that the equilibrium still applies and the constants R_g, R_r, and K are unaffected. We may therefore write, by analogy to Eq. (3-10),

$$R_g = R_r = k_r n_n p_n = K \qquad (3\text{-}12)$$

Combining Eqs. (3-10) and (3-12) with (3-6) gives

$$n_n p_n = n_i p_i = K' \qquad (3\text{-}13)$$

Obviously, since $n_n > n_i$ by definition, it follows that p_n is less than p_i. The analogy to the solubility product principle is straightforward. If, by virtue of the addition of an impurity, the concentration of electrons is increased above the intrinsic value in an n-type semiconductor, the concentration of holes must decrease to a new value below the intrinsic value in order that their concentration product remain constant; and the competing reactions, Eqs. (3-7) and (3-9), proceed at the same rates.

A simple relationship can easily be derived to relate the constant p-n product to the impurity concentration and the intrinsic conductivity. Consider a sample of silicon at room temperature containing both acceptors and donors. From our discussion of compensation in a preceding section, we know that the donors and acceptors will neutralize each other, producing no charge carriers until the one in excess has completely compensated the impurity present at the lower concentration. Let us assume that the donors are in excess and that there are additional electrons in the conduction band as a result. In order that the p-n product remain constant, recombination of hole-electron pairs must take place to reach a new equilibrium value. The concentration of the electrons in the conduction band (n_n) will be given by

$$n_n = N_D + n_i - X \qquad (3\text{-}14)$$

where N_D = concentration of uncompensated ionized donors
$\quad n_i$ = concentration of intrinsic electrons corresponding to temperature of sample
$\quad X$ = concentration of carriers which must recombine to maintain a constant p-n product

Similarly, the concentration of holes p_n is given by

$$p_n = p_i - X \qquad (3\text{-}15)$$

where p_i = intrinsic concentration of holes at sample temperature involved

Obviously, the value of X in both the foregoing cases is the same. If we now multiply the two expressions and equate them to the p-n product for the temperature of the sample, we have

$$(N_D + n_i - X)(p_i - X) = K' \tag{3-16}$$

This is a simple quadratic equation in one unknown. An analogous expression results for a p-type sample.

Upon substituting Eq. (3-15) into Eq. (3-16) and recognizing that N_D is much larger than $n_i - X$ for doping levels of more than 1 apb impurity (5×10^{13} atoms per cm³), we see that

$$p_n = \frac{K'}{N_D} \tag{3-17}$$

For silicon at room temperature, having a p-n product (K) of 4×10^{21} cm⁻⁶ and using $N_D = 5 \times 10^{13}$ atoms per cm³, the minority carrier concentration p_n becomes 8×10^7 per cm³, which justifies our simplifying assumption.

REFERENCES

1. R. D. Middlebrook: "Introduction to Junction Transistor Theory," pp. 67, 292, John Wiley & Sons, Inc., New York, 1957.
2. G. L. Pearson and W. H. Brattain: History of Semiconductor Research, *Proc. IRE*, **43:** 1794–1806 (1955).
3. F. J. Morin and J. P. Maita: Electrical Properties of Silicon Containing Arsenic and Boron, *Phys. Rev.*, **96:** 28–35 (1954).
4. J. N. Shive: "Properties, Physics and Design of Semiconductor Devices," p. 294, D. Van Nostrand Company, Inc., Princeton, N.J., 1959.

4: Fermi Statistics and Fermi Level

4-1. THE FERMI PROBABILITY FUNCTION

It is now possible to compute the number of charge carriers in the valence and conduction bands of a semiconductor in the temperature range where there is complete impurity ionization (above $-70°C$ for silicon). To do this we need to know the p-n product at the temperature involved and the impurity concentrations (both p- and n-type).

Although we can thus determine the carrier densities, we have no information about the energy levels which are occupied by the electrons. (Holes are really unoccupied energy levels in the valence band and are therefore part of the electron distribution picture.) So far we know only that the electron distribution among the permitted energy levels, including the vacancies (holes), must be consistent with the Pauli exclusion principle.

Information about the electron energy distribution will be needed when we discuss the p-n junction. We have seen that the addition of an n-type impurity to a semiconductor increases the electron concentration in the conduction band with energy levels above the forbidden gap. Conversely, a p-type impurity increases the number of unoccupied energy levels in the valence band below the forbidden gap. If an n-type semiconductor were brought into close contact with a p-type specimen, it might be assumed that there would be a continuous flow of electrons from the apparent high-energy region in the n-type material to the region of lower energy in the p-type specimen. This does not occur even when the semiconductor single crystal is continuous across the boundary between the two regions (p-n junction). To reconcile this apparent anomaly, we must assume that the average electron energies in the two regions are equal.

As an aid to understanding this and other phenomena, a statistical approach will be employed. We shall consider first the electron energy distribution in a metallic conductor (say, copper), which represents the simplest case. Next we shall apply the same approach to the intrinsic, the single-impurity, and the compensated semiconductor in turn. This discussion will enable us to reconcile the apparent difference in average electron energy between n- and p-type semiconductors and to produce a model of a p-n junction having a uniform average energy throughout the specimen at a given temperature. We begin by recalling the band energy model, which is based on a series of allowed levels each capable of containing two electrons only (because of the Pauli exclusion principle) and arranged in bands of fixed energy values with forbidden bands or gaps between. Each lower band is filled with electrons, in pairs, until we reach the uppermost or highest-energy band or bands, where there are no longer enough electrons to completely fill all the available energy states.

Based on this model, the phenomenon of electrical conduction (and therefore transistor action) is concerned with electrons which are free to move under the influence of an applied field. Such electrons are always found associated with bands which are incompletely filled* or which lose electrons because of changes in temperature, external stimulus, etc. These bands constitute the transition region, which is defined as that part of the energy spectrum where there is a finite mathematical probability that the energy levels will not all have their full quota of electrons. Stated in other terms, we shall be concerned with electron states and electron energies which are capable of being influenced by external stimuli of reasonable proportions, and where the effects can be observed as changes in electrical properties, chiefly conductivity.

Electrons corresponding to these energy levels are free in that they can be caused to move and thereby conduct a current. Our concept of free electrons in this connection, however, must be broad and must include, at one extreme, valence electrons of an insulator which normally fill the valence band, with a wide energy gap to be traversed before reaching the conduction band. These electrons can, if the temperature and applied field are high enough, jump to the conduction band and produce a measurable conductivity. Conversely, we must remember that at low temperatures all semiconductors suffer a decrease in the number of free charge carriers and essentially become insulators. Finally, in metals, the conductivity changes less drastically with temperature, neither approaching zero at very low temperatures nor exhibiting the very large changes at higher temperatures characteristic of semiconductors.

We shall henceforth deal with the behavior of free electrons, or those in the transition region, for all solids. It should be mentioned that this region is quite narrow by comparison with the total energy spectrum and that the dividing line between filled and unoccupied bands is relatively

---

* For a brief qualitative treatment of conduction in filled bands, see Appendix.

sharp (at ordinary temperatures). The approximate boundary line between regions of predominately occupied and predominately empty energy levels is known as the *Fermi energy level*, or *Fermi level*, in honor of the coauthor of the statistical system which governs the behavior of free electrons. For temperatures greater than absolute zero, the probability that an electron has an energy E equal to the Fermi energy E_F is ½. This is a natural consequence of the Fermi-Dirac distribution law, which expresses the probability that an electron will have any energy E. In this and subsequent discussion the Fermi equation and its physical implications will be examined, with particular emphasis on semiconductors.

It has been shown experimentally that free electrons in semiconductors and metals obey the rules of Fermi-Dirac statistics. By applying these rules to an array of electrons $\mathfrak{N}$ distributed over G energy states, it is possible to obtain an expression for $\mathfrak{D}$, the number of independent ways in which the electrons can be distributed among the energy states [1].

The quantity $\mathfrak{D}$ then becomes an index of the order or disorder of the system, and it is identical with the quantity in the Boltzmann definition for the entropy of distribution of objects:

$$S = k \ln \mathfrak{D} \tag{4-1}$$

where S is the entropy and k is Boltzmann's constant, which has been determined for this case on the basis of the perfect-gas laws.

If we impose the rules of Fermi-Dirac statistics, which state that the objects (electrons) are indistinguishable and noninteracting and that only one is permitted per energy state (from the Pauli exclusion principle), we obtain for $\mathfrak{D}$

$$\mathfrak{D} = \prod \frac{G_i(G_i - 1) \cdots (G_i - \mathfrak{N}_i + 1)}{\mathfrak{N}_i!} \tag{4-2}$$

and

$$S = k \ln \left[\prod \frac{G_i(G_i - 1) \cdots (G_i - \mathfrak{N}_i + 1)}{\mathfrak{N}_i!} \right] \tag{4-3}$$

Equation (4-3) states that the more ways there are of rearranging a system of objects, the more disordered it will be, and hence the larger the entropy.

It is also a basic premise of thermodynamics that, in any system free to seek equilibrium, the entropy tends to become a maximum. We therefore wish to maximize S and also to impose two other conditions on the overall system, based on our concept of a free-electron model. This requires that the total number of electrons $\mathfrak{N}$ be constant and also that the overall or internal energy of the electrons and therefore of the system be fixed.

Mathematically,
$$\sum_j \mathfrak{N}_j = \mathfrak{N} \tag{4-4}$$
and
$$\sum_j \mathfrak{N}_j E_j = U \tag{4-5}$$

where U is the internal energy and E_j is the energy possessed by the group containing $\mathfrak{N}_j$ objects.

To apply Lagrange's method of undetermined multipliers, we rewrite the above, using $\ln \mathfrak{D}$ for convenience,

$$\ln \mathfrak{D} = \sum_i [\ln G_i + \ln(G_i - 1) + \cdots + \ln(G_i - \mathfrak{N}_i + 1)]$$
$$- \sum_i (\ln 1 + \ln 2 + \cdots + \ln \mathfrak{N}_i) \tag{4-6}$$

$$\mathfrak{N} - \sum_j \mathfrak{N}_j = 0 \tag{4-7}$$

$$U - \sum_j \mathfrak{N}_j E_j = 0 \tag{4-8}$$

By differentiating and letting $\mathfrak{N}_i$ increase by unity, we obtain

$$\frac{\partial(\ln \mathfrak{D})}{\partial \mathfrak{N}_i} = \ln(G_i - \mathfrak{N}_i) - \ln(\mathfrak{N}_i + 1) \tag{4-9}$$

$$\frac{\partial}{\partial \mathfrak{N}_i}\left(\mathfrak{N} - \sum_j \mathfrak{N}_j\right) = -1 \tag{4-10}$$

and
$$\frac{\partial}{\partial \mathfrak{N}_i}\left(U - \sum_j \mathfrak{N}_j E_j\right) = -E_i \tag{4-11}$$

From these expressions, we have

$$\ln \frac{G_i - \mathfrak{N}_i}{\mathfrak{N}_i + 1} - \lambda_1 - E_i \lambda_2 = 0 \tag{4-12}$$

where λ_1 and λ_2 are lagrangian multipliers. If

$$\mathfrak{N}_i \gg 1 \tag{4-13}$$

$$\frac{\mathfrak{N}_i}{G_i} = \frac{1}{\exp(\lambda_1 + E_i \lambda_2) + 1} \tag{4-14}$$

To evaluate λ_2, it is recalled that

$$\mathfrak{D} = f(\mathfrak{N}_j) \tag{4-15}$$

Then
$$d \ln \mathfrak{D} = \sum_j \left(\frac{\partial}{\partial \mathfrak{N}_j} \ln \mathfrak{D}\right) d\mathfrak{N}_j \tag{4-16}$$

and from Eqs. (4-9) and (4-12)

$$\frac{\partial}{\partial \mathfrak{N}_i} \ln \mathfrak{D} = \lambda_1 + E_i \lambda_2 \qquad (4\text{-}17)$$

From
$$dS = k\, d(\ln \mathfrak{D}) \qquad (4\text{-}18)$$
we may write
$$dS = k \sum_j (\lambda_1 + E_j \lambda_2)\, d\mathfrak{N}_j \qquad (4\text{-}19)$$

However, since $\mathfrak{N}$ is fixed,
$$\sum_j d\mathfrak{N}_j = 0 \qquad (4\text{-}20)$$

and
$$dS = k\lambda_2 \sum_j E_j\, d\mathfrak{N}_j \qquad (4\text{-}21)$$
$$= k\lambda_2\, dU \qquad (4\text{-}22)$$

from Eq. (4-5).

From the thermodynamic definition
$$dU = T\, dS - P\, d\mathfrak{V} + \mu\, d\mathfrak{N} \qquad (4\text{-}23)$$

If $\mathfrak{N}$ and $\mathfrak{V}$ are constants and μ is the Gibbs potential* of any constituent present in amount $\mathfrak{N}$ (number of moles, in chemical terms),

$$\left(\frac{\partial U}{\partial S}\right)_{\mathfrak{V},\mathfrak{N}} = T \qquad (4\text{-}24)$$

and therefore
$$\lambda_2 = \frac{1}{kT} \qquad (4\text{-}25)$$

Now, we assign a value to λ_1,
$$\lambda_1 = -\frac{E_F}{kT} \qquad (4\text{-}26)$$

and
$$\frac{\mathfrak{N}_i}{G_i} = \frac{1}{\exp\,(E_i - E_F)/kT + 1} \qquad (4\text{-}27)$$

The right side of Eq. (4-27) is called the Fermi function $P(E_i)$. We now identify G_i as the number of energy states in a small region dE about E and $\mathfrak{N}_i$ as the number of electrons whose energy lies in a small region dE about E. When we pass to differentials and generalize by dropping the subscript i on E, this becomes

$$d\mathfrak{N} = P(E)\, dG \qquad (4\text{-}28)$$

and
$$P(E) = \frac{1}{\exp\,(E - E_F)/kT + 1} \qquad (4\text{-}29)$$

The function $P(E)$ determines the number of the available states dG (with energies in the interval $E \pm dE$) which are occupied by electrons.

---

* The Gibbs potential μ is identically the Lewis and Randall partial molar free energy $\bar{\mathfrak{F}} = (\partial \mathfrak{F}/\partial \mathfrak{N})_{T,P}$ for one species.

These electrons, whose number is given by $d\mathfrak{N}$, will also of necessity have energies lying in the same differential energy range. Since we know from our model that only one electron may be accommodated per state, the maximum value of $P(E)$ must be unity, in which case $d\mathfrak{N} = dG$. In addition, we intuitively predict that there must be energy levels and available states therein which are not occupied, so that $d\mathfrak{N}$ may become equal to zero. In this case, since we postulate that dG does not vanish, the function $P(E)$ must approach zero.

If we consider the meaning of the Fermi expression in terms of a single electron, it is apparent that $P(E)$ expresses the probability that an electron will be found in a state of energy E. Note that there is nothing in the function $P(E)$ to tell us that an energy state exists at the chosen energy. As we shall see, the Fermi function may predict that the probability is 100 per cent for a state whose energy level would lie in a forbidden gap in the band structure. However, since in this case $dG = 0$, no electron will be found at such an energy level. We shall find it necessary to return to the interpretation of $P(E)$ as a probability function and to examine its meaning in the case of real metals and semiconductors in later sections.

So far we have said little about the reference energy E_F, which is known as the Fermi energy or Fermi level. It has the significance of a reference value for a system, and it is a measure of average energy of the free electrons in the transition region for a given temperature T. We shall consider further interpretations in terms of our model of a metal or semiconductor after relating this statistically derived quantity to the more familiar thermodynamic functions.

The Helmholtz function is given by

$$d\mathcal{Q} = dU - T\,dS \qquad (4\text{-}30)$$

From Eqs. (4-19), (4-25), and (4-26) we may write

$$d\mathcal{Q} = dU - Tk \sum_j (\lambda_1 + E_j \lambda_2)\, d\mathfrak{N}_j \qquad (4\text{-}31)$$

$$= dU - Tk \sum_j \left(\frac{-E_F}{kT} + \frac{E_j}{kT}\right) d\mathfrak{N}_j \qquad (4\text{-}32)$$

$$= dU + E_F \sum_j d\mathfrak{N}_j - \sum_j E_j\, d\mathfrak{N}_j \qquad (4\text{-}33)$$

We now remove the restriction that $\mathfrak{N}$ is fixed, corresponding to a system in which the number of free electrons may vary (as with temperature). The second term on the right does not vanish, but the first and third terms cancel owing to Eq. (4-5). Then,

$$d\mathcal{Q} = E_F\, d\mathfrak{N} \qquad (4\text{-}34)$$

or

$$\left(\frac{\partial \mathcal{Q}}{\partial \mathfrak{N}}\right)_{v,T} = E_F \qquad (4\text{-}35)$$

We may also write Eq. (4-30) in terms of the Gibbs potential, or electrochemical potential,

$$d\mathcal{G} = -P\,d\mathcal{U} - S\,dT + \mu\,d\mathcal{N} \qquad (4\text{-}36)$$

Therefore,
$$\left(\frac{\partial \mathcal{G}}{\partial \mathcal{N}}\right)_{\mathcal{U},T} = \mu \qquad (4\text{-}37)$$

and
$$\mu = E_F \qquad (4\text{-}38)$$

Finally, we may also show the relationship between the Gibbs function (free energy $\mathfrak{F}$) and E_F.

$$d\mathfrak{F} = -S\,dT + \mathcal{U}\,dP + \mu\,d\mathcal{N} \qquad (4\text{-}39)$$

and
$$\left(\frac{\partial \mathfrak{F}}{\partial \mathcal{N}}\right)_{P,T} = \mu = E_F \qquad (4\text{-}40)$$

Also, note that if T, P, $\mathcal{U}$, and U are constant, as in the Fermi free-electron gas, at equilibrium, we have [see Eq. (4-23)]

$$\left(\frac{\partial \mathcal{G}}{\partial \mathcal{N}}\right)_{\mathcal{U},T} = \left(\frac{\partial \mathfrak{F}}{\partial \mathcal{N}}\right)_{P,T} = -T\left(\frac{\partial S}{\partial \mathcal{N}}\right)_{U,\mathcal{U}} = E_F \qquad (4\text{-}41)$$

In the above expressions $\mathcal{N}$ symbolizes the number of electrons in the system. It will be found more convenient, in later work, to speak of n, the concentration of electrons per cubic centimeter. It is therefore worth mentioning that if the volume of the system is constant, any change in $\mathcal{N}$ is also a change in n, and Eq. (4-41) could be written

$$\left(\frac{\partial \mathcal{G}}{\partial n}\right)_{\mathcal{U},T} = \left(\frac{\partial \mathfrak{F}}{\partial n}\right)_{P,T} = -T\left(\frac{\partial S}{\partial n}\right)_{U,\mathcal{U}} = E_F \qquad (4\text{-}42)$$

From quantum physics we may write an explicit expression for E_F in terms of the electron concentration n. If we then express this for absolute zero, we obtain

$$E_F = \frac{\hbar^2}{2m}(3\pi^2 n)^{2/3} \qquad (4\text{-}43)$$

where $\hbar$ = Planck's constant divided by 2π
 m — electron mass

In the case of a metal, silver, this gives a value of 5.5 ev. The result is a good illustration of several fundamental facts. At absolute zero all electrons are at their lowest possible energy levels. However, because of the Pauli exclusion principle, even at absolute zero levels up to a certain value of energy will be filled in order to accommodate the electrons. Above that energy, all levels will be empty. This level is, therefore, the Fermi level at absolute zero. The energy of the electron at this level is 5.5 ev, and not zero as would be the case without the exclusion principle. Further, the Fermi level is here the top of a region of filled levels and can go no lower. This concept is helpful in assigning physical meaning to E_F or μ.

We have seen that the Fermi distribution function indicates the probability that a level of energy E is occupied by an electron. This implies that there are empty energy levels within the range covered by the probability function. Since our definition of the transition region states that all energy levels below it are filled, the Fermi function for these low-lying levels must have the value of unity (100 per cent). In the transition region the probability of occupation changes in a smooth mathematical function from 100 per cent to zero as the energy levels change from nearly all occupied to practically all vacant. Above the transition region are many more states, all vacant (indicated by the zero value of the Fermi function).

The quantity E_F, or Fermi level, is the energy about which the probability function described above is symmetrical. It is also, as previously stated, the energy level which has a probability of 0.5 (50 per cent) of being occupied. In addition, if the transition from full to empty is not abrupt, the Fermi level may be said to define the energy below which there are as many unoccupied states as there are occupied states above. This definition will be reassessed for impurity semiconductors in a later section, but it is valid for metals and intrinsic semiconductors. We have also shown that the Fermi level represents the average energy of the free electrons in the transition region at any temperature T. The voltage equivalent of this Fermi energy V_F is sometimes called the electrochemical potential referred to these free electrons.

Inspection of the Fermi distribution expression [Eq. (4-29)] will show that, for values of E much less than the reference value E_F, the value of $P(E)$ approaches unity (because the value of the exponential term approaches zero). When E equals E_F, the function has a value of $\frac{1}{2}$ (because the value of the exponential to the zero power is equal to unity). Finally, when E is much greater than E_F, the function will approach zero (because the value of the exponential is approaching infinity).

4-2. DEPENDENCE ON TEMPERATURE

Figure 4-1 shows a number of typical curves as derived for several temperature levels from 0 to 1500° Kelvin over an energy range of plus or minus 0.6 ev with the Fermi level taken as the zero of energy for all temperatures. In the illustration the probability (0 to 1) that a given energy level will be occupied is plotted as a function of the energy level involved. All these probability curves intersect at the point where the probability is $\frac{1}{2}$, the Fermi energy level. The shape of this probability curve for any temperature is independent of the energy at which the Fermi level occurs, since the probability depends only on the energy difference between the Fermi level and the level being investigated. These curves are for a solid, and they show the set of energy levels in the band defined by $E = E_F \pm 0.6$ ev.

Let us consider the curve in Fig. 4-1 corresponding to 0°K. Theoretically, at this temperature there are no unfilled levels below the Fermi level and no electrons in an excited state. The Fermi level therefore occupies a position on the energy scale coincident with the uppermost filled level as in the earlier example. The probability of an electron being

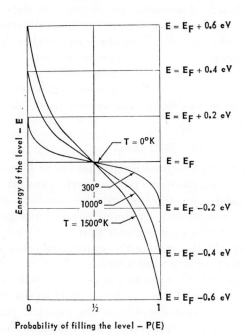

FIG. 4-1. Temperature dependence of the Fermi distribution function. (*After J. N. Shive, courtesy of D. Van Nostrand Company, Inc.* [2].)

excited above this energy level is 0, and the probability of filled bands below this point is 1. There is no physical significance to the horizontal line in this case.

When the temperature rises, electrons are excited into upper levels and leave behind an equal number of vacancies, with the Fermi level still marking the point at which the probability of a filled level is ½. With a further increase in temperature, the energy range in which electrons may be found above the Fermi level is extended, as is the energy range where we may expect to find vacancies below the Fermi level.

We have already mentioned that the Fermi function is a statement of probability concerning the transition region and that it may indicate finite probabilities for electrons in states which are not allowed by the band theory. Semiconductors are a good example of this situation, since the Fermi probability function has a real value over the entire transition region, which includes the valence band, forbidden gap, and conduction band. Therefore, in order to describe the electron distributions for a semiconductor, one must know not only the Fermi probability function but also the density of allowed states at each energy level. In succeeding sections we shall observe the manner in which the energies of the band edges in a semiconductor are shifted with respect to the Fermi energy level under the influence of impurity concentrations, variations in the temperature, etc. Finally, it has been shown that the Fermi energy represents an average electrochemical potential, or Gibbs free energy, for all the electrons in the transition region. Even when the Fermi level corresponds to an energy level within a forbidden gap, as in semiconductors, it retains its value as an index of the average energy of the electrons in the transition region.

4-3. CONSTANCY OF THE FERMI LEVEL

Another important concept with reference to the Fermi energy level is that it is constant throughout any physical system at a given temperature. (A system is defined as a complete crystalline structure such as a sample of a metal or semiconductor.) A simple physical analogy would be two containers of water with their contents at different levels but connected by an open pipe. In this system the water would flow back and forth through the connecting pipe until a constant level was established.

As will be seen later in the operation of p-n junction devices, the application of an electric field can have a profound effect on the relative positions of the Fermi levels on opposite sides of the junction, just as if a pump were operating in the pipe between the water tanks in the example given above.

4-4. METALS

Figure 4-2 shows the electron energy levels as they might appear in a real metal in a typical Fermi distribution. The partially filled band of energy levels is depicted. Although only a fraction of the total number of available energy levels can be shown, we have schematically illustrated the numerical equivalence of occupied states above and vacancies below the Fermi level.

4-5. INTRINSIC SEMICONDUCTORS

For intrinsic semiconductors the Fermi level will be found at the center of the forbidden gap. Reference to Fig. 4-3 will show that this must be true, since by definition an intrinsic semiconductor has as many conduction electrons as holes. At ordinary temperatures the electrons occupy

Fermi Statistics and Fermi Level 43

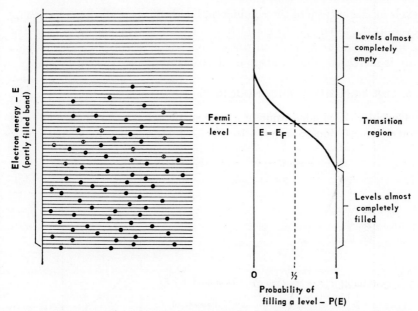

FIG. 4-2. The partly filled energy levels of a metallic conductor and its Fermi function. (*After J. N. Shive, courtesy of D. Van Nostrand Company, Inc.* [3].)

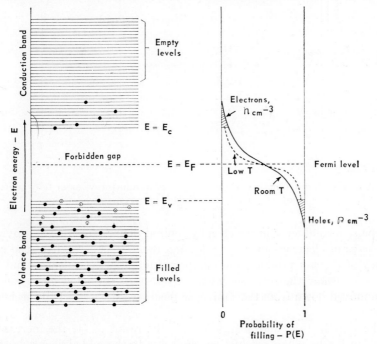

FIG. 4-3. The Fermi function of an intrinsic semiconductor. (*After J. N. Shive, courtesy of D. Van Nostrand Company, Inc.* [4].)

levels at the bottom of the conduction band, whereas all the holes occupy levels at the top of the valence band. This situation continues to be true at all temperatures for an intrinsic semiconductor, since as the temperature is increased, the numbers of electrons and holes increase equally.

4-6. ONE–IMPURITY SEMICONDUCTORS

Consider the single-impurity n-type semiconductor having a small concentration of donor atoms. In Fig. 4-4 we picture schematically an

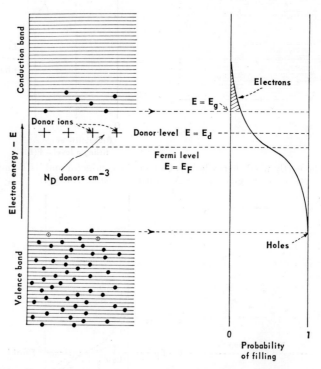

FIG. 4-4. The Fermi function of an n-type semiconductor (donors fully ionized). (*After J. N. Shive, courtesy of D. Van Nostrand Company, Inc.* [5].)

n-type semiconductor in which fully ionized donor atoms have provided an excess of electrons in the conduction band. As a result of the constancy of the p-n product, the number of holes in the valence band has fallen far below the intrinsic value. Note that the band energy levels have moved down from their intrinsic position with respect to the Fermi level.

The shift in the band energy level is the result of (1) the increase in occupied levels in the conduction band due to the added electrons from the ionized donors and (2) the decrease in vacancies in the valence band due to the limitation set by the constancy of the $p\text{-}n$ product. Therefore, the probability of finding a filled energy level near the lower end of the conduction band has increased because of the ionized donors, while the probability of finding filled levels in the valence band has also increased to keep the $p\text{-}n$ product constant. Since the shape of the Fermi distribution curve is fixed for a given temperature, the increased probabilities

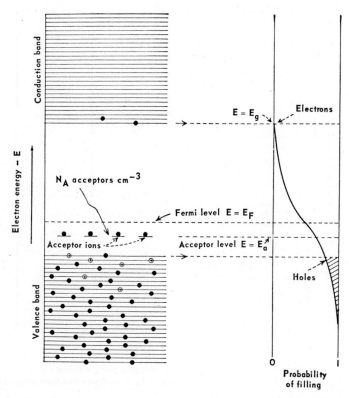

FIG. 4-5. The Fermi function of a p-type semiconductor (acceptors fully ionized). (*After J. N. Shive, courtesy of D. Van Nostrand Company, Inc.* [6].)

result in a shift of the entire band structure. There is no change in the average energy of the system (if the temperature remains constant). The degree of shift of the band levels is not linear but exponential with respect to donor concentration. That is, the change in band energy levels becomes proportionately less as the donor concentration increases.

An analogous argument holds for the p-type semiconductor. Figure 4-5 shows the shift in band energy levels toward the Fermi level that accompanied the addition of acceptor impurities. At the low concentrations normally encountered, the impurity ions themselves (donors or acceptors) have no effect on the position of the Fermi level.

4-7. THE COMPENSATED SEMICONDUCTOR

Last to be considered is the case of a semiconductor containing both donor and acceptor impurities. This is not only the most general case in

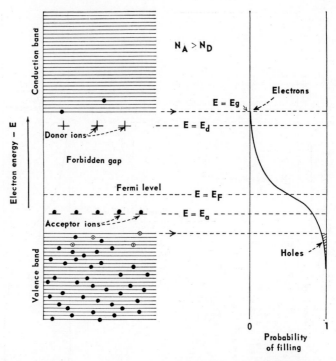

FIG. 4-6. The Fermi energy distribution for a compensated semiconductor containing excess acceptors. (*After J. N. Shive, courtesy of D. Van Nostrand Company, Inc.* [7].)

the application of Fermi statistics but the one most frequently encountered in practice. For silicon, which has not yet been prepared in intrinsic purity, this is the usual situation; and the material behaves as if it were a one-impurity semiconductor having a concentration of impurity equal to the net difference.

In Fig. 4-6 such a two-impurity sample is pictured schematically. The acceptors outnumber the donors, and the sample resembles a *p*-type semiconductor with band energy levels positioned with respect to the Fermi level as though they were influenced only by the excess of acceptors over donors.

REFERENCES

1. A recent discussion of this material may be found in A. Nussbaum: "Semiconductor Device Physics," chap. 2, Prentice-Hall, Inc., Englewood Cliffs, N.J., 1962.
2. J. N. Shive: "The Properties, Physics and Design of Semiconductor Devices," p. 307, D. Van Nostrand Company, Inc., Princeton, N.J., 1959.
3. Ref. 2, p. 305.
4. Ref. 2, p. 308.
5. Ref. 2, p. 321.
6. Ref. 2, p. 314.
7. Ref. 2, p. 324.

5: Symmetrical p-n Junctions

5-1. INTRODUCTION

We have now considered the fundamental properties of a semiconductor in sufficient detail to arrive at an understanding of the mechanisms which govern the behavior of junction devices. In the next sections we shall discuss the formation of a theoretical p-n junction from isolated sections of p- and n-type semiconductor (silicon), the various electrical currents which flow in the vicinity of the junction at thermal equilibrium, and the effect of external variations in electric field and temperature on these currents. We shall also consider how the bulk properties of the semiconductor affect the flow of current in the region of the junction.

Before discussing the p-n junction, we must define a term frequently used in device terminology: *bias*. This refers simply to the application of a voltage gradient across a semiconductor device. When this voltage is applied to a p-n junction in the direction in which current flows readily, it is commonly termed a *forward bias*. Conversely, when the voltage is applied in the opposite direction (in which current flows with difficulty) the device is said to be *biased in the reverse direction*. In the initial discussion to follow, we shall consider a p-n junction under zero bias (no applied voltage).

5-2. FORMING A p-n JUNCTION

If two samples of equally doped silicon, one p-type and one n-type, could be brought into intimate physical contact, we might be able to observe the instantaneous phenomena that take place in the formation of a p-n junction. This physical approach has not been successful, and junctions must be formed either during crystal growth or by impurity addition to the single crystal after it exists. In neither of these processes is it possible to observe what happens during the formation of the junction. In addition, since a junction is believed to reach equilibrium in a time of the order of a millimicrosecond, we are forced to examine an existing *p-n* junction and theorize concerning its formation. The theories developed to date are able to explain most of the observed experimental data, but complete understanding has not been achieved.

If we were able to form a *p-n* junction by bringing a sample of *p*-type silicon into intimate contact with a sample of *n*-type silicon, the initial situation might be as shown in Fig. 5-1. The uncircled symbols represent free carriers. The plus signs represent free holes, and the minus signs represent free electrons, both of which are due to ionization of donor and acceptor impurities. The circled symbols correspond to ionized impurities which are frozen in the crystal lattice (Chap. 3).

We now assume that certain processes must take place instantaneously in the vicinity of a junction (if it could be formed as we have postulated) and that the result of these interactions is a structure composed of five

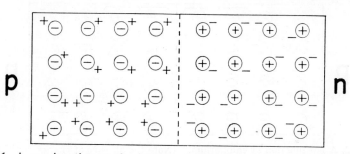

FIG. 5-1. A *p-n* junction at the instant of formation (before recombination and diffusion).

distinct regions each having a different concentration of charged particles. We then find that, based on this qualitative picture, we are able to predict the observed energy levels for the electrons in the vicinity of the junction.

The first interaction between the *p*- and *n*-type regions consists of a migration of electrons into the *p*-type region and holes into the *n*-type region. This migration, in the absence of an external field, is due to the difference in concentration of free electrons and holes on opposite sides of the junction. This movement of electrons and holes in the absence of an electric field, called *diffusion*, is analogous to the spread of ink dropped on a blotter.

When the charge carriers cross over the boundary region (junction), they cause an increase in the minority carrier concentration of the region which they enter. Since the majority carrier concentration in this region is fixed by the number of ionized impurity atoms which are present, the constancy of the *p-n* product is immediately violated. Therefore, in order to restore equilibrium conditions, recombination of holes with electrons takes place until the equilibrium value is again established.

The annihilation of holes in the n-type region and electrons in the p-type region results in an electrical imbalance due to the presence of ionized donors and acceptors, respectively, which are no longer neutralized by the presence of charged particles of the opposite sign. This means that on the left, or p-type side, there is a net deficiency of positive holes and a negative charge results. Conversely, on the n-type side, there is a deficiency of negative electrons and a positive charge develops.

As this process continues, the gradual accumulation of charge exerts a strong electrostatic repulsion for free carriers from the other side of the junction. It therefore tends to oppose further migration of electrons into the p-type region and holes into the n-type region. The migration persists until the space charge formed by the uncompensated ionized

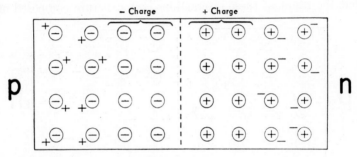

FIG. 5-2. A p-n junction at equilibrium. (Note the absence of charge carriers in the junction region.)

impurity atoms builds up sufficiently on both sides of the junction to prevent diffusion from the opposite side. When equilibrium has been achieved, no further net migration of charged particles takes place, and the free carriers on each side of the junction are repelled (Fig. 5-2).

A second process which contributes to the characteristics of a p-n junction resembles compensation as discussed previously. However, in this case the reverse of what we normally call compensation is assumed to take place. Owing to the space-charge effect, electrons are repelled from the p-type impurity ions close to the junction and are recaptured by the n-type ions which exert an attraction because of their positive charge. This process results in a narrow region of effectively intrinsic silicon immediately adjacent to the junction. (Although intrinsic silicon is defined as silicon containing no impurities, the impurities are now electrically inactive; and it is convenient to refer to this region as the intrinsic region of the p-n junction [1].)

The intrinsic region is believed to spread out on both sides of the junction until it becomes broad enough to impede further transfer of electrons owing to its high ohmic resistance. In other words, it acts like an insulator of high dielectric strength and resistivity. Experimentally, the exact nature of this region cannot be explored, but it is consistent with theory and has been found necessary in certain quantitative treatments of current flow in and near junctions.

Finally, at some distance from the actual junction, and beyond the regions of high space charge, the semiconductor remains unchanged in character and behaves like normal p- or n-type material, as discussed in preceding sections.

We have thus defined five distinct regions characteristic of the formation of a p-n junction. From normal p-type material we passed through

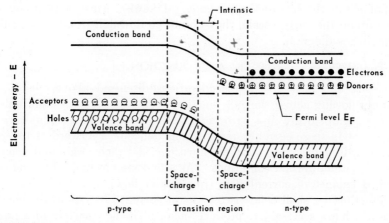

FIG. 5-3. Detailed band energy structure for an unbiased p-n junction in thermal equilibrium.

a region of high negative space charge into a narrow intrinsic layer and thence into a region of high positive space charge before reaching normal n-type material. In order to proceed further, we must know something about the energy levels or band structure of this junction region. If we now picture the junction in terms of a band energy diagram, it is important to remember the significance of the Fermi level, which is a measure of the average electron energy in any system. It will be recalled that the presence of acceptor impurities caused the band energy levels to move up with respect to the Fermi level and that the presence of donor impurities caused the band energy levels to move down. However, when specimens of unlike type are formed into a p-n junction, the concept of constancy of the Fermi level must still apply, and it therefore follows that the Fermi level is the same on both sides of the junction and that the energy bands are displaced with respect to each other. If the

52 Semiconductor Junctions and Devices

Fermi level is taken as a common reference,* it will also follow that the amount of shift in the band structures will be in proportion to the concentrations of acceptors and donors in the p- and n-type regions, respectively. In the present example we are concerned with a symmetrical, or equally doped, junction, and the energy shifts are of equal magnitude.

The resulting band diagram for a p-n junction, Fig. 5-3, shows the shifts imposed by the Fermi level concept and the manner in which the donor and acceptor impurities have compensated each other in the center of the junction so that this region behaves much as if it were intrinsic. The position of the bands on each side is consistent with the way the majority carriers prefer to move and the way these carriers behave near a junction. Electrons prefer to move down on the energy scale; consequently, they do not climb the hill to the p region. Likewise, since holes prefer to move up on the energy scale, they do not go downhill from the p to the n side in the diagram.

5-3. EQUILIBRIUM CURRENTS IN A p-n JUNCTION [2]

Despite the space-charge considerations discussed above, there are currents flowing across the p-n junction even when it is at equilibrium at room temperature and with zero bias. Because of the importance of changes in these currents under the influence of a forward or reverse bias, as will be discussed later, we must consider their origin and magnitude under equilibrium conditions. As a first step in understanding the detailed balance of current flow (free carrier flow) in the vicinity of a p-n junction, we shall define two types of current in terms of the conditions which are necessary to produce them. The flow pattern at a junction will then be considered in these terms and in terms of majority and minority carriers.

A flow of charged mobile particles in a conductor can be produced either by diffusion or direct action of an electric field. Diffusion, as was mentioned earlier, results from a concentration gradient. In a semiconductor such a situation can arise in two ways. First, a local increase in carrier concentration may result from the injection of carriers by electrical or other means, after which the carriers spread out in all directions with a decreasing concentration as a function of distance from the point of origin. Second, a similar diffusion can take place at a p-n junction during its formation, as postulated in Sec. 5-2, where the concentration gradient is across the interface between two samples of opposite type and therefore of widely different carrier concentrations.

* The choice of a reference energy level from which to evaluate changes within a junction is largely a matter of convenience, since only differences are involved. When it becomes necessary to compare electron energies in noncontinuous phases, the matter of an absolute reference or zero energy level requires a more elaborate discussion (Chap. 17).

A diffusion current is therefore the result of a concentration gradient and the charges on the carriers. It is important to recognize that diffusion takes place regardless of the charge on the carrier. Neutral particles must obey the same law under a concentration gradient, and we detect a current only when the particles carry an electric charge. The magnitude of the current is therefore proportional only to the concentration and concentration gradient at a given time and location.

Two additional facts must be emphasized in discussing diffusion currents: (1) the diffusion current is independent of applied field and (2) it is not at all affected by repulsive electrostatic forces between the charged particles. These observations reinforce the statement that such currents depend only on concentration, since we would normally expect charged particles to be influenced by both electric fields and mutual repulsion. A physical example of this may be realized if we picture the diffusion of holes in an n-type sample. Here each positively charged hole is surrounded by a much more concentrated atmosphere of negatively charged particles, or electrons. The qualitative result is that the holes cannot sense the charges of other holes, nor can an external field of normal magnitude exert any influence on them, since the negatively charged medium in which they find themselves is an excellent buffer against such relatively puny effects. The holes therefore diffuse by a concentration-dependent mechanism only.

The second type of current is the drift current, which is more familiar. This is a current composed of charged-particle flow (holes or electrons) under the influence of an electric field or potential gradient. Typical examples are the flow of electrons in a wire and holes through a p-type semiconductor, where the magnitude of the current is proportional to the number of charge carriers (concentration, but not concentration gradient) and the electric potential applied. If we now consider our p-n junction in the absence of an applied field, we expect no drift current through the p- or n-type bulk material at some distance from the junction. In this expectation, we are correct.

There is, however, a space charge of positive and negative ionic charges, which can have the effect of an applied electric field over the very small distance involved at a p-n junction, and drift currents of a highly localized and special nature are indeed present.* First, these drift currents are

* In this discussion the currents due to thermally generated electron-hole pairs are treated as drift currents. Although some authors treat these carrier flows as diffusion currents and derive quantitative expressions for them which are consistent with observed behavior, the theory is equally capable of following the interpretation used herein. In addition, the qualitative physical interpretation is more consistent, since the field produced by the space charge would tend to move them as indicated, whereas the concentration gradient is opposed to motion in the directions chosen. In either case, the important fact that their number is a temperature-controlled quantity suffices to explain the reverse saturation current phenomena for the p-n junction.

not dependent on external fields applied to the sample containing the junction, since any reasonable external field is insignificant compared to the enormous fields associated with the space-charge region ($>5 \times 10^5$ volts per cm) [3]. Also, the carriers which move under the influence of this field are minority carriers at the point of origin, and they are propelled across the junction against the concentration gradient into a region of high concentration. These currents are very small and effectively constant (determined by the impurity concentrations of the material of the junction and therefore the space charges, among other quantities). They can therefore be ignored in many cases except for the effect of temperature, which, by influencing the number of free carriers, can exert a profound effect on drift currents and thereby on the properties of the junction.

Coming now to the p-n junction, we find both types of current, with respect to both holes and electrons, flowing across the junction. However, since the junction is at thermal equilibrium, with no external field and no external current path, the net algebraic sum of the four currents must be zero. Considering the electrons first, we find that a few on the n-type side will acquire enough thermal energy to surmount the space-charge barrier and will then flow, as a diffusion current, across the junction into the p-type region, where they disappear through recombination. Simultaneously, electron-hole pairs are constantly being generated by thermal activity throughout the sample. When such a pair is generated in or near the space-charge region, the electron is propelled by the field from the p-type to the n-type side, against the concentration gradient. Exactly analogous reasoning gives rise to two hole currents, one a diffusion current and one a drift current.

By convention, it is usual to refer to the drift currents as I_g, since they have their origin in thermal generation of hole-electron pairs and since they are affected only by thermal changes in the sample. Therefore, the drift current I_g is always shown opposing the concentration gradient at a p-n junction, moving electrons from the p- to the n-type side and holes in reverse. Also by convention, the diffusion current, which moves with the concentration gradient, is known as I_f, since this is found to be the forward direction of current flow across a junction when it is externally biased. For the present, it is enough to recall that the subscript for forward indicates flow in the direction of the concentration gradient, as is proper for a diffusion current. Figure 5-4 shows these currents in schematic form. In each case the length of the arrow is proportional to the magnitude of the current and, for the unbiased junction at thermal equilibrium, the sum of the four vectors must be zero.

Before closing this discussion of the equilibrium-current patterns typical of a p-n junction, it is instructive to review the pattern of carrier flow with emphasis on the role of minority carriers. Inspection of the mechanisms detailed above, along with the concept of the constant p-n product, will illustrate the principle that the junction device is controlled by minority carrier behavior. The utility of this understanding will become apparent if it is realized that only a very limited number of examples of majority carrier action are yet known, and those in highly specialized structures.

It should be obvious that whenever a hole or electron crosses a p-n junction, it perforce changes from majority to minority carrier or vice versa. Thus, holes, which are majority carriers in a p-type region, may

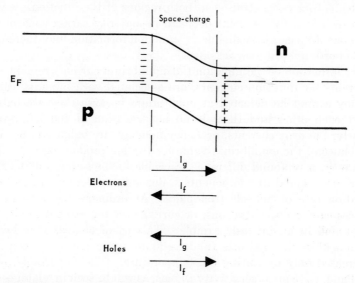

FIG. 5-4. Equilibrium currents in an unbiased symmetrical p-n junction. (*After N. B. Hannay, by permission* [4].)

diffuse into an n-type region, where they become minority carriers. Such behavior was observed during the formation of the p-n junction and in setting up the I_f current component for the equilibrium-current analysis. On the other hand, electrons in or near the p-type region (minority carriers) may be carried against the concentration gradient by the field due to the space charge into the n-type region, where they become majority carriers. This constitutes the I_g, or drift-current, component in the equilibrium situation.

Although the drift-current components I_g involve progression of a minority carrier against the concentration gradient and are therefore the reverse of usual minority carrier behavior, a quantitative examination of the origin and magnitude of these drift currents will illustrate their relative importance compared to I_f components.

We first review the process of junction formation in which carriers diffuse from their region of high (majority) concentration to the opposite side of the junction, where they are minority carriers and cause an increase in the concentration of that species. A violation of the p-n product results, and recombination occurs until equilibrium is reestablished. Note that the reverse process is also proceeding, so that each side of the junction is losing majority carriers in two ways: (1) by diffusion across the barrier and (2) by recombination with the influx of minority carriers from the other side. Both processes are minority carrier effects, since the violation of the p-n product is the result of increased minority carrier concentration in both regions of the junction. (As will be shown below, the per cent decrease in majority carrier concentration due to out diffusion is negligible.) Junction formation may therefore be regarded as a minority carrier effect.

After formation of the junction the equilibrium currents remain. The I_f currents, or diffusion currents, are associated with minority carriers, but they cannot be detected at equilibrium because they are balanced against each other and there is no net violation of the p-n product. Similarly, the drift currents, or I_g components, are balanced and no net disturbance of the equilibrium value for the p-n product occurs. There is, however, a profound difference traceable to the sources of the I_f and I_g currents. The drift components I_g depend completely on the thermal generation rate of hole-electron pairs. At ordinary temperature, this produces carrier concentrations, or currents, of the same order as those to be found in an intrinsic sample. This is, of course, very low and independent of external bias (under equilibrium conditions), and it can be increased only by raising the temperature. The I_f currents, on the other hand, have great sensitivity to bias, as will be seen in a later section, and have virtually unlimited supplies of carriers available to diffuse across the junction. Once across the space-charge barrier, they become minority carriers and produce changes as profound as in the initial formation of the junction. Therefore, the role of the minority carriers is also dominant in current flows at a junction after it has been formed.

It is of interest to conclude this discussion by anticipating a future situation. Suppose, instead of the equilibria prevailing after the formation of a p-n junction, we inject minority carriers, say, holes, across a junction in such a way that no compensating reverse process takes place. If our p-n product is 10^{21} cm^{-6} (a reasonable value for silicon) [5] and the concentration of electrons in the n-type region is 10^{13} cm^{-3} (corresponding to high-purity material), the initial concentration of minority carriers (holes) will be 10^8 cm^{-3}. If we now inject holes up to 10^{10}, we change this concentration by a factor of 100, or in terms of an increment, from 1×10^8 to 100×10^8. To maintain neutrality, the corresponding change in the electron density need be only from $100{,}000 \times 10^8$ to $100{,}099 \times 10^8$, or about one part in 1,000.*

If the injection process continues, there will be competition between the disturbing force and the tendency for the p-n product to reassert itself, but it is obvious from the above that the contribution by the majority carriers to a change in overall sample conductivity is, and remains, negligible. This dominant role played by minority carriers is a basic property of junction structures and one which will be frequently invoked in understanding the function of devices.

5-4. ENERGY, FIELD, AND CHARGE DENSITY [6]

We have mentioned space-charge density and the electric field associated therewith in explaining certain aspects of p-n junction behavior on a qualitative basis. We now wish to look more closely at these quantities and to derive expressions which will permit us to evaluate the magnitude of the field at the junction. From a consideration of the band energy diagram, Fig. 5-3, it is obvious that the energy of an electron occupying a fixed level at, say, the lower edge of the conduction band undergoes a change on the energy scale in passing from the p- to the n-type region of the semiconductor. Numerically this change is equal, in electron volts, to the magnitude of the shift of a band edge, and it can be measured from any reference such as the bottom of the conduction band in the n-type region, the Fermi energy level, or any other datum line. This potential difference for an electron is known as the barrier potential or potential-energy change brought about by the formation of the p-n structure. This localized potential barrier is similar to the contact potential at a thermocouple junction, and, as in the latter case, it cannot produce useful energy if the temperature of the system is constant. By the use of Poisson's equation, we can relate the potential difference to the electric field and finally to the space charge.

Before considering these quantities and their usefulness in explaining p-n junction behavior, it is necessary to reconcile the constant Fermi energy level concept with a different electron potential energy in the p- and n-type portions of the junction and adjoining regions. It has been emphasized that the Fermi energy level is an average value for all the free electrons which define the transition region and that it is invariant throughout the system, despite differences in potential (and therefore in electron energy) arising during junction formation. To resolve the

---

* In this calculation, the p-n product has been exceeded, since

$$(100 \times 10^8)(100{,}099 \times 10^8) = 100.099 \times 10^{21}$$

To restore equilibrium, 99×10^8 hole and electron recombinations must occur (cf. Sec. 3-7).

apparent inconsistency, we recall the assumptions made in setting up the model for treatment by Fermi-Dirac statistics. The electrons in the transition region were treated as uncharged particles and were assumed to be free of the restraint normally associated with the potential valleys corresponding to the positive atomic nuclei. This assumption is parallel to that which stated that they did not interact, despite their negative charge and the normal repulsion forces which might be expected. They were, of course, constrained to the volume of the system, and for a given temperature the total internal energy was also fixed. Beyond this, they were entirely free to distribute themselves among the allowed energy states in accordance with the Pauli exclusion principle and the band energy structure for the material involved, solely on the basis of their kinetic energies.

This simplified picture serves very well for the normal case of a metal or intrinsic semiconductor, and it is extended without serious error to cover a relatively broad range of doping levels (up to the onset of degeneracy at about 10^{17} atoms per cm^3). The success of the simple model reassures us that we were justified in using an array of neutral particles. The effects of the periodic lattice potential and the mutual repulsion due to the charge on the electrons are, in other words, very small compared to the kinetic-energy component. Under other circumstances, such as junction formation, degeneracy, and surface and contact effects, however, a more refined model is required so that interactions between the electrons and potential fields due to space charges, external potential gradients, etc., may be accounted for. Since we are concerned with the effect of a barrier or other form of internal voltage difference, we must include the potential-energy term due to the interaction of a potential gradient with a system of charged particles. If we refer to the expression for the internal energy of the system

$$dU = T\,dS - P\,d\mathcal{U} + \mu\,d\mathfrak{N} \tag{4-23}$$

we recognize that μ, the Gibbs potential, is restricted to kinetic energy only. To account for potential energy, we introduce a new variable Q, the total electrostatic charge on the particles, into the thermodynamic equation for the internal energy. By definition, the potential is a function of U determined by

$$\left(\frac{\partial U}{\partial Q}\right)_{S,\mathcal{U},\mathfrak{N}} = V \tag{5-1}$$

From this we obtain

$$dU = T\,dS - P\,d\mathcal{U} + \mu\,d\mathfrak{N} + V\,dQ \tag{5-2}$$

where V = potential at a point in the system, volts
Q = total electrostatic charge on the particles, coulombs

Note that in a system where the potential may be ignored, or the particles carry no charge, Eq. (5-2) is identical with Eq. (4-23).

It is also true that
$$dQ = -q\, d\mathfrak{N} \tag{5-3}$$
where the minus sign reflects the sign of the charge on the electron. Then,
$$dU = T\, dS - P\, d\mathfrak{v} + \mu\, d\mathfrak{N} - qV\, d\mathfrak{N} \tag{5-4}$$
or
$$dU = T\, dS - P\, d\mathfrak{v} + (\mu - qV)\, d\mathfrak{N} \tag{5-5}$$

We now define $\bar{\mu}$ to include both the kinetic-energy term μ and the potential-energy term $-qV$, in electron volts,
$$\bar{\mu} = \mu - qV \tag{5-6}$$
and Eq. (5-5) becomes
$$dU = T\, dS - P\, d\mathfrak{v} + \bar{\mu}\, d\mathfrak{N} \tag{5-7}$$

From this it can easily be shown that Eq. (4-41) is valid for $\bar{\mu}$ and that
$$\left(\frac{\partial \mathfrak{a}}{\partial \mathfrak{N}}\right)_{T,\mathfrak{v}} = \left(\frac{\partial \mathfrak{F}}{\partial \mathfrak{N}}\right)_{P,T} = -T\left(\frac{\partial S}{\partial \mathfrak{N}}\right)_{U,\mathfrak{v}} = E_F = \bar{\mu} \tag{5-8}$$

The physical meaning of $\bar{\mu}$ is still that of the electrochemical potential, and it is numerically identical to μ in previous discussions. However, in situations where potential effects must also be considered, $\bar{\mu}$ is the generalized form which recognizes that part of the total energy is a potential term.

From a statistical point of view, the concept of a potential-energy term is frequently written [7]
$$E_F = E_{POT}(x) + f\left(\frac{n(x)}{g}\right) = \text{const} \tag{5-9}$$
where $n(x)$ is the concentration of electrons in a given energy range as a function of position and g is the density of states in that energy range.

In this expression $E_{POT}(x)$ is the potential-energy term as a function of position along the x direction (in a one-dimensional model) and $f[n(x)/g]$ is a concentration factor expressing the distribution of electrons among the allowed levels. The equation describes the interdependent changes in the potential energy and the purely kinetic-energy distribution of electrons in a crystal. Both terms are written as functions of position, since these potential changes are usually local.

It is beyond the scope of this work to delve further into explicit relationships for $\bar{\mu}$ or $f[n(x)/g]$ where the simple kinetic-energy model fails. The effect of these terms is to maintain the Fermi energy constant (at constant temperature) despite changes in $E_{POT}(x)$ as a function of distance in moving from the p-type to the n-type region around a p-n

junction. Thus, if at a position $x = x_n$ on the n-type side the $E_{POT}(x)$ becomes more negative as a result of closer binding to the lattice, the presence of impurity ions, or the formation of a junction, the distribution (or population) term based on kinetic energy will increase the average concentration of electrons in higher energy levels to overcome the decrease and maintain E_F constant.

If we now ignore the population term in the expression for the Fermi energy level, we may consider only the potential change seen by an electron, and from this we may derive expressions for the field strength and space-charge density in passing from the p- to the n-type regions, using a one-dimensional approximation and considering only the x distance through the junction. Both Poisson's equation (in one dimension)

$$\frac{d^2V}{dx^2} = -\frac{\rho(x)}{\epsilon} = -\frac{\rho(x)}{\kappa\epsilon_0} \tag{5-10}$$

and Coulomb's law

$$F = \frac{q^2}{4\pi\epsilon x^2} = \frac{q^2}{4\pi\kappa\epsilon_0 x^2} \tag{5-11}$$

may be derived from one of Maxwell's equations. In Eq. (5-10) the potential V is expressed in volts, the x coordinate is in meters, and the charge density ρ is in coulombs per cubic meter. In Eq. (5-11) the force F is in newtons (1 newton = 10^5 dynes), the like charges q are in coulombs, and the charge separation x is in meters.

Equations (5-10) and (5-11) are written in the mks system of units. In both expressions ϵ is the permittivity (sometimes called the absolute dielectric constant or capacitivity) and is defined as the force between two charged conductors measured in a homogeneous insulator or dielectric. The corresponding quantity in vacuum is ϵ_0, and the two are related by $\kappa = \epsilon/\epsilon_0$, where κ is the normal, dimensionless dielectric constant (16 for germanium and 12 for silicon). In this system of units, ϵ_0 has the units of farads per meter and a numerical value of 8.85×10^{-12}. (This is equivalent to 8.85×10^{-14} farads per cm, which is the value used in most semiconductor work. The units are then those of the mks system, with centimeters in place of meters.) Since the farad is a unit of capacitance, it is defined in Sec. 5-6 as the ratio between charge, in coulombs, and potential difference, in volts.

The electric field strength $\mathcal{E}$ is the link between the potential V, in volts, and the force F, in newtons, and it is expressed as volts per meter:*

$$\mathcal{E} = \frac{F}{q} = \frac{dV}{dx} \quad \text{volts per m} \tag{5-12}$$

* Without reverting to formal dimensional analysis, the following simple argument will show that the units in the above expressions are consistent. Coulomb's law

Symmetrical p-n Junctions

In semiconductor junction devices we may identify the charge density $\rho(x)$ in terms of the donor and acceptor concentrations. Thus

$$\rho(x) = qN_D \qquad (5\text{-}13)$$

in the donor region and

$$\rho(x) = qN_A \qquad (5\text{-}14)$$

in the acceptor region, where q is the electronic charge (1.6×10^{-19} coulomb) and N_D and N_A are the net donor and acceptor concentrations, respectively. The three functions, potential, field strength, and charge density, are shown in Fig. 5-5, where V_0 is the potential difference, in volts, between the n- and p-type regions.

If we apply the above relationships to the narrow junction region with potential differences of the magnitude normally associated with junction devices, we shall encounter field strengths of the order of hundreds of thousands of volts per centimeter. This high field is important on the reverse cycle of rectifiers and in many switching operations in other devices. When we consider this and the extremely high charge densities associated therewith, it is not difficult to understand why free electrons

---

defines the unit of electrostatic charge:

$$F = \frac{q^2}{4\pi\kappa_0 x^2} \quad \text{or} \quad \frac{(\text{kg})(\text{m})}{\text{sec}^2} = \frac{(\text{coulomb}^2)(\text{volt})(\text{m})}{(\text{m}^2)(\text{coulomb})} \qquad (a)$$

Potential or emf is defined by

$$\text{Work} = q(V_2 - V_1) \quad \text{or} \quad \frac{(\text{kg})(\text{m}^2)}{\text{sec}^2} = (\text{coulomb})(\text{volt}) \qquad (b)$$

Work is also a force acting through a distance, $W = (\text{force})(x)$, and force is therefore work/distance. From (b)

$$F = \frac{W}{x} \quad \text{or} \quad \frac{(\text{kg})(\text{m})}{\text{sec}^2} = \frac{(\text{coulomb})(\text{volt})}{\text{m}} \qquad (c)$$

which is identical with Coulomb's law in (a).

Poisson's equation states

$$\frac{d^2V}{dx^2} = -\frac{\rho(x)}{\kappa\epsilon_0} \quad \text{or} \quad \frac{\text{volt}}{\text{m}^2} = \frac{(\text{coulomb})(\text{volt})(\text{m})}{(\text{m}^3)(\text{coulomb})} \qquad (d)$$

Canceling, $\qquad\qquad\qquad\quad \dfrac{\text{volt}}{\text{m}^2} = \dfrac{\text{volt}}{\text{m}^2} \qquad (e)$

Finally, $\qquad \varepsilon = \dfrac{F}{q} = \dfrac{dV}{dx} \quad \text{or} \quad \dfrac{(\text{kg})(\text{m})}{(\text{coulomb})(\text{sec}^2)} = \dfrac{\text{volt}}{\text{m}} \qquad (f)$

and rearranging, $\qquad \dfrac{(\text{kg})(\text{m})}{\text{sec}^2} = \dfrac{(\text{coulomb})(\text{volt})}{\text{m}} \qquad (g)$

which is again identical with Coulomb's law in (a).

or holes do not cross the junction region in any significant numbers except under outside stimulus. We have seen in the preceding section that there are very small drift and diffusion currents resulting from thermal activation of charge carriers which cross the junction, but that these are manyfold smaller than any useful signal.

The existence of such high field strengths and space-charge concentrations in an otherwise neutral system explains in part why semiconductor

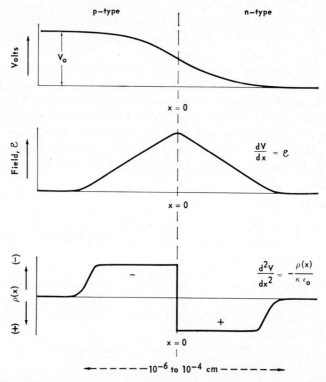

FIG. 5-5. Potential, space-charge density, and field distributions. (*After J. N. Shive, courtesy of D. Van Nostrand Company, Inc.* [8].)

devices can be so small and still perform the many functions of the much larger vacuum tubes.

5-5. JUNCTION WIDTH

The actual width of the junction or the space-charge region between normal p- and n-type silicon in a device plays an important role in the device performance. In general, junction widths vary from 10^{-6} to 10^{-4} cm. When we consider typical devices later, it will be helpful to know how the width of a junction depends on composition, bias, etc.

We have already seen that, during the formation of a *p-n* junction, the majority carriers in the two types of silicon involved annihilate each other by recombination to form regions of positively and negatively charged ions in the *n*- and *p*-type metal, respectively, separated by the narrow "intrinsic" zone. The ions which produce the space charges are fixed in the lattice at regular intervals and at a concentration proportional to the original doping level of the *p*- and *n*-type material. (The actual density of these local impurity sites is low, if we remember that doping is significant at a few atoms of impurity per 10^9 silicon atoms in the host lattice.) In order that overall charge neutrality of the system be maintained, the number of exposed, or unneutralized, impurity ions on the plus side of the junction (*n*-type side) must exactly equal the number on the negative or *p*-type side. Since these ions are fixed in the lattice at average separations inversely proportional to their concentration, the volume of the space-charge regions will vary and, for a given cross section, the junction width (usually taken across the outer edges of the depletion or space-charge regions) will be a function of composition.

In view of the above it is obvious that in a highly purified region containing a very small number of impurities the depletion process will penetrate farther from the center of the junction than would be the case for highly doped material. By this mechanism, the junction is effectively wider when higher-purity materials are used on one or both sides. A sharp, narrow junction results if highly doped material is used for both *p*- and *n*-type regions.

The width of the unbiased junction may be predicted from a knowledge of the impurity concentration profile as a function of distance through the junction. It is also necessary to know the potential-barrier height, which is dependent only on the concentration of ionized impurities in the *n*- and *p*-type regions. It may be shown that V_0, the barrier height, is represented by [9]

$$V_0 = \frac{kT}{q} \ln \frac{n_n p_p}{n_i^2} \qquad (5\text{-}15)$$

which, if we assume 100 per cent impurity ionization, becomes

$$V_0 = \frac{kT}{q} \ln \frac{N_D N_A}{n_i^2} \qquad (5\text{-}16)$$

In these equations, V_0 is measured in volts, Boltzmann's constant k is in electron volts per degree, q is in coulombs, and the other quantities express concentrations, as before.

An explicit statement for V_0 having thus been obtained, the junction width may be quantitatively determined for two well-defined cases. In the first of these, the step junction, conductivity type changes sharply from n- to p-type at the junction. Such a structure is approximated by processes such as alloying, to be described later. The second case involves the assumption that the change from n- to p-type is linear, continuous, and more gradual across the junction. This is a simplification of the structure to be found in grown or diffused junctions.

The derivation of the above expressions is deferred to the next section, since the value of junction width is used in the calculation of junction capacitance. Further, in the case of junction capacitance we wish to consider the effect of bias, which, as will be seen, may also change junction width. It should also be mentioned that in most practical cases neither of the above assumptions regarding impurity gradients in the junction is entirely valid and junction width and capacitance calculations must be approximate.

In commercial practice the width of the junction can be critical, particularly when capacitance or high-frequency response is important. By a suitable choice of materials, doping or growing techniques, etc., the fabricator can learn to control this structural parameter rather closely. However, the properties required in the phases beyond the edge of the junction may dictate the impurity level which is necessary so that the device will function. In most devices we shall see that the junction is a composite consisting of a highly doped phase on one side and a lightly doped phase on the other. Frequently, the highly doped material is present to provide adequate electrical conductivity. In almost every case the composition of both phases represents a compromise between the desired electrical conductivity through the bulk and the desired characteristics of the junction itself.

Finally, as will be found particularly useful in connection with charge-controlled devices, the width of the junction is a function of the bias voltage. In this case the applied field may either oppose or reinforce the existing space charge, causing readjustment in the width of the depletion layer and therefore of the junction. The mechanism here may be understood if it is recalled that the space-charge regions cause the electrons (or holes) to retreat from the junction until the diffusion tendency balances the barrier voltage. If an external field is now applied by biasing the junction, the clouds of free carriers are subject to an additional voltage gradient and will either move farther away from or closer to the center of the junction, depending upon the direction of the field compared to that already present at the barrier. Equilibrium will again be restored when the diffusion tendency equals the net sum of the combined fields. As might be expected, this effect is also dependent on the impurity concentrations of the n- and p-type regions (or barrier height, which is directly related) and the shape of the concentration gradient as a function of linear distance. Since, in many practical cases, the concentrations are nonuniform in the junction region, a precise determination of effective depletion layer width for a device under bias frequently involves a rather laborious calculation.

The externally measurable parameter most closely related to the effective junction width is the capacitance of the junction. In the next section we shall present several simple formulas relating the above-mentioned variables to the capacitance.

5-6. JUNCTION OR BARRIER CAPACITANCE

One of the most important ways in which junction devices differ in performance from the conventional vacuum tube is with respect to the high capacitance associated with the p-n junction. By comparison, the vacuum tube capable of performing rectification or amplification has a very low capacitance between its active elements.

The capacitance of a junction device results from the space-charge regions on both sides of the intrinsic layer of semiconductor in the junction. We have already observed that the space-charge densities can be extremely high and that the insulating layer between them is exceedingly thin. This is analogous to a metal-plate capacitor, where the highly charged plates correspond to the space charges and the air gap between corresponds to the semiconductor intrinsic layer. However, in the p-n junction, the space-charge regions are even thinner than the metal foil in the capacitor and the width of the insulating (intrinsic) layer is also small, depending upon the manner in which the junction was produced (diffused, alloyed, grown, etc.). In addition, the magnitude of the space-charge density is a function of the doping level, and, as we have seen, highly doped material will produce a narrow space-charge region whose charge density will be extremely high.

Capacitance in the p-n junction must be considered when circuits are designed. A device maker has certain means at his disposal to control this effect, but these methods of capacitance control will frequently affect other parameters in the operation of the junction device. For example, increased doping may result in a narrower space-charge region, but it may simultaneously affect the peak inverse voltage and minority carrier lifetime. The space-charge density, which is directly proportional to the capacitance, may in part be neutralized by a proper selection of the bias across the junction. However, the selection of bias is frequently a function of the device application and cannot be varied outside relatively narrow limits.

By making certain assumptions concerning the impurity concentration gradient, useful relationships between capacitance and other junction parameters may be derived [10]. We recall that the total capacitance is defined as the ratio between charge and potential difference, given in farads.

$$C_T = \frac{Q}{V} \quad \text{farads} \tag{5-17}$$

where Q = total charge stored in the depletion region, coulombs
V = electric potential, volts

For the usual case of a unit area

$$C = \frac{C_T}{A} = \frac{Q}{AV} \quad \text{farads per cm}^2 \tag{5-18}$$

where A represents, in this case, 1 cm².

To relate the general definition of capacitance to the charge density, we consider a volume bounded by two planes of unit area A separated by a distance x. If the charge density is uniform and equal to $\rho(x)$, the total charge Q of the system is related to the charge density by

$$Q = \rho(x)Ax \tag{5-19}$$

The field between the planes is given by

$$\mathcal{E} = \frac{V}{x} \tag{5-20}$$

and the specific capacitance C is

$$C = \frac{C_T}{A} = \frac{Q}{AV} = \frac{A\rho(x)x}{A\mathcal{E}x} = \frac{\rho(x)}{\mathcal{E}} \tag{5-21}$$

Since, from Coulomb's law,

$$\mathcal{E} = \frac{Q}{A\epsilon} \tag{5-22}$$

Eq. (5-21) becomes, where $\rho(x) = Q/Ax$,

$$C = \frac{\rho(x)\epsilon A}{Q} = \frac{Q\epsilon A}{QAx} = \frac{\epsilon}{x} \quad \text{farads per cm}^{2*} \tag{5-23}$$

* The case of the parallel-plate capacitor is treated in the same way except that the total charge is concentrated on two plates, of area A, separated by a distance x. Then,

$$C = \frac{Q}{VA} \tag{a}$$

and from Eqs. (5-20) and (5-22)

$$\mathcal{E} = \frac{Q}{A\epsilon} = \frac{V}{x} \tag{b}$$

Then

$$V = \frac{Qx}{A\epsilon} \tag{c}$$

and

$$C = \frac{\epsilon}{x} \quad \text{farads per cm}^2 \tag{d}$$

In terms of the dielectric constant, Eq. (5-23) becomes

$$C = \frac{\kappa \epsilon_0}{x_T} \quad \text{farads per cm}^2 \tag{5-24}$$

where x_T is the total width of the junction, or depletion region, taken across the boundaries of the space-charge volumes, as in the preceding discussion. Equation (5-23) shows the origin of the units (farads per cm) in which ϵ is normally expressed.

Since we know the value of ϵ_0, we may, if we assume that silicon is nearly intrinsic throughout a junction, use the value of 12 for the dielectric constant κ. To calculate the junction capacitance, we need an expression for x_T in terms of the parameters which control the width.

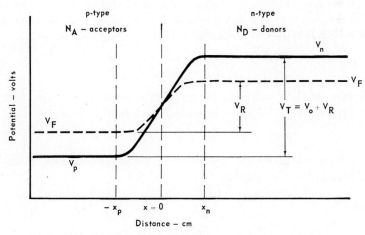

FIG. 5-6. Potential-distance plot for reverse-biased junction.

We have already seen that junction width depends on the doping levels of the p- and n-type material from which the junction was made, the concentration gradient as a function of distance x through the junction, and the potential difference, which will be the net sum of the barrier potential V_0 and the applied bias. (For the present, we shall restrict our discussion to the case of reverse bias, or an applied voltage which reinforces the barrier potential and results in depletion layer widening.)

The situation may be represented schematically as shown in Fig. 5-6. In this case the potential differences, in volts, are plotted vertically instead of the more customary energy levels, since we are concerned with barrier height and reverse bias, both usually expressed in volts. Note that the Fermi voltage V_F has been shifted at the junction by an amount V_R, corresponding to the reverse bias. For the sake of symmetry, the original barrier height V_0 is split and appears as two voltage increments whose sum is equal to $V_T - V_R$. The total potential difference at the junction is V_T, which equals $V_0 + V_R$ and also $|V_n - V_p|$, where V_n and V_p are the potentials (in volts) of the n region and p region, respectively, referred to any convenient common reference level such as the top of the valence band. Note also that, in the absence of a bias voltage, $|V_n - V_p|$ would simply equal V_0, the barrier height.

Figure 5-6 does not attempt to represent any particular gradient of impurity concentration in the junction. The potential curves are shown varying smoothly, as a compromise between a linear and a step configuration. If we now assume a step junction and identify x_p and x_n as the limit of penetration of the depletion layer into the p- and n-type material, respectively, we see that the junction width is given as

$$x_T = |x_p| + x_n \tag{5-25}$$

For a step junction we may also write

$$\rho(x) = -qN_A \quad \text{for } -x_p < x < 0 \tag{5-26}$$
and
$$\rho(x) = qN_D \quad \text{for } 0 < x < x_n \tag{5-27}$$

which express the conditions for an abrupt transition at the midpoint of the junction. Outside the junction, the following must be true:

$$\rho(x) = 0 \quad \text{for } x < -x_p \text{ and } x > x_n \tag{5-28}$$

To obtain values for the potential V, we substitute Eqs. (5-26) and (5-27) into Poisson's equation

$$\frac{d^2 V_p}{dx^2} = \frac{qN_A}{\kappa\epsilon_0} \qquad p \text{ region} \tag{5-29}$$

$$\frac{d^2 V_n}{dx^2} = -\frac{qN_D}{\kappa\epsilon_0} \qquad n \text{ region} \tag{5-30}$$

By integrating twice and introducing constants of integration, we get

$$\frac{dV_p}{dx} = \frac{qN_A x}{\kappa\epsilon_0} + k_p \tag{5-31}$$

$$V_p = \frac{qN_A x^2}{2\kappa\epsilon_0} + k_p x + k'_p \tag{5-32}$$

and
$$\frac{dV_n}{dx} = -\frac{qN_D x}{\kappa\epsilon_0} + k_n \tag{5-33}$$

$$V_n = -\frac{qN_D x^2}{2\kappa\epsilon_0} + k_n x + k'_n \tag{5-34}$$

If $x = 0$, $V_p = V_n$, since voltage is continuous and therefore $k'_n = k'_p$. Furthermore, from the first integrations, giving dV/dx, we know that at $x = 0$ the field is continuous and, therefore, $k_n = k_p$.

Finally at $x < -x_p$ and $x > x_n$, the fields must be zero (outside the depletion layer boundary). We may therefore write, from Eqs. (5-31) and (5-33),

$$k_p = k_n = \frac{qN_A x_p}{\kappa\epsilon_0} = \frac{qN_D x_n}{\kappa\epsilon_0} \qquad (5\text{-}35)$$

and
$$N_A x_p = N_D x_n \qquad (5\text{-}36)$$

an obvious statement of charge equality.

If we now use Eq. (5-35) to obtain values for k_p' and k_n' in Eqs. (5-32) and (5-34), we have

$$V_p = \frac{qN_A}{2\kappa\epsilon_0} x^2 + \frac{qN_A x_p}{\kappa\epsilon_0} x + k_p' \qquad (5\text{-}37)$$

and
$$V_n = -\frac{qN_D}{2\kappa\epsilon_0} x^2 + \frac{qN_D x_n}{\kappa\epsilon_0} x + k_n' \qquad (5\text{-}38)$$

Subtracting, to obtain V_T, the total potential difference from $-x_p$ to x_n, gives

$$V_T = V_0 + V_R = |V_n - V_p| = \frac{q}{2\kappa\epsilon_0}(N_D x_n^2 + N_A x_p^2) \qquad (5\text{-}39)$$

and from Eq. (5-36), $N_A = N_D(x_n/x_p)$. Therefore

$$V_T = \frac{qN_D}{2\kappa\epsilon_0} x_n(x_n + x_p) \qquad (5\text{-}40)$$

As we have seen, in a useful junction $N_A \gg N_D$ and therefore $x_n \gg x_p$. Therefore,

$$x_T = x_n + |x_p| \approx x_n \qquad (5\text{-}41)$$

and
$$x_T = \sqrt{\frac{2\kappa\epsilon_0 V_T}{qN_D}} \quad \text{cm} \quad n\text{-type} \qquad (5\text{-}42)$$

Similarly,
$$x_T = \sqrt{\frac{2\kappa\epsilon_0 V_T}{qN_A}} \quad \text{cm} \quad p\text{-type} \qquad (5\text{-}43)$$

Finally, by substitution in Eq. (5-24),

$$C = \sqrt{\frac{q\kappa\epsilon_0 N_D}{2V_T}} \quad \text{farads per cm}^2 \quad n\text{-type} \qquad (5\text{-}44)$$

and
$$C = \sqrt{\frac{q\kappa\epsilon_0 N_A}{2V_T}} \quad \text{farads per cm}^2 \quad p\text{-type} \qquad (5\text{-}45)$$

where the correct expression to use depends on the condition of inequality used to obtain $x_T = x_n$ in Eq. (5-41). If the n-type region is of higher resistivity, x_T approaches x_n and Eq. (5-44) holds. If the p-type region is of high resistivity, Eq. (5-45) will govern. Thus it has been shown that for a step junction the capacitance per unit area varies as $V_T^{-\frac{1}{2}}$, where $V_T \geq V_0$. It can also be shown that the capacitance decreases as the square root of the resistivity.

Analogous reasoning for the grown or diffused junction, whose concentration gradient may be represented as a linear function in x, shows that the capacitance per unit area varies as $V_T^{-1/3}$, thus suggesting a means of investigating the structure of the junction by observing the change of capacitance with changing reverse bias. Unfortunately, with increasing voltage the difference disappears. However, the range of most interest is usually that in which the voltage dependence is changing, and detailed analyses of the junction structure are available. Note that this dependence on voltage arises from the variation in junction width, and it is not observed for a classic fixed-plate capacitor.

If the junction is biased in the forward direction $(V_T < V_0)$, there is a sharp increase in capacitance, owing to the decrease in the voltage and therefore the narrowing of the depletion layer as well as the flow of carriers, which invalidates the above model based on impurity ions alone. Analysis of this case is not within the scope of this effort, but the effect of the rapid rise in capacitance is an important consideration in the high-frequency behavior of devices.

Finally, it should be noted that the calculation of junction width, deferred until this discussion, follows directly from Eqs. (5-42) and (5-43) depending on the type of the high-resistivity portion of the junction. Such a calculation is valid for values of $V_T \geq V_0$ up to a nonequilibrium condition of reverse bias.

5-7. THE EFFECT OF TEMPERATURE ON A p-n JUNCTION

As previously noted, when semiconductors are heated, they ultimately reach a temperature at which they cease to function as semiconductors and behave more like metals. Silicon approaches this intrinsic condition at about 150°C. At all higher temperatures the impurity contribution to the conductivity is trivial, since it is overwhelmed by thermally generated hole-electron pairs. On the other hand, from -70 to 150°C the conductivity of silicon is dominated by its electrically active impurities and the intrinsic contribution is negligible.

If we examine a p-n junction in the light of the above facts, we can easily predict its behavior with increasing temperature. Since a junction consists of an intrinsic region bounded by two highly concentrated space-charge regions which are due to the impurity ions and not the silicon itself, we may treat the entire junction as a single-impurity semiconductor. At about the same temperature as for a silicon sample having no junction, the thermal generation of hole-electron pairs will become controlling. The space-charge regions and the intrinsic layer will disappear, since the thermal generation of hole-electron pairs will provide excess carriers which will rapidly neutralize the ions and the entire junction will become conducting to the same degree as the bulk material. Under these conditions the device has degenerated into a current carrier.

The foregoing paragraph describes the situation when the temperature is well into the intrinsic range, the extreme case. At lower temperatures, as the temperature is increased, junction devices gradually lose efficiency even though impurity conductivity still dominates. This is due to two thermal effects: (1) the statistical increase in the energies of the electrons (and holes) in the conduction and valence bands, which permits more electrons (and holes) to surmount the potential barrier, and (2) the increased generation of hole-electron pairs within the region of the junction, which also contributes to current flow. These effects lead to greatly increased reverse current in a rectifier and current leakage in switching devices, resulting in decreased efficiency in both cases. All semiconductor devices employing a junction are subject to this temperature limitation.

One of the principal reasons for the pronounced interest in silicon is its performance in devices which can operate at temperatures substantially above room temperature and still retain their efficiency. It is possible to design devices for high-temperature operation only if the semiconductor itself does not become intrinsic or near-intrinsic. The only other large-volume commercial elemental semiconductor, germanium, begins to become intrinsic at about 75°C. Many silicon devices function at very nearly top efficiency at temperatures at which the corresponding germanium devices fail to function at all.

5-8. BREAKDOWN IN p-n JUNCTIONS

There is another important aspect of p-n junctions in semiconductors which will contribute to our comprehension of useful junction applications. The nonequilibrium condition which results from increasing reverse bias and is manifested by a sharp increase in reverse current is known as *breakdown*. Later we shall see that there are cases in which the onset of this behavior is a limitation on the usefulness of a device, while in other cases the phenomenon is exploited to make voltage regulators. In all cases the current becomes many orders of magnitude higher than the normal reverse or drift-current components I_g, and the effective ohmic resistance of the junction falls almost to zero with a relatively small increase in bias. (There is, of course, still the resistance of the p and n regions beyond the junction, which is unchanged and is still a part of the circuit in all practical cases.)

The mechanism of breakdown in junction devices is therefore of great practical importance and theoretical interest, since, for example, breakdown limits the reverse voltage that can be applied to rectifiers and transistors. In such practical situations, breakdown is usually considered to be the result of bulk properties as well as the properties of the junction; but most of the experimental work and much of the theory have concerned only junction behavior, since it dominates the overall picture.

For purposes of this discussion we shall consider only those phenomena within the depletion region. If we now consider the junction in terms of a parallel-plate capacitor, we can understand why breakdown occurs. For a unit area we know that increasing voltage will have two effects which tend to offset each other. The field $\mathcal{E}$, in the depletion region is defined as dV/dx, and we know that if V_T increases, the junction width also increases. However, if we refer to Eq. (5-16), we see that the barrier height is constant and is a function only of impurity concentrations (which in turn govern the space-charge density) for a given temperature. Therefore, since $V_T = V_0 + V_R$, as we have already seen, V_T increases directly with V_R for a given junction. The junction width, on the other hand, increases only as $V_T^{1/2}$ [Eq. (5-42)]. Therefore, since the charge density is unchanged, the net effect is an increase in $\mathcal{E}$ directly proportional to $V_R^{1/2}$. Obviously, at some point the dielectric in the capacitor will fail, and by analogy, we find the intrinsic junction region also breaking down.

Two kinds of breakdown in junctions were originally postulated. In both cases the theoretical approach was an effort to explain the detailed mechanism of high-voltage junction failure in applications such as rectifiers and transistors, where high-voltage capability was an index of merit. Although it has since been learned that only one of the mechanisms applies to this group of voltage breakdowns in actual practice, the other explanation cannot be dismissed, since there is another broad region in which it may well be significant, as will be seen later in the discussion of zener diodes. We shall therefore look first at these models starting with field emission, or zener breakdown [11,12], which was first believed to be the correct physical picture.

A mechanism based on internal field emission was originally called zener breakdown because of the early work of Dr. Clarence Zener in this area. The phenomenon is analogous to cold field emission from metals. This postulates a spontaneous rupture of covalent bonds involving normal lattice atoms in the immediate vicinity of the junction as a consequence of the high field (both barrier potential and reverse bias across the junction depletion width). This disruption of bonds produces hole-electron pairs in large numbers which increase the equilibrium currents due to thermally generated carriers (I_g components). The current thus generated becomes very large; it is limited only by external circuit resistance, as in a gas discharge. For silicon the fields required are of the order of 1.5×10^6 volts per cm, which are theoretically quite reasonable for highly doped junctions. The Zener effect was therefore considered to be the principal one involved in the breakdown of semiconductor diodes.

However, the zener mechanism implies a critical field and should therefore result in the breakdown voltage being proportional to the volume resistivity, but this is not observed. Also, predictions concerning the slope of the breakdown portion of the current-voltage reverse curve are infrequently fulfilled. Experimentally, simple zener behavior according to the classic model is rarely seen. The existence of this breakdown mechanism is not really in question. It can, and undoubtedly does, exist, but there are other mechanisms which obscure zener behavior.

An alternative explanation identifies avalanche breakdown [3,13], which usually occurs in high-voltage cases before the Zener effect can become operative. Avalanche breakdown is believed to be responsible for practically all such failures (except possibly in very thin junctions). The avalanche mechanism is analogous to the Townsend breakdown in a gas tube, where the field becomes high enough to cause current-carrying electrons to ionize the gas molecules by collision, producing more ions and electrons in a multiplication, or chain reaction, sequence.

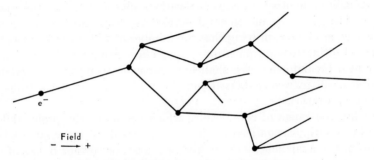

Fig. 5-7. Ionization in avalanche breakdown.

Figure 5-7 pictures the avalanche effect schematically. In this diagram we consider the electrons and holes comprising the saturation current of a diode, and it is conjectured that these electrons become accelerated to higher and higher drift velocities as the voltage across the diode increases. Eventually, these electrons possess enough energy to ionize other carriers; that is, they collide with electrons, for example, bound to the silicon atoms and impart enough energy to free them from the parent atom, creating new electron-hole pairs. These electrons in turn may be accelerated in the field to ionize other pairs. This cumulative process leads to an extremely rapid increase of current when the voltage approaches values at which the multiplication takes place.

It should be mentioned that the junctions to which the preceding discussion originally applied were designed for higher voltages. They were therefore made of relatively pure material, so that substantial reverse voltages were sustained prior to breakdown. In addition, it is chiefly in such structures that simple avalanche breakdown is observed. Lower voltages and less pure materials result in different mechanisms, which are less easily defined than high-voltage avalanche.

The characteristics of avalanche breakdown are fortunately rather distinctive, and it is not difficult to identify this behavior. We have already mentioned that the process is to be expected for high-voltage junctions made of relatively pure materials. In a quantitative sense these terms may be taken to mean voltages above the 20- to 50-volt range and doping levels well under the degenerate range, that is, less than 10^{17} impurities per cm^3. Other distinguishing behavior involves reverse-current multiplication, prebreakdown noise, and visible radiation [14].

Reverse-current multiplication before avalanche goes out of control is a logical consequence of the mechanism described above (Fig. 5-7). This means that there will be an increase in the reverse current (exceeding the normal I_g currents) as a function of bias before breakdown. Experimentally [13], this is measured by injecting a constant number of carriers (electrically or photoelectrically) at various bias levels. As the reverse voltage is raised, the incremental current increase from each injection pulse becomes larger, proving that the number of added carriers has been increased owing to multiplication. This effect cannot be seen in a true zener field breakdown. We shall have occasion to return to avalanche multiplication in reverse-biased junctions when we consider transistor current-gain ratios.

Finally, the range of reverse bias which results in the prebreakdown effects of multiplication also produces two secondary, but characteristic, indications of avalanche. Prebreakdown noise, or current pulses of constant amplitude occurring at voltages just below the critical value, result from ionizing events due to a few high-energy electrons which cause momentary small and nonsustaining pulses. Furthermore, in a darkened room, if the external edge of the junction is exposed, tiny sparks, or microplasmas, will often be observed at points of surface lattice imperfection. These occur at the same time as some of the pulses due to noise, and they are part of that behavior.

No equivalent noise or microplasma is observed for zener or other types of breakdown. In some cases the junction may emit visible radiation, but from a large area and continuously. The effects outlined above are therefore quite sufficient to identify avalanche breakdown whenever it occurs.

Avalanche-type breakdown is used in some devices, which will be described later, to perform switching actions. In these cases the breakdown must be reversible and not damage the junction. In most other structures full breakdown is more or less catastrophic, and permanent damage may result. In these cases the prebreakdown current noise is a useful test to establish the voltage rating of such a unit without the chance of destruction.

We have now considered the avalanche and field emission or Zener interpretations of reverse breakdown in *p-n* junctions. From a practical point of view, the avalanche model covers a very large fraction of normal device behavior, while the true Zener effect is rarely, if ever, identifiable in these junctions. In the discussion thus far, the breakdown effect has been an undesirable limitation on the overall performance of semiconductor structures such as rectifiers and transistors. (The avalanche switching device is, of course, an exception.)

On the other hand, the potentialities of controlled breakdown as a voltage-regulating or -limiting function was recognized early. To be practical, such junctions must be capable of breakdown and recovery for a very large number of cycles. Also, the breakdown current flowing in the reverse direction must be limited either within the device itself (by volume resistivity of material outside the junction region) or by limiting series resistances. Furthermore, such limit devices could be useful over a very wide voltage range—from a volt (or even less) to hundreds of volts. For such applications the term "breakdown" no longer carries the implication of a limitation or shortcoming of a junction, but becomes a controlled property around which the device is designed.

Commercial voltage-limiting diodes, known as zener diodes, became available early in the development of device technology. The name "zener diode" was originally used because this explanation of breakdown was believed to be universally correct. Today the term is still in use to cover the breakdown mechanism. The range and properties of zener diodes have been extended from the relatively pure semiconductor junctions considered above, where the avalanche mechanism is the governing factor, down to highly doped degenerate structures where a third, entirely different mechanism is postulated. Throughout the range, problems of permanent damage or irreversibility, current limitation, and reproducibility of the breakdown voltage have been overcome by redesign and very careful control of fabricating techniques.

This technology has advanced to a high degree of sophistication, so that not only is the reverse breakdown a controlled constant, but the current-voltage characteristic is also subject to adjustment. It is therefore possible to pass from a high resistance to an almost complete short circuit abruptly, with very small change of reverse bias, or to proceed gradually through a transition region, offering the possibility of voltage regulation.

Diodes prepared for such applications in the low-voltage range (less than about 5 volts) are believed by some to be controlled by an entirely different breakdown mechanism known as tunneling. Briefly, this is a phenomenon whereby charge carriers overcome a potential barrier between states of equal energy. In other words, if electrons on one side of the barrier are at the same energy as empty states on the far side, a quantum-statistical fraction of these electrons will appear in the empty states per unit time at any given temperature. Such effects become appreciable when the doping levels are high, resulting in narrow junctions, high space-charge densities, and a plentiful supply of carriers. All of these increase the probability of tunneling and therefore, with increased bias, large current flow [15,16].

The physical picture of this process is rather difficult, since there is no classical equivalent to the appearance of electrons on the far side of the barrier without an increase in electron energy. Ordinary thermal excitation cannot account for the magnitude of the tunneling currents. No attempt will therefore be made to derive explicit expressions for this behavior. For present purposes, the experimental evidence will suffice, although we shall return to the subject of tunneling from the point of view of band energy systems in a later chapter. Meanwhile, we shall have occasion to concern ourselves with only one breakdown mechanism of the three described in this section, since the avalanche mode is the principal mechanism seen in the great majority of useful device types.

REFERENCES

1. W. Shockley: Transistor Electronics: Imperfections, Unipolar and Analog Transistors, *Proc. IRE*, **40**: 1303–1311 (1952).
2. R. D. Middlebrook: "Introduction to Junction Transistor Theory," pp. 86–92, John Wiley & Sons, Inc., New York, 1957.
3. K. G. McKay: Avalanche Breakdown in Silicon, *Phys. Rev.*, **94**: 877 (1954).
4. N. B. Hannay: Semiconductor Principles, in N. B. Hannay (ed.), "Semiconductors," p. 46, Reinhold Publishing Corporation, New York, 1959.
5. J. N. Shive: "The Properties, Physics and Design of Semiconductor Devices," p. 473, D. Van Nostrand Company, Inc., Princeton, N.J., 1959.
6. A. Nussbaum: "Semiconductor Device Physics," pp. 67, 68, Prentice-Hall, Inc., Englewood Cliffs, N.J., 1962.
7. E. Spenke: "Electronic Semiconductors," pp. 292ff., McGraw-Hill Book Company, New York, 1958.
8. Ref. 5, p. 347.
9. A. B. Phillips: "Transistor Engineering," p. 96, McGraw-Hill Book Company, New York, 1962.
10. Ref. 9, pp. 108–124.
11. C. Zener: A Theory of the Electrical Breakdown of Solid Dielectrics, *Proc. Roy. Soc. (London)*, **A145**: 523 (1934).
12. K. B. McAfee, E. J. Ryder, W. Shockley, and M. Sparks: Observations of Zener Current in Germanium p-n Junctions, *Phys. Rev.*, **83**: 650, 651 (1951).
13. K. G. McKay and K. B. McAfee: Electron Multiplication in Silicon and Germanium, *Phys. Rev.*, **91**: 1079–1084 (1953).
14. L. B. Valdes: "The Physical Theory of Transistors," p. 204, McGraw-Hill Book Company, New York, 1961.
15. J. R. Tillman and F. F. Roberts: "An Introduction to the Theory and Practice of Transistors," pp. 120, 121, John Wiley & Sons, Inc., New York, 1961.
16. R. P. Nanavati: "An Introduction to Semiconductor Electronics," p. 111, McGraw-Hill Book Company, New York, 1963.

6: Biased p-n Junctions

6-1. BIASED SYMMETRICAL JUNCTION (NO INJECTION)

We have discussed the formation, structure, and properties of a p-n junction in silicon, its behavior at room temperature and elevated temperatures, and the phenomenon of breakdown. We shall now consider junction behavior under operating forward and reverse bias. The junction is said to be biased in the forward direction when the n-type region is connected to the negative terminal of a battery. Under these conditions electrons will tend to flow from the negative terminal of the battery into the n-type region of the junction. Conversely, holes will flow from the positive terminal of the battery into the p-type region. When junctions are reverse-biased, electrons flow into the p region and holes flow into the n region. It will be recalled that holes or vacancies move in the direction opposite to electron flow, which is a convenient concept in explaining current flow in junction devices. The same result would be achieved if only electrons were considered.

In the following sections we shall consider symmetrical junctions in which both the p- and n-type bulk regions contain silicon of relatively high purity (less than about 25 apb net impurity). In junctions in which one or both bulk regions contain silicon of lower purity (in the range of about 100 apb) a phenomenon known as *injection* may take place. This refers to a greatly increased flow of one or both charge carriers across the junction, and it is a nonequilibrium situation. Discussion of these effects will be deferred to a later section, and for the present we shall consider the simpler junction.

Figure 6-1 is a schematic representation of current flow in a p-n junction at thermal equilibrium and with forward and reverse bias applied. Both bulk regions are of high purity. A close inspection of this diagram confirms one of the most important properties of all p-n junctions in semiconductors, namely, that they act as rectifiers because they permit much larger currents to flow in the forward than in the reverse direction. In Fig. 5-4 we saw that there is no net current flow on the part of either electrons or holes in the absence of an applied voltage. The equal and opposite flows of electrons and holes under these circumstances represent very minute currents, and electrical neutrality is preserved at all times. In the forward direction, as shown in Fig. 6-1, both electrons and holes are crossing the junction region under the influence of a forward voltage from

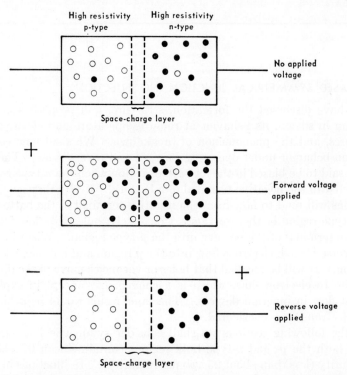

FIG. 6-1. Schematic diagram of symmetrical *p-n* junction under bias. (*After W. C. Dunlap, by permission* [1].)

a region of high concentration to a region of low concentration. In the reverse direction, both species of carriers are being pulled away from the junction region and one might assume, therefore, that no current flows in the circuit. This assumption is not quantitatively true, however, since there are secondary effects, already observed in the junction at thermal equilibrium, which give rise to small but important reverse currents. In the next two sections we shall explain with the aid of band diagrams the various current flows which take place under forward and reverse applied voltages.

6-2. SYMMETRICAL p-n JUNCTION UNDER REVERSE BIAS

When a *p-n* junction is subjected to reverse bias, there is a tendency for the mobile majority carriers to be attracted away from the junction so that the equilibrium space-charge barrier is, in effect, augmented. The minority carriers will, of course, be attracted by the bias field but will be prevented from reaching the junction by the space charge. When we considered the current flows in the case of the unbiased junction, we observed a small but continuous drift current of minority carriers arising

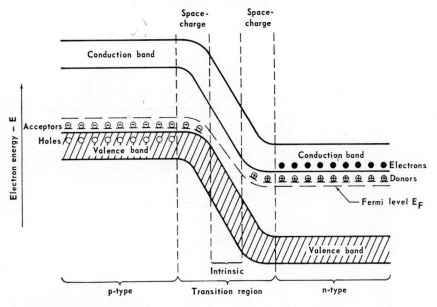

FIG. 6-2. Detailed band energy structure for a symmetrical reverse-biased *p-n* junction in thermal equilibrium.

from thermal generation in the junction region. This was nullified by an equal and opposite diffusion current of each type of carrier made up of majority carriers of high instantaneous statistical energy. The situation for the reverse-biased junction may be explained on the basis of these same currents.

Consider the band energy diagram for a reverse-biased junction shown in Fig. 6-2. The height of the potential hill has been increased by an amount which is proportional to the applied voltage. The Fermi level is no longer horizontal and reflects the nonequilibrium condition imposed by the applied potential. It is therefore more difficult for a hole in the *p*-type region or an electron in the *n*-type region to acquire the necessary energy by thermal excitation to surmount the potential barrier at the junction. This results in a sharp reduction in the forward diffusion current due to majority carriers.

On the other hand, minority carrier flow, which arises from the continual thermal generation of hole-electron pairs, remains unchanged if the temperature of the sample is the same as that in our example at thermal equilibrium under no bias. Since the minority carriers flow downhill under the influence of the potential barrier, and since their number is a

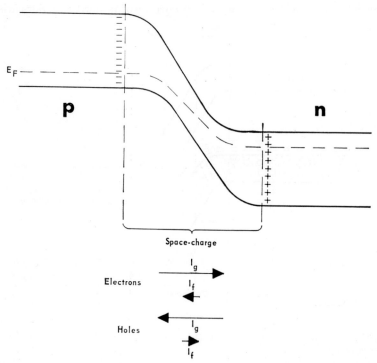

Fig. 6-3. Equilibrium currents in a reverse-biased symmetrical *p-n* junction. (*After N. B. Hannay, by permission* [2].)

function of temperature only, the current I_g carried by them is unaffected by the reverse bias. Shown schematically and by analogy with Fig. 5-4, this situation is represented in Fig. 6-3.

The arrows corresponding to the forward current flow have been shortened owing to the increased potential barrier, whereas the arrows representing the minority reverse current I_g remain unchanged in spite of the applied voltage. Since the minority carrier drift current is independent of the applied voltage, the reverse current of any biased *p-n* junction is a constant value and is unchanged with increasing bias, up to the point of breakdown, as previously discussed.

6-3. SYMMETRICAL *p-n* JUNCTION UNDER FORWARD BIAS

We have discussed the *p-n* junction under reverse bias when almost no current flows and when it therefore acts like a very high resistance or like a rectifier on the reverse half cycle. By analogous reasoning we shall now consider the *p-n* junction under forward bias, when we shall expect a great increase in current flow. Figure 6-4 shows the band diagram of a *p-n* junction under forward bias. The height of the potential barrier has been decreased, and we now find the Fermi level raised in the *n*-type region. This is a nonequilibrium situation attributable to the applied

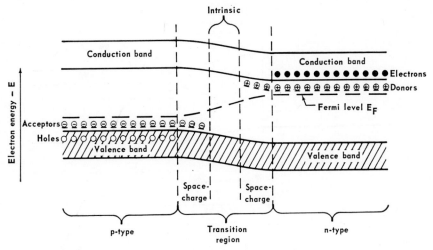

FIG. 6-4. Detailed band energy structure for a symmetrical forward-biased *p-n* junction in thermal equilibrium.

field. As in the discussion of reverse bias above, we shall hold the temperature constant.

Once again we shall consider the flow of minority and majority carriers. Figure 6-5 may now be compared with Fig. 5-4, the unbiased junction at thermal equilibrium. With the lowering of the potential barrier due to the forward bias it has become much easier for the majority carriers to surmount the barrier as a result of thermal activation. This is shown by a lengthening of the arrows indicating the amount of diffusion current carried by the majority carriers in the forward direction.

As before, with the temperature of the system unchanged, the rate of generation of hole-electron pairs remains constant. Since these minority carrier currents are independent of potential, they remain unchanged for the forward-biased junction as well. The sum of these current components is therefore greatly increased over the equilibrium value, and the apparent resistance of the junction diminishes.

As in the case of the equilibrium junction, the majority carriers move by diffusion under the influence of a concentration gradient, and not in proportion to the applied voltage. The forward bias simply lowers the level of energy they must attain by thermal excitation so that they may diffuse past the junction. The diffusional mechanism for the flow of majority carriers is the same in all cases.

The foregoing statements concerning diffusional flow of majority carriers, the drift flow of thermally generated minority carriers, and the relationship between these flows and the applied field or bias are valid for the special symmetrical type of *p-n* junction we are here considering. It

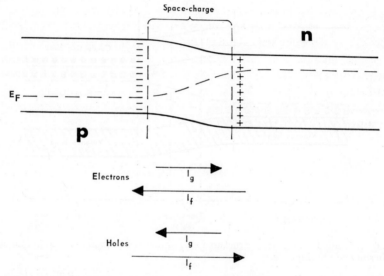

Fig. 6-5. Equilibrium currents in a forward-biased symmetrical *p-n* junction. (*After* N. B. Hannay, by permission [2].)

should be recalled that this junction is composed of high-purity materials in both regions and is subject to a forward bias of only about one volt and reverse biases within the range of equilibrium behavior. Later we shall observe the behavior of junctions composed, at least in part, of materials having higher impurity concentrations and biased to a greater degree. (In the case of reverse breakdown of a junction under a sufficiently high field we have already seen that the current at breakdown can be sharply influenced by small changes in bias.)

6-4. THE RECTIFIER EQUATION

We have now considered the equilibrium behavior of a *p-n* junction under forward and reverse bias. We have assumed that the bulk material on each side of the junction is highly purified. We have restricted ourselves to relatively small applied potentials and imposed the condition of thermal equilibrium at room temperature for all examples. Under these conditions, we have observed that current flow in the forward case is the direct result of increased diffusion at the junction and is not linearly related to the voltages which are applied.

In the case of reverse bias the barrier height effectively stopped the diffusion of majority carriers but had no influence on the drift of minority carriers. Since the number of minority carriers resulting from thermal generation of hole-electron pairs is a constant for a given temperature and since the height of the potential drop through which they fall does not affect the current produced by them, we find that the reverse current through the junction quickly reaches a constant value known as the saturation current. This saturation current remains constant under increasing negative bias until breakdown in the junction occurs. The voltage corresponding to this breakdown is known as the *peak inverse voltage*, or *PIV*.

All the current-carrying properties of a *p-n* junction, except reverse breakdown, may be mathematically expressed by a simple equation known as the rectifier equation:

$$I = I_s \left[\exp\left(\frac{qV}{kT}\right) - 1 \right] \qquad (6\text{-}1)$$

where the total current I in either the forward or reverse direction is expressed in terms of the reverse saturation current I_s, the charge on the electron q, the applied bias voltage V, Boltzmann's constant k, and the absolute temperature T, in degrees Kelvin. This function is plotted as Fig. 6-6a, where the solid lines correspond to the actual values derived from the rectifier equation with the exception of the line labeled "body resistance limiting." At the point where this line intercepts the plot of the rectifier equation, a dotted transition curve has been shown. In actual practice the forward voltage drop as a function of current will, at this point, depart from the rectifier equation and be governed entirely by the bulk resistance of the silicon outside the junction region. The dotted line shown at a reverse bias of about 150 volts illustrates the reverse breakdown of a typical *p-n* junction. In Fig. 6-6b similar data for an actual rectifier have been plotted in a more conventional way. Note, however, that the scales on the ordinate and abscissa have changed markedly on passing the origin.

Inspection of these plots of the rectifier equation permits us to comment on some of the problems facing the device manufacturer. In the forward direction it is desirable to pass as large a current as possible with a minimum voltage across the junction. If the body resistance becomes limiting and the voltage drop increases steeply as shown, there will be power loss in the device resulting not only in lower efficiency but in the generation of heat which, if not removed, can soon destroy the rectifying property of the *p-n* junction. Removing this heat can often impose serious design problems. To minimize this effect, the device maker endeavors to

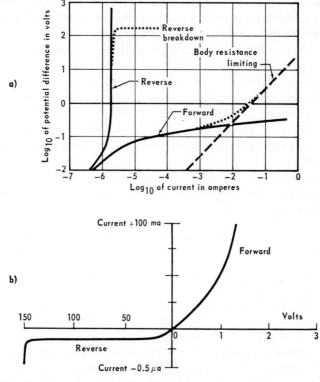

Fig. 6-6. Rectifier characteristic for symmetrical *p-n* junction with small signal. (*After J. N. Shive, courtesy of D. Van Nostrand Company, Inc.* [3].)

increase the bulk conductivity by doping, without impairing other desirable characteristics such as peak inverse voltage and lifetime of minority carriers, and to make the current path as short as possible by reducing the thickness of the bulk layer between the junction and the forward contact.

On the other hand, a high peak inverse voltage is desirable if the rectifier is to handle large quantities of power at reasonable currents. To achieve a high peak inverse voltage, the junction must be theoretically perfect, heating must be held to a minimum, and highly uniform high-resistivity materials should be used on each side of the junction. It is apparent that the device maker is forced to compromise in his design and utilize all his skill in fabrication.

6-5. UNSYMMETRICAL JUNCTIONS WITH INJECTION

Up to this point we have confined ourselves to a consideration of p-n junctions formed between high-resistivity bulk regions. Under these conditions the p-n junction cannot handle significant amounts of power per unit area without generating heat and otherwise damaging the structure. Devices fabricated with such symmetrical junctions belong to the group known as *small-signal varieties*. While they have usefulness in

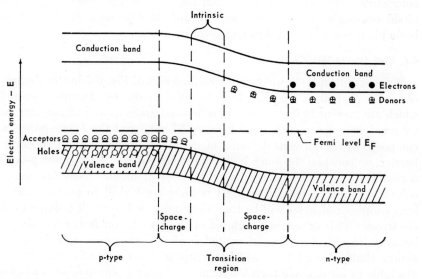

Fig. 6-7. Detailed band energy structure for an unsymmetrically doped, unbiased p-n junction in thermal equilibrium.

certain applications, their ability to handle higher current is so limited that it has been necessary to modify them by the process known as injection.

An emitter or injection type of junction differs from the junctions we have been considering in that one side is heavily doped with respect to the other. A typical example of such an unsymmetrical configuration would involve a junction between 20-ohm-cm n-type silicon and 0.2-ohm-cm p-type silicon, as shown schematically in Fig. 6-7. It will be noticed that the Fermi level in this junction, shown with no applied bias, is much nearer the acceptor level than the donor level. The space-charge region on the heavily doped p-type side is narrower than on the n-type side because of the higher concentration of ionized impurities.

86 Semiconductor Junctions and Devices

Because of the practical utility of the unsymmetrical or emitter junction, which is said to inject mobile carriers across the barrier, it is of interest to examine it under conditions of zero, forward, and reverse bias, as was done for the symmetrical small-signal case. Unfortunately, the configuration does not yield to a simple exact analysis as was possible in our previous example. In addition, when under forward bias, a nonequilibrium condition exists, which renders the calculation still more difficult. We shall therefore confine ourselves to a physical picture which cannot be explicitly interpreted in theoretical terms but which will serve to introduce the principles of emitter junction operation and its current-voltage characteristics. In this chapter we shall apply these considerations to the diode or single-junction rectifier. It will later be seen that the same basic phenomena are vital to transistor operation.

6-6. THE UNBIASED CASE

We first consider the unsymmetrical junction of Fig. 6-7 in the absence of bias. As before, we shall be interested in the equilibrium currents which are present in the absence of a thermal gradient and whose sum, as we previously discovered, must be zero. Looking at the I_f components, composed of the forward diffusion of majority carriers across the energy barrier, we consider the much higher impurity concentration in the p-type region and the consequently larger number of mobile holes (assuming complete impurity ionization). It is therefore logical to assume that a larger number of holes will be thermally activated per unit volume than electrons. This is another example of the population factor mentioned before. The total number of excited carriers is the product of the probability that a higher level will be occupied and the number of carriers available to fill it. We therefore expect a larger I_f component for holes than for electrons in this case. Figure 6-8 shows an example of the unsymmetrical situation in a familiar format.

If the above argument with respect to the I_f components is valid, we must now establish an equilibrium condition of zero net current flow for the unbiased junction at thermal equilibrium by adjustment of the I_g components. We know, however, that the source of the mobile carriers which make up these drift currents is the thermal generation of hole-electron pairs in the junction zone. Obviously, as many holes as electrons are thermally generated, and the available mobile carriers for the two I_g current components are equal.

To justify the postulated difference between the hole drift component and the electron drift component, we shall introduce the concept of *collector efficiency*. This refers to the ability of a junction, under reverse field or bias, to "collect" minority carriers which are driven to it from a region beyond the junction and to add them to the majority carriers already present. In our previous discussion of junction currents at equilibrium we were dealing with a symmetrical junction, and it was a reasonable assumption that the collector efficiencies were equal for both the n- and p-type regions adjacent to the space-charge layers, regardless of their absolute values.

In the present unsymmetrical case we postulate a difference in this efficiency based on the unsymmetrical field and space-charge distribution. The highly impure p-type region has a narrow negative space-charge layer of very high concentration, so that a hole (minority carrier

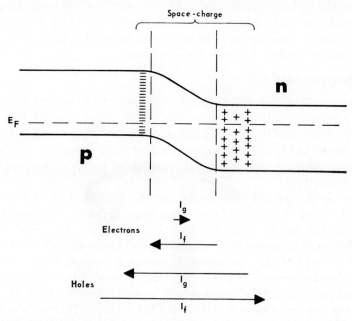

FIG. 6-8. Equilibrium currents in an unbiased, unsymmetrical p-n junction. (*After* N. B. Hannay, by permission [2].)

from the n-type material) is strongly and effectively attracted and collected. On the other side of the junction, however, the positive space charge is more diffuse and the attraction for an electron is smaller, with the result that some thermally generated electrons are lost to other processes (recombination with a hole, loss of energy due to collisions, etc.) before they are collected. Therefore, the I_g components are not equal, and their inequality balances out the I_f difference.

Finally, it should be mentioned that the length of the current component arrows in Fig. 6-8 is an indication of another property of unsymmetrical junctions. In the case at hand, with a highly doped p-type region, both I_f and I_g for holes are relatively large compared with those

88 Semiconductor Junctions and Devices

quantities for electrons, a fact of significance for the forward-biased case. Also, the I_g currents are small in comparison with the I_f components. This is logical, since the rate of thermal generation is normally small and a function of temperature only, whereas the diffusion currents are proportional to concentration, which in this case is much higher in the p-type region than our previous example.

It should also be apparent that the actual magnitude of the current components cannot be truly represented by arrows of nominal lengths, such as those of Fig. 6-8. Further, there is no clear-cut dividing line between the truly symmetrical case of our earlier example and the highly unsymmetrical case presented here. In practice, every junction device represents a compromise and can best be understood in terms of a physical picture somewhere between these extremes.

6-7. FORWARD BIAS

From the foregoing, it will be relatively easy to deduce the properties of an unsymmetrical junction under conditions of applied bias. We shall, however, find it necessary to introduce several new concepts, each of which will add to our physical picture and make it easier to understand single and multiple unsymmetrical junctions of both injecting (emitter) and collector types.

If an unsymmetrical junction having the configuration of Fig. 6-7 is forward-biased, there is, as expected, a large increase in I_f current components while the I_g components remain the same. In a typical case, for a difference in doping level of 100 to 1, the forward current will consist almost entirely of the one species of majority carriers, holes. In fact, better than 99 per cent of the current will be carried by holes across the junction into the lightly doped n-type region. Arriving here, the holes will be so numerous as to literally flood the region. This will result in a greatly increased conductivity in the n-type region near the junction, decreasing as the holes move away and are annihilated by electrons. This influx of holes creates a nonequilibrium condition exceeding the p-n product (since the holes have a finite life expectancy before random recombination destroys them) and creating additional positive space charge in the n-type region. To counter this imbalance, electrons from the external circuit flow into the n-type region, neutralizing the space charge and accelerating the recombination process. This effect, however, also results in a nonequilibrium situation in that the influx of electrons also exceeds the p-n product.

The net effect of this entire injection process under forward bias is to maintain a very high conductivity in the normally high resistivity n-type region, causing large currents to flow for a very modest forward bias. Without injection, the high-resistivity n-type region would behave much as it did in the symmetrical case, obeying the rectifier equation and introducing a limiting body resistance as the current increased. Finally, the heavily doped p-type region on the emitter side of the junction already offers very small resistance to the flow of holes (or electrons), so that the overall bulk resistivity of the junction and associated normal regions is reduced many times compared to the symmetrical case.

It is beyond the scope of this work to attempt an evaluation of all the theoretical and practical considerations which enter into the design of an emitter-type junction. The examples discussed are to be regarded as typical but not at all inclusive, and many variations will be found in practice. In many cases advantage is taken of otherwise undesirable effects to produce devices of special character and application. A qualitative sampling of some of these will be presented later.

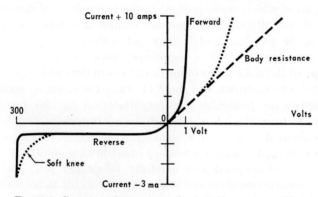

FIG. 6-9. Current-voltage curves for injection-type rectifier.

The current-voltage behavior of a forward-biased junction with injection is depicted in Fig. 6-9. Note the radical differences in the voltage scales for forward and reverse bias. In the forward case, which is a typical commercial power diode, we see that the current rises rapidly with slight increases in voltage until, at about one volt, it has reached the rated output of the unit (10 amp). For comparison, a typical small-signal forward characteristic is shown. (Note that the current scale is different from that of Fig. 6-6, with which it may be compared.) The significant advantage of injection in terms of forward drop and current-carrying ability may be clearly seen by inspecting the two forward current-voltage curves.

6-8. REVERSE BIAS

In the reverse-biased condition we encounter much the same situation as before. The barrier height increases and current components I_f fall to zero, despite the normally increased I_f corresponding to majority carrier hole flow from the low-resistivity p-type region. In practical cases the reverse bias required to wipe out the I_f due to the larger hole concentration is negligible compared to other effects. As before, there is a saturation current I_s made up of the components I_g. Once again, the magnitude of the reverse bias does not affect the carriers which create I_s, and, to a first approximation, one might expect I_s for the unsymmetrical case to be the same as I_s for the high-resistivity symmetrical junction. In both cases the current carriers are thermally generated at a rate dependent only on the temperature. Actually, this is not a bad assumption for most unsymmetrical junctions.

There are a number of secondary effects, such as bulk resistivity of the material making up the junction (often the lightly doped side is made as near intrinsic as practical), lifetime of minority carriers, and physical thickness, all of which are important in the technology of device fabrication. In almost all cases the necessary compromise represents the optimum economic solution, and it may not realize the highest potential performance. However, it is significant that the reverse saturation currents are of the same order of magnitude as in the ideal high-resistivity symmetrical configuration, and the PIV values are equally good.

To complete the picture of the unsymmetrical rectifier junction, the reverse characteristic of Fig. 6-9 will be of interest. The saturation current is about 0.5 ma until a breakdown voltage (PIV) is reached at about 300 volts. At this point a sharp breakdown is shown; it is characteristic of a well-designed and carefully fabricated junction rectifier. This characteristic reverse curve is frequently difficult to achieve because of fabrication problems. The dotted curve labeled "soft knee" is in some degree typical of most commercial devices. The effect of a soft knee is to permit a rise in the saturation reverse current at a voltage below that for which the rectifier was designed. This increased current on the reverse cycle produces heat and would soon destroy the device. Accordingly, manufacturers derate (downgrade) such a rectifier and sell it as a device rated only at some lesser voltage at which this degradation has not yet taken place (about 200 volts in this example).

Rectifiers employing the injection or emission principle are also known as *conductivity-modulated power rectifiers*. The origin of this term is obvious from a consideration of the conduction mechanism.

REFERENCES

1. W. C. Dunlap: "Introduction to Semiconductors," p. 150, John Wiley & Sons, Inc., New York, 1957.
2. N. B. Hannay: Semiconductor Principles, in N. B. Hannay (ed.), "Semiconductors," p. 46, Reinhold Publishing Corporation, New York, 1959.
3. J. N. Shive: "The Properties, Physics and Design of Semiconductor Devices," p. 354, D. Van Nostrand Company, Inc., Princeton, N.J., 1959.

7: Single-junction Devices

7-1. INTRODUCTION

By far the most important commercial single-junction application is the rectifier. Rectifiers include small-current units as well as those employing injection across the junction to permit high levels of power to be handled. We have actually been talking about rectifiers in the preceding sections dealing with the detailed structure and electrical properties of the p-n junction, since all junctions in semiconductors are inherently rectifying. To make such a p-n junction into an operable rectifier, we merely attach leads to the p and n regions and protect the active element from outside contamination or damage by suitable encapsulation.

There is another class of single-junction devices, known as point-contact diodes, which may serve as rectifiers. They are now mainly of historic interest, since they have largely been replaced by junction units, although there is still a small demand for point-contact types. Since their operation is not as well understood theoretically as junction devices, we shall confine our present attention to the theory of junction structures. Later, we shall review some of the explanations of the mechanism involved in the operation of point-contact diodes as a separate class. (This classification will include the so-called "bonded rectifier," which involves a welded connection between an aluminum wire and an n-type semiconductor body. Although such rectifiers might be regarded as alloy-junction types, their behavior in a circuit places them most logically with the point-contact class.)

7-2. JUNCTION FABRICATION

Much of the foregoing discussion was based on a hypothetical structure formed by bringing into intimate contact two semiconductor bodies of different type and assuming that this technique would allow a junction to develop. This concept was valuable, since it permitted us to consider the formation of a p-n junction under ideal conditions and to study the individual steps which are believed to lead to the equilibrium configuration of a p-n junction but which in actuality take place almost instantaneously.

Since this theoretical method of formation is physically impossible, it is desirable to examine the various procedures which are employed in the fabrication of junctions in actual commercial practice. Most of

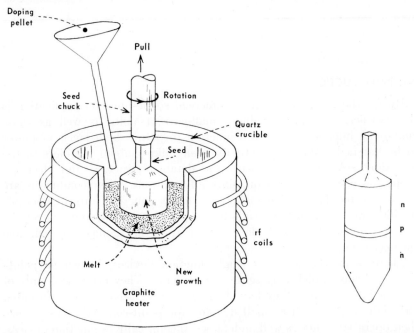

FIG. 7-1. An n-p-n junction produced by the Czochralski crystal-pulling process. (*After J. N. Shive, courtesy of D. Van Nostrand Company, Inc.* [7].)

these fabrication techniques are in current use, the choice among them being dictated by the type of device to be produced. In some cases where two or more methods would be satisfactory, the selection is dictated by economics and/or past experience of the fabricator.

7-2.1. The Grown Junction. This was one of the pioneer techniques used for producing large-area junctions. It consists of successive doping of the melt from which the crystal is being grown by alternate inoculations of p- and n-type impurities. Figure 7-1 illustrates a typical crystal-growing setup with provision for alternate doping. Because of the variables inevitably associated with crystal growing in general [1], control of the doping operation and hence the location of the junction in the crystal is difficult. The success of the method depends largely upon the design of the furnace and the degree of skill acquired by the operator. In addition, if several junctions are to be grown in a given crystal, the impurity concentration must be increased with each addition of the alternate-type dopant, usually resulting in a decrease in resistivity in each successive zone. Several other factors introduce additional degrees of variability into the process. For example, the impurity concentration in the melt is being continually increased, over and above any dopants which are purposely added, by normal segregation as the growth proceeds. (This increase in impurity concentration is further enhanced by the fact that the liquid volume is being continually decreased by the amount of solid crystal frozen from it.) In addition, there may be changes in the impurity concentration in the melt owing to volatilization of the impurities and solution of impurities as a result of attack on the crucible.

One variation of this grown-junction procedure utilizes the alternate doping technique on a crystal grown by the floating-zone process. Figure 7-2 is a schematic representation of this procedure. To be successful, the rod being zone-refined must emerge as a single crystal after one doping pass. If additional passes are made, large differences in the segregation coefficients of the impurities involved will produce anomalous results [2,3].

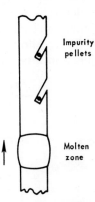

FIG. 7-2. Addition of impurities during crystal growth by the floating-zone technique. (*After M. Tanenbaum, by permission* [3].)

For all of the above reasons production of junctions by growth techniques with alternate doping can be successfully applied only when the resistivity specifications are not critical and when the quality of the semiconductor is low enough that the attendant waste does not render it uneconomical.

7-2.2. Rate-grown Junctions. This method takes advantage of the variation in effective segregation coefficient occurring with changes in conditions of crystal growth or, more specifically, changes in the rate at which the crystal is rotated and withdrawn from the melt [4,5]. A typical example of this procedure involves a melt doped with boron and antimony. Because the effective segregation coefficient of antimony increases with pulling speed, whereas that of boron remains essentially constant, a sudden decrease in pulling speed serves to drop the antimony concentration in the solid below that of boron, and n-p-n-p structures with extremely thin p-type layers are formed. Regions only a few tenths of a mil thick have been produced by this method. In addition, the rate-growth process may be repeated a number of times throughout the crystal pull. As many as 50 n-p-n-p structures have been produced in a single crystal [6].

The success of this technique depends upon accurate knowledge of (1) the manner in which the effective segregation coefficients of the dopants vary with pull conditions and (2) the concentration of p- and n-type impurities so that the change in effective segregation coefficients will result in the change of type necessary to produce the layered structure. This requires a very precise adjustment of the impurity concentration in the initial melt. Finally, a knowledge of the shape (and preferably a means of controlling the shape) of the growth interface is required in order that the resultant crystal may be efficiently cut into individual junction elements.

7-2.3. Alloyed or Fused Junctions. In this technique of forming a junction a pellet of doping material, either the pure element or a high-

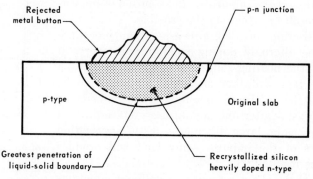

FIG. 7-3. A p-n junction produced by alloying. (*After J. N. Shive, courtesy of D. Van Nostrand Company, Inc.* [12].)

content silicon alloy of that element, is placed on the surface of the prepared silicon wafer. This assembly is then heated in a vacuum or inert atmosphere until the dopant pellet liquefies and alloys itself to the silicon wafer. Upon subsequent freezing, most of the dopant is rejected by the body of the silicon, leaving a layered structure involving a p-n junction. The exact location of the junction depends on the time and temperature of heating. The junction may occur in an area of the wafer which was never liquefied but into which the dopant has diffused. It may also occur at the interface between the liquid and solid phase or, indeed, back in the refrozen pellet [8,9].

Because of these variables, if it is desired to produce an n-p-n layered structure in which the p region is to be as thin as possible (a desirable characteristic for transistors, as will be shown later), the alloying cycle must be precisely controlled and the uniformity of materials from wafer to wafer must be extremely constant. Figure 7-3 illustrates a typical alloy junction.

7-2.4. Melt-back Junctions. In this method of producing, say, n-p-n structures an n-type single crystal containing the proper amounts of both p- and n-type impurities is cut up into elements of the proper size for the finished device. Each element is heated on one end until a molten drop is produced. It is then cooled under controlled conditions so that normal segregation removes most of the n-type impurity from the portion next to the solid rod and thereby produces a thin p-type region. Complete solidification of the molten zone is then permitted under normal freezing conditions. The n-type impurity which was rejected during the early freezing adds to the concentration of n-type impurities in the refrozen n-type region [10,11].

Successful operation of this technique requires a knowledge of the segregation coefficients of the impurities involved and a careful adjust-

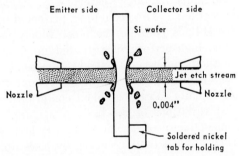

Fig. 7-4. Formation of a surface-barrier transistor element. (*After J. W. Tiley and R. A. Williams, by permission* [13].)

ment of the concentrations of these impurities so that the initial portion of the molten zone which freezes results in a p-type composition. In actuality, this process is analogous to the production of n-p-n semiconductors by the rate-grown process, since once again the transition in type is accomplished by careful adjustment of the product of impurity concentration and effective segregation coefficient. This process, therefore, suffers from the weaknesses already described for the rate-grown junction.

7-2.5. Surface-barrier Junctions. During the development of transistor technology, it became apparent that in any n-p-n or p-n-p structure it was desirable that the central region be physically as thin as possible. In 1953 the Philco Corporation made an important contribution toward accomplishing that objective. Their technique involves a process known as *jet-etching*. When the wafer (p- or n-type) has been reduced by grinding to the minimum practical thickness, it is placed between two opposed coaxial streams of jet solution, as shown in Fig. 7-4, and the etching current is turned on. The etching proceeds until the thickness of the wafer in the center has been reduced to about 0.2 mil (0.0002 in.). At this point the current is reversed and plating from the solution begins.

The metal which is to be plated on the thin center section of the wafer is of opposite type to the silicon. For example, for an n-type silicon wafer the etching and plating solution would be based on indium or some other p-type impurity. The electroplating is then allowed to proceed, using the wafer and the plating solution jet as current carriers, until an optimum thickness of plate has been achieved. Leads are then soldered to the plated surface, and the heat required to attach the leads serves to form the junction between the wafer and the plated areas. The device derives its name from the fact that the junctions or barriers are essentially on the surface of the silicon [13].

Because these areas are extremely small, the junctions so produced are a hybrid between point-contact and large-area junctions. Since they are closer to the large-area junction in performance, however, they are considered in the large-area-device category.

7-2.6. Diffused Junctions. This technique, perfected by the Bell Telephone Laboratories [14], is perhaps the most widely used method for making reproducible, high-performance graded junction devices. The simplicity of the method and ease of control recommend its use wherever possible. Its success is limited only by the purity and uniformity of the semiconductor and dopants and the ambient conditions.

There are several variations of the basic procedure. They all start with a wafer of semiconductor of specified impurity level and type. The doping agent is then applied to the surface by electroplating, painting (applying a solution of the impurity and evaporating the solvent), or by exposing the surface to a gaseous form of the dopant. Area masking can be easily achieved for any of these techniques. The wafer is then subjected to a carefully controlled time-temperature cycle during which the dopant diffuses into the solid to produce a junction at a uniform distance from the surface. The diffused layer is now of a different type than the base material. At the conclusion of the diffusion operation excess dopant is removed from the surface of the semiconductor. By diffusing into both sides of the original wafer, it has been possible to produce base regions less than 0.05 mil in thickness.

7-2.7. Epitaxial Junctions. A relatively recent technique [15-17] has become commercially important owing to the emphasis on size reduction, both as an advantage where space and weight must be minimized and because small physical dimensions, and particularly the thickness of certain active regions in multijunction structures, enhance high-frequency performance. Although we are not concerned with miniaturization, we shall later discuss the problems of high-frequency response.

Essentially, epitaxial growth refers to the formation of new layers of crystalline solid material on a well-oriented single crystal in such a manner that the new, or expitaxial, layer perpetuates the crystalline structure of the original substrate. The advantages are twofold. First, the thickness of the new growth may be controlled to narrow limits and much thinner layers may be reproducibly achieved (growth rates on germanium of 100 μ per day have been reported). Second, by appropriately doping the new growth, the type, resistivity, and resistivity gradient are all under continuous direct control. This is not true of diffusion processes, for example, where the dopant concentration gradient follows the fixed distribution laws for the process and wherein variations are possible only within limits.

Experimentally, the growth of epitaxial layers for semiconductors proceeds by a gas-solid reaction in which the single crystal substrate, heated to the desired temperature (1000 to 1250°C) is exposed to a gas phase containing a volatile compound of the semiconductor plus volatile compounds of the doping agent at known concentrations. Most workers employ an atmospheric pressure system using a very dilute stream of $SiCl_4$ in hydrogen gas with doping vapor added. This gives best control of crystallinity and avoids the difficulties of vacuum techniques.

Although commercial devices are now being produced on a large scale, there are some advanced configurations which suffer from difficulties with diffusion effects during growth and resulting variation in the exact location of the junction.

7-3. RECTIFIERS

We have made a detailed study of the rectifying properties of the p-n junction and have now described seven techniques available for obtaining such junctions, but we have not described commercially available rectifiers. Of the methods used to fabricate junctions, diffusion, alloying, junction growth, and epitaxial techniques are used for the manufacture of rectifiers. In the following paragraphs we shall compare the properties of and applications for rectifiers prepared by these four techniques.

The first junction diodes to be developed were made from crystals containing grown junctions. However, a disadvantage frequently encountered in grown junctions involves the shape of the radial profile of the junction within the crystal. Since this profile is never flat, it imposes limitations on the economics of the process as well as design problems due to the variation of the junction profile within the slice. Practically, therefore, grown-junction rectifiers are small in cross section and are consequently sharply limited in their power ratings. Furthermore, since injection-type junctions are limited for practical reasons to one per crystal pull, injection rectifiers starting with a grown junction would be unworkable from a cost standpoint. This limitation results from the fact that grown-junction crystals are made either by a series of alternate light dopings with p- and n-type impurities or by rate-growing techniques. Both the p- and n-type regions which are produced are of relatively high resistivity material, and as a result, the junctions cut from these crystals are not suited to the injection process.

Since current ratings for grown-junction rectifiers are sharply limited, the rectifiers can be used only in applications where power is not a consideration. Otherwise, these devices exhibit excellent characteristics with PIV's as high as 1,000 volts and with forward currents in excess of 100 ma at 1 volt applied potential. Because of their excellent electrical characteristics, millions of these units are in service in a variety of uses where high-power capabilities are not necessary [18].

Alloyed or fused rectifiers suffer from the same junction area limitation. This is due to the high surface tension of the alloying agents and their failure to wet the surface of the semiconductor. When the alloying material melts, it forms a small ball on the surface of the semiconductor which limits the resultant junction area [19]. In addition, the junction will frequently show cracks and strains after the leads have been soldered into position, owing to the unequal coefficients of thermal expansion between the solder and the base material. Another limitation characteristic of alloyed-junction rectifiers is a lack of response at frequencies above a megacycle. In most other respects these rectifiers are the equivalent of good grown-junction devices.

By means of special techniques, alloyed rectifiers employing several square centimeters of junction area have been obtained with germanium as the base material. Such units can pass hundreds of amperes in the forward direction. However, it is necessary to water-cool these devices, and they are exceedingly expensive and sensitive to fabrication techniques [20]. With the development of the diffused-junction silicon rectifier to be described in the next paragraph, work along these lines has been virtually suspended.

There is no theoretical limit on the area of a diffused junction; and diffused junctions can be made to inject, since the diffusion can be continued until any concentration of impurity has been attained. From a practical point of view the present diffused-junction rectifiers are limited in area primarily by the perfection of the crystal from which they are made. Most attention is currently on devices fabricated from silicon, owing to the temperature limitation of germanium. Injection-type rectifiers made from silicon can easily carry forward currents of several hundred amperes, but thus far they are somewhat limited in their voltage rating.

Theoretically, it is only necessary to increase the thickness and purity of the high-resistivity material on one side of the junction to reach almost any desired voltage rating in silicon, up to the limit of the intrinsic resistivity. Practically, questions of crystal perfection, loss of lifetime during processing, and such fabrication problems as edge leakage have thus far imposed a limit of about 4,000 volts PIV per junction [21,22].

In summary, there will continue to be a demand for low-power grown-junction and alloy-junction rectifiers. At the same time, however, diffused-junction silicon units for power rectification will be steadily improved and may eventually take the place of most other types of power rectifiers (such as copper-oxide, selenium, germanium, and vacuum-tube).

Epitaxially grown diodes are not yet in general use except in special p-i-n structures to be discussed later, since the more expensive technique is not competitive in most areas. However, for high-frequency operation and special applications, they have a definite place. It is also probable that costs will come down as the technology advances and their use will eventually become relatively widespread.

7-4. CONTROLLED RECTIFIERS

For the sake of completeness, and to alert the reader to some of the pitfalls of commercial terminology, we shall make brief mention of a special device now being widely used under the general name of *controlled rectifier*. Devices of this type are sometimes called triodes, since by analogy to vacuum tubes there are three external leads, or by various trade names such as Westinghouse's Trinistor. They will also sometimes be described as controlled junction rectifiers with or without mention of a third, or control, electrode. Actually, the structure contains three junctions and four regions, and its detailed operation is rather complicated. In one type of construction these devices are controlled power rectifiers with some of the characteristics of the familiar gas thyratron of vacuum-tube origin. In another configuration they function as high-speed transistor switches but without power-handling ability, in which case they are connected as tetrodes, but with three junctions. Finally, in yet another application, they are known as switching diodes, and they are provided with only two leads despite the three junctions. Again their action is like that of a thyratron, but without the grid lead to control the onset of conduction.

Since we have not yet considered multijunction structures, a discussion of these and other three-junction configurations will be deferred until the classic case, that is, the two-junction transistor, has been presented. It must be recognized that the variety of multijunction structures is tremendous, and it would be hopeless to attempt to cover them all. However, the principles to be illustrated herein should provide a basic understanding of single- and double-junction devices and some facility in analyzing the behavior of more complex arrangements.

Unfortunately, there is one area about which we can do little. As the art of the semiconductor has sprung up, there have been inconsistencies in nomenclature such as were discussed above. There are many of these now in existence, and many more are likely to arise. However, if the reader is forewarned and looks a little more closely, he should soon be able to recognize the "p-n-p-n diode" switch for the semantic anomaly it is and be able to cope with it.

7-5. ZENER DIODES

We have already met this group of devices in connection with the discussion of reverse breakdown in p-n junctions (Sec. 5-8). In examining the mechanisms of reverse breakdown, we identified avalanche, zener, and tunneling. Of these the first, or avalanche, may be clearly recognized from a number of characteristics which can be observed experimentally. It is usually encountered in high-voltage breakdown (greater than 20- to 50-volt range, depending on the junction configuration) and with relatively pure materials.

Zener, or field emission, breakdown is not easy to recognize, and it is believed not to exist to any appreciable extent in the higher-voltage ranges. The low-voltage range, on the other hand, is believed to involve tunneling, a phenomenon which we defined in terms of the quantum-statistical probability that electrons of a given energy will be found in empty states of the same energy on the far side of a potential barrier. It was stated that this range of voltage was from 0 to 5 volts reverse bias. Nothing was said about the prevailing mechanism in the range from 5 volts reverse bias up to a value between 20 and 50 volts where avalanche takes over.

Since zener diodes (so called by virtue of function, not mechanism) are available from the range of one volt to several hundred volts, it is of interest to inquire more closely into the mode of breakdown over the entire low-voltage range. The situation is not entirely clear. Some authors make assumptions which lead to a true zener field emission mechanism for the entire preavalanche mode, while others find that tunneling is more accurate, particularly below the range 2 to 5 volts, and that from there up to avalanche there may well be a combination of Zener and tunneling effects [23–26].

The source of the argument can be appreciated if we recall the basic conditions which are believed to lead to the tunnel mechanism. First, the junction is heavily doped, resulting in a narrow, sharp, step junction with very high space-charge densities and steep field gradients. Second, there is a plentiful supply of carriers (electrons), meaning low-resistivity regions. Then, with the tunneling probability high, reverse bias can easily produce a tunneling condition.

With this in mind, we construct a band energy diagram for two heavily doped phases making up a p-n junction. By analogy with the emitter side of the unsymmetrical junction of Fig. 6-7, we may represent the situation as the upper picture, Fig. 7-5a. Here we see the Fermi level near the band edges (and therefore the impurity ion levels) and further indicate a region of empty states in the valence band on the p-type side

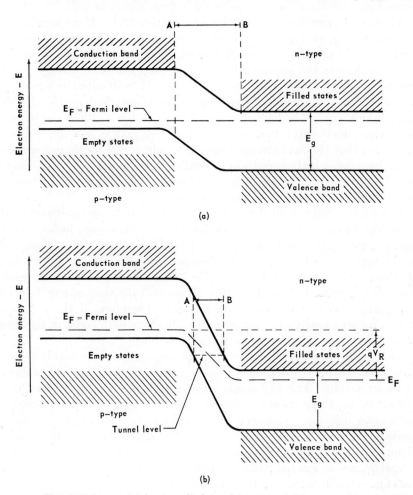

FIG. 7-5. Low-voltage zener diode at (a) zero and (b) reverse bias.

and filled states in the conduction band on the n-type side. Notice also that there is only a small energy gap between these two sets of states. If a reverse bias is now applied, the bands shift by an amount qV_R and cause the filled states in the n-type region to line up with the empty states in the p-type region.

This fulfills the last requirement for tunneling, since we now have occupied and empty states on opposite sides of the barrier at the same energy. A current will now flow by the tunneling mechanism, and it will be far larger than the equilibrium reverse current. Note also that if the reverse bias were greater than V_R, the situation would still produce tunneling, only then from the p- to the n-type region as well.

There is, however, an alternative explanation [27]. If we considered the conditions for true zener field emission, we would simply require that the field reach a value high enough to ionize lattice atoms in the junction despite the opposing barrier and then overcome energy losses due to collisions. We now recall that the field built into the structure is proportional to the space-charge densities (dependent on the doping level) which define the barrier height as well as the width of the barrier for a step junction (cf. Sec. 5-5). If a reverse bias is now applied, we have seen that the field strength increases only as $V_R^{1/2}$ owing to junction width increase. If the sum V_T of the built-in potential V_0 and the reverse bias V_R produces a field of sufficient intensity, we could get zener breakdown despite the increase in junction width.

If we now express field strength in terms of distance per unit potential drop, high fields will appear as short distances; and if we now look again at Fig. 7-5, we could regard the distance A-B as a measure of field strength. In the upper, or unbiased, case the distance is large, but after application of reverse bias the distance A-B in the lower drawing is much reduced, indicating increase in field strength. (Note that the vertical gap E_g does not change even in the depletion region.) Therefore, if the distance A-B is reduced enough to correspond to a field strength equal to the true zener field, we could postulate a zener mechanism.

The choice between mechanisms is not easy, since these highly doped junctions are nearly degenerate and quantitative estimations appear to be uncertain. For any single zener diode, particularly in the low-voltage range, there are often second-order effects which make a rigorous calculation difficult. However, in the next section we shall consider a diode so degenerate in both regions that the space-charge width is only about 100 A. In this structure a trivial voltage signal produces a considerable current reaction, and questions of Zener versus tunneling effects are eliminated in favor of a tunneling mechanism.

Therefore we could perhaps use this as a limiting case for zener structures and conclude that the very highly doped low-voltage units operate by a mechanism which is principally tunneling. Then, at some point (several authors quote the 5-volt region) avalanche breakdown intervenes and is an appreciable part of the behavior. It is then only logical to predict that the avalanche effect will become increasingly significant with increasing bias.

Unfortunately, there is no general agreement about the preavalanche mechanism. Some authorities believe that true zener or internal field-effect breakdown is at least partly responsible at all such voltages (except perhaps the very low range). Other authors substitute the zener mechanism entirely (with no mention of tunnel effect at all) over an unspecified range, starting with the lowest working voltage for zener diodes up to the onset of avalanche.

It is therefore clear that zener diodes are not at all simple in theory; and if we also consider the variety of current-voltage characteristics which are built into commercial units, it is not surprising that no simple, first-order analysis is valid for the entire group. Notwithstanding, the variety and low cost of these devices have produced a wide market.

In closing, it might be mentioned that one of the largest uses for zener diodes is in transistor circuitry of the small-signal type at voltages in the range of 3 to 18 volts. In this range it is found that there are two competing temperature coefficients, one associated with incipient avalanche and the other with zener or tunneling breakdown. Near 6 volts they tend to balance and the devices are almost temperature-independent, a fact of great practical value [28].

Nothing has been said about the forward behavior of a zener diode. As might be expected from the doping level, there is virtually no voltage drop. It is also worth mentioning that power levels in excess of 100 watts are very unusual. Because of capacitance problems the areas are as small as possible, so that, except for special structures, the power levels are normally below the 10-watt range.

7-6. TUNNEL OR ESAKI DIODES

These fascinating devices are named for Dr. L. Esaki [29,30], who first observed a form of rectification by tunneling in very highly doped junctions. In the tunnel diode we also find an important example of the negative-resistance phenomenon. An esaki diode is composed of heavily doped p- and n-type germanium or silicon. The doping level is far higher than any we have previously met, and it is termed *degenerate*. This means that the Fermi level is close to or in the valence or conduction band, as the case may be. (In quantitative work, there are various considerations which describe the exact position of the Fermi level for the degenerate case. These involve the energies of the ionized donors and acceptors, the total impurity level, etc.) For purposes of this discussion we shall simplify the energy band diagram and show only the band edges, the Fermi level, and the effect of bias. For a doubly degenerate diode structure, we may represent the band energy diagram for zero bias as shown in Fig. 7-6.

Because of the heavy doping the Fermi level is below the edge of the valence band on the p-type side and above the edge of the conduction band in the n-type region. In addition, because of the very high doping level the space-charge density is high, the depletion regions are sharp, and the intrinsic layer is very thin (10^{-6} to 10^{-7} cm). These conditions are recognized as basic requirements giving rise to the phenomenon of tunneling, a mechanism which calls for an electron to make a direct transition across the gap from an occupied level to an unoccupied level at the same energy. In the structure as drawn, however, there are very few equivalent energy levels which are filled and vacant on opposite sides of the depletion layer, and only a small tunnel current would be expected. At zero bias, therefore, as shown in Fig. 7-6, there will be all

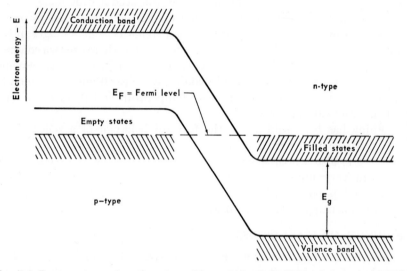

FIG. 7-6. Degenerate p-n junction at zero bias. (*After L. B. Valdes, by permission* [31].)

four normal junction currents of which we have previously spoken plus two tunneling effects, and the sum of all six will be zero.

At increasing forward bias, the Fermi level on the n-type side will rise as shown by the three diagrams of Fig. 7-7. In the case of Fig. 7-7a the filled n-type states have come partly opposite the empty p-type states and one-way tunneling dominates, producing a rapid increase in forward current for a small voltage. This rise is much steeper than a normal forward characteristic for an emitter junction. In Fig. 7-7b the forward bias has increased until there is the maximum possible coincidence between the filled states in the conduction band and the empty states in the valence band. For this condition the tunnel current reaches a maximum, and it is still much larger than the normal forward current. As the forward bias increases further, the condition of Fig. 7-7c is reached and, to a first approximation, tunneling ceases. Note, however, that the tunnel current has decreased while the forward bias has increased. The characteristic has therefore gone through a region of negative resistance.

Finally, as the forward bias continues to increase, the normal forward diffusion current appears and the characteristic reverts to a positive-

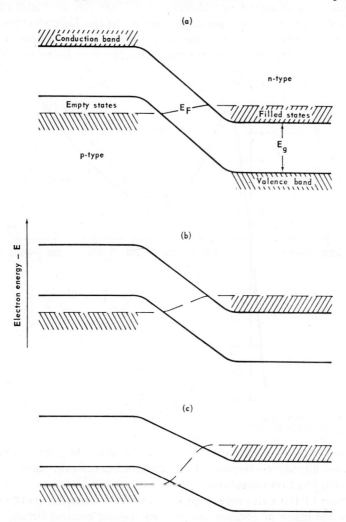

FIG. 7-7. Forward bias on degenerate p-n junction. (*After L. B. Valdes, by permission* [31].)

resistance behavior, with a total current composed of the dwindling tunnel contribution and the increasing forward-diffusion components. This entire current-voltage characteristic is shown in Fig. 7-8. In this plot the points corresponding to conditions a, b, and c of Fig. 7-7 are indicated.

Note that the currents and voltages are very small, less than 0.5 volt and 5 ma. On a conventional diode plot the region of interest would not be seen. Note also that there is a region of negative resistance in which the forward current drops from about 5 to 1 ma while the voltage increases from about 80 to 250 mv. A further characteristic of these junctions is a very high capacitance (owing to the high charge density and very narrow space-charge region), so that the area must be kept very small in practice. The reverse behavior may be predicted from

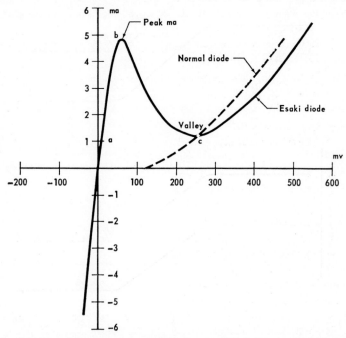

FIG. 7-8. Current-voltage plot for esaki diode. (*After G. W. A. Dummer and J. W. Granville, by permission* [32].)

our knowledge of zener diodes. Since the zero bias condition of an esaki unit already corresponds to a zener in breakdown, there is almost no reverse junction resistance.

Because of the extremely high doping levels required many of these devices are made of germanium or a compound semiconductor such as gallium arsenide, although some manufacturers offer silicon esaki diodes. Despite the limitations of low current and voltage and high capacitance, these devices are widely used as amplifiers at frequencies in the gigacycle range or as very fast switching elements. (In the negative-resistance region, there is a change from 100 to 10 ohms with a voltage decrease of less than 0.2 volt.) Finally, it should be mentioned that the high doping level makes the diode relatively insensitive to temperature, contamination, and fabrication errors.

7-7. PHOTOCELLS

In the early discussion of p-n junctions it was mentioned that electron-hole generation within the barrier could be produced by external stimuli such as an electric field, heat, or radiation. In a photocell this sensitivity to radiation is put to practical use. In this application the resistivity of the bulk material must be so chosen that the photocell fits into the circuit for which it is designed. Usually the element consists of a bar of semiconductor with a junction normal to its axis and with leads attached to the ends. To avoid contamination, the sensing element is usually imbedded in clear plastic. In operation, the junction is normally reverse-biased. This condition makes it most sensitive to generated hole-electron pairs in the junction region. The change in conductivity as a result of carrier generation is a measure of the amount of radiation which falls on the junction. Each junction is sensitive to only a limited range of wavelengths, and photocells are therefore made of a variety of base materials of which elemental semiconductors are only a small portion. Silicon is most sensitive in the infrared range and finds greatest utility in that area. A schematic diagram of such a photocell is shown in Fig. 7-9.

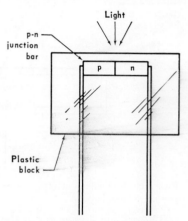

FIG. 7-9. A p-n junction photocell. (*After J. N. Shive, courtesy of D. Van Nostrand Company, Inc.* [33].)

7-8. SOLAR CELLS* [34,35]

A solar cell is prepared by exposing wafers of n-type silicon having a resistivity in the 1 ohm-cm range to the vapors of boron trichloride at high temperatures (1000 to 1200°C). As the result of the diffusion process, a p-type skin is formed over the entire surface. One connection is made to the p-type skin, and another is made to the substrate by means of a window which was masked over during the diffusion. When this element is exposed to visible radiation, for example, sunlight, the radiant energy generates hole-electron pairs in the p-type skin. The excess minority carriers, in this case electrons, diffuse to the junction

* Adapted with permission from J. N. Shive, courtesy of D. Van Nostrand Company, Inc. [36].

where, as we have seen before, they are readily collected and enter the n-type substrate. Once they are in the n-type substrate, they must occupy energy levels higher than the electrons already present at thermal equilibrium and are therefore capable of performing work. If the external circuit is now completed, the electrons will flow out of the n-type substrate through the external circuit and return to the p-type skin. The process will continue as long as the radiant energy is present.

Figure 7-10 shows such a solar cell element. Under open-circuit conditions, the voltage generated by such a cell is approximately 0.5 volt. If short-circuited, the cell will deliver about 25 ma per cm² if exposed directly to the noonday sun. Under these conditions the voltage will drop to about 0.3 volt.

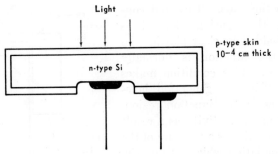

FIG. 7-10. Bell solar battery. (*After J. N. Shive, courtesy of D. Van Nostrand Company, Inc.* [36].)

The critical dimension in the fabrication of a solar cell is the thickness of the p-type skin. If the skin is too thick, the minority carriers generated by the radiation will recombine before being collected at the junction. On the other hand, if the skin is too thin, lateral resistance will impede proper collection.

Silicon is the currently accepted optimum material for solar cells because of the higher energy gap, which produces a higher voltage than any other practical semiconductor. The spectral range to which silicon is responsive extends from the near infrared through the visible. Since the highest intensity of solar radiation lies in the visible green region, there are no serious sensitivity problems when using silicon for solar radiation. Other semiconductors, having smaller forbidden gaps, function with higher efficiency in the far infrared than silicon, despite their lower open-circuit voltage. This is true of germanium under normal incandescent lighting.

Recently, in connection with space exploration, the opposite configuration (n-type skin on p-type body) for solar cells has become popular. Although less efficient, the structure has greater resistance to damaging high-energy radiation and radioactive decay products.

REFERENCES

1. J. Czochralski: Ein neues Verfahren zur Messung der Kristallisationsgeschwindigkeit der Metalle, *Z. Physik. Chem.*, **92**: 219 (1918).
2. W. G. Pfann: "Zone Melting," pp. 89–97, John Wiley & Sons, Inc., New York, 1958.
3. M. Tanenbaum: Semiconductor Crystal Growing, in N. B. Hannay (ed.), "Semiconductors," p. 124, Reinhold Publishing Corporation, New York, 1959.
4. R. N. Hall: p-n Junctions Produced by Rate Growth Variation, *Phys. Rev.*, **88**: 139 (1952).
5. R. N. Hall: Segregation of Impurities During the Growth of Germanium and Silicon Crystals, *J. Phys. Chem.*, **57**: 836 (1953).
6. W. C. Dunlap: Methods of Preparing p-n Junctions, in L. P. Hunter (ed.), "Handbook of Semiconductor Electronics," 2d ed., pp. 7–11, McGraw-Hill Book Company, New York, 1962.
7. J. N. Shive: "Properties, Physics and Design of Semiconductor Devices," p. 459, D. Van Nostrand Company, Inc., Princeton, N.J., 1959.
8. J. S. Saby and W. C. Dunlap: Impurity Diffusion and Space-charge Layers in "Fused-impurity" p-n Junctions, *Phys. Rev.*, **90**: 630 (1953).
9. J. J. Ebers: Alloyed Junction Transistor Development, *Bell Lab. Record*, **34**: 8 (1956).
10. R. N. Hall: Melt-back Transistors, paper presented at AIEE-IRE Semiconductor Device Research Conference, Philadelphia, June, 1955.
11. J. I. Pankove: Transistor Fabrication by the Melt-Quench Process, *Proc. IRE*, **44**: 185 (1956).
12. Ref. 7, p. 191.
13. J. W. Tiley and R. A. Williams: Part II, Electrochemical Techniques for Fabrication of Surface-barrier Transistors, *Proc. IRE*, **41**: 1706–1708 (1953).
14. F. J. Biondi (ed.): "Transistor Technology," vol. 3, chaps. 1, 3, and 8, D. Van Nostrand Company, Inc., Princeton, N.J., 1958. (C. D. Thurmond, Impurity Control and Junction Formation in Silicon and Germanium; C. S. Fuller, Diffusion Techniques; C. J. Frosch, Silicon Diffusion Technology; M. B. Prince, Diffused p-n Junction Silicon Rectifiers; H. S. Veloric and K. D. Smith, Silicon Diffused Junction "Avalanche" Diodes.)
15. H. C. Theuerer: Epitaxial Silicon Films by the Hydrogen Reduction of $SiCl_4$, *J. Electrochem. Soc.*, **108**: 649 (1961).
16. A. Mark: Single Crystal Silicon Overgrowths, *J. Electrochem. Soc.*, **108**: 880 (1961).
17. E. G. Bylander: Kinetics of Silicon Crystal Growth from $SiCl_4$ Decomposition, *J. Electrochem. Soc.*, **109**: 1171 (1962).
18. W. C. Dunlap: "Introduction to Semiconductors," p. 341, John Wiley & Sons, Inc., New York, 1957.
19. Ref. 14, chap. 7B. (G. L. Pearson and B. Sawyer, Silicon p-n Junction Alloy Diodes.)
20. Ref. 18, p. 344.
21. E. M. Pell: Ion Drift in an n-p Junction, *J. Appl. Phys.*, **31**: 291–302 (1960).
22. H. S. Veloric and M. B. Prince: High-voltage Conductivity-modulated Silicon Rectifier, *Bell System Tech. J.*, **36**: 975–1004 (1957).
23. R. P. Nanavati: "An Introduction to Semiconductor Electronics," pp. 108–115, McGraw-Hill Book Company, New York, 1963.
24. J. R. Tillman and F. F. Roberts: "An Introduction to the Theory and Practice of Transistors," pp. 117–123, John Wiley & Sons, Inc., New York, 1961.

25. L. B. Valdes: "The Physical Theory of Transistors," pp. 203–209, McGraw-Hill Book Company, New York, 1961.
26. L. Pincherle in "Encyclopaedic Dictionary of Physics," vol. 7, pp. 475, 476, The Macmillan Company, New York, 1962.
27. Ref. 25, pp. 207, 208.
28. G. W. A. Dummer and J. W. Granville: "Miniature and Microminiature Electronics," p. 41, John Wiley & Sons, Inc., New York, 1961.
29. L. Esaki: New Phenomenon in Narrow Germanium p-n Junctions, *Phys. Rev.*, **109**: 603, 604 (1958).
30. Ref. 25, pp. 347–350.
31. Ref. 25, p. 348.
32. Ref. 28, p. 45.
33. Ref. 7, p. 147.
34. D. M. Chapin, C. S. Fuller, and G. L. Pearson: A New Silicon p-n Junction Photocell for Converting Solar Radiation into Electrical Power, *J. Appl. Phys.*, **25**: 676 (1954).
35. M. B. Prince: Silicon Solar Energy Converters, *J. Appl. Phys.*, **26**: 534 (1955).
36. Ref. 7, pp. 156, 157.

8: Simple Transistors or Two-junction Devices

8-1. INTRODUCTION

In this chapter we shall discuss the operation of simple two-junction transistors and we shall use an amplifier as our example. The selection of this particular device is dictated by its simplicity and usefulness. In addition, its functioning is readily explainable in terms of the junction theory already presented. Point-contact equivalents of this structure will be treated in a later chapter.

8-2. THE n-p-n STRUCTURE AND ITS OPERATION

We shall begin with two unconnected junctions and refresh our memory as to their behavior under bias. We shall then combine these two junctions in such a fashion as to form an n-p-n junction transistor. Let us consider an unsymmetrical n-p junction in which the n portion has a resistivity in the neighborhood of 0.2 ohm-cm and the p portion has a resistivity in the neighborhood of 20 ohm-cm. With such a junction, for very small forward bias a large current will flow by the process known as injection.

We now wish to introduce a new concept in connection with the bias. If the bias voltage were varied, we would expect to see a corresponding variation in the current. It is therefore possible to superimpose an alternating voltage on the bias voltage. As long as the alternating potential does not equal the bias voltage in the opposite direction, there will be no time at which current will be interrupted. If the a-c voltage were from a normal 60-cycle supply, the current through the junction would also vary at 60 cycles but would always flow in the forward direction. We have thus superimposed an alternating component on the current flowing through the junction.

Let us now consider an unsymmetrical p-n junction whose p-type region has a resistivity of 20 ohm-cm and whose n-type region has a resistivity of 100 ohm-cm. Since both resistivities are high, we would not expect a significant flow of current under reverse bias. However, it will be recalled that a reverse-biased junction promotes the flow of thermally generated minority carriers across the junction to the majority side. Thus, if it were possible to inject electrons close to the junction on the p-type side, they would fall through the potential gradient and become majority carriers on the n-type side. If this injection were continued and a circuit were completed to the end of the n-type region, a flow of current in the form of electrons would result.

This current would be flowing in the same direction as the reverse current which is characteristic of the reverse-biased junction. It has been shown that the rate of thermal generation in the junction normally limits the reverse current, regardless of reverse bias, up to breakdown. However, by postulating injection close to the junction, we utilize the field of the space charge and effectively increase the reverse current flow. This phenomenon of increased reverse current flow in the p-n reverse-biased junction just described is at the root of all transistor action. It is only necessary to accomplish injection on the p-type side of the junction close enough to the junction that the electrons are repelled by neither the space-charge effect nor the applied reverse bias, but instead are collected and added to the reverse current.

We recall our first unsymmetrical forward-biased n-p junction and the mechanism of current flow by injection across the junction. If we could substitute for the contact on the p-type end of this junction a region very close to a p-n junction under reverse bias, we would in effect be injecting electrons close enough to the second junction that they would cross the junction owing to the potential gradient and emerge from the opposite end of the n-type portion.

Fortunately, we have several techniques which will permit us to dope a p-type bar at each end and control the resistivity of the doped portions at any chosen values. We shall therefore dope a 20-ohm-cm p-type bar at each end but at different doping levels of n-type impurity. One end we shall elect to dope to a resistivity of 0.2 ohm-cm, n-type. The other end we shall lightly dope to produce a (compensated) resistivity of 100 ohm-cm, n-type. We now see that the p-type region in the middle is common to both junctions and that the n-type regions have the same resistivities as the n-type regions in the individual junctions previously discussed.

Let us now reproduce the conditions of bias which we previously described. A small d-c source is connected between the p-type region and the 0.2-ohm-cm n-type region in such a way as to produce a forward bias. For the second junction we connect a d-c source between the central p-type region and the 100-ohm-cm n-type region. This source will be connected in such a way as to reverse-bias this junction. We now have the configuration shown in Fig. 8-1.

In this configuration of the two junctions we have seen that electrons supplied to the low-resistivity n-type material flow through the body of this region and arrive at the first junction in such a way that they are injected past the space-charge region and easily flow, as a result of the diffusion gradient, into the p-type region. These electrons, which are now in the p-type region, are collected by the reverse-biased junction and added to the saturation current flowing under reverse bias. This results in an enhanced current flow in the reverse direction, from the central p-type region through the n-type region and the external circuit.

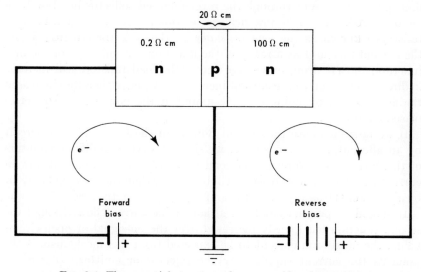

Fig. 8-1. The essential structure of an n-p-n biased transistor.

Quantitatively, a very small potential applied in the forward direction to the first junction is capable of sending large numbers of electrons to the second junction, where they are incorporated into the reverse current flow. For example, the forward-biased junction, under a potential of 0.5 volt, may sustain a current of 100 ma. This current is composed primarily of electrons flowing through the low-resistivity region, which has an effective overall resistance of about 5 ohms. In the case of the second junction, by virtue of its reverse bias and its high-resistivity n-type region, a current of about 0.5 ma results from an impressed potential of 500 volts. The resistance of this circuit is therefore 10^6 ohms. (These voltage-current values for the reverse-biased second junction are valid only when the first junction circuit is open.)

When the first junction (forward-biased) circuit is closed, the resultant

114 *Semiconductor Junctions and Devices*

electron flow across this junction causes a marked change in the electron flow across the reverse-biased junction owing to injection and collection. Assuming 100 per cent efficiency (which will be very nearly true) this "reverse current" will now amount to 100.5 ma flowing through the circuit under the influence of the full reverse bias of 500 volts. The effective resistance of the reverse-biased circuit has been reduced from 10^6 to about 5,000 ohms. (Since the current flow across the reverse-biased junction has apparently increased by a factor of about 200, the effective resistance of this circuit has decreased by the same factor.)

The above description has given us the clue to transistor action. A very small signal impressed on the forward-biased emitter junction has multiplied the current through the reverse-biased collector junction by a factor of 200. To make practical use of this increase, we shall insert a resistance known as the *load resistance* in the collector circuit. When the current increases as a result of the injection (or emission) of the first junction, the IR drop, or voltage across the load resistor, will increase in direct proportion to the current increase. We have therefore increased the current and by doing so have amplified the power in the reverse-biased circuit.

If we now modulate the current flowing in the forward-biased circuit by an alternating-current source, we shall observe a similar modulation of the current flow through the load resistor. This modulated flow of current through the load resistor will be greater than the no-signal current by a factor of 200 as a result of the amplifying effect of the second junction.

It should be pointed out that, although the current flow through the second junction has greatly increased over the no-signal or open-circuit value, the maximum current to be expected from this mechanism is the same as the current supplied by the injector or emitter. Therefore, taking both junctions into account, the maximum current amplification is a factor of 1. The true amplification is a power increase by virtue of the resistances of the emitter and collector circuits—an increase of I^2R, or power. We shall later see that there are other circumstances under which current amplification is produced, but it is a good general rule to consider the transistor as a power-amplifying device and to regard it, in general, in the light of the above example.

Figure 8-2 shows the fully developed circuit described above. The alternating signal source, the load resistance, and the voltmeter across the load resistance have been added to the basic circuit shown in Fig. 8-1. Figure 8-2 also introduces the nomenclature for the various regions of a transistor. These terms have their origin in the earliest point-contact devices, which will be described later. Their significance for the junction transistor, however, is completely analogous. Hence, the forward-biased junction which injects electrons into the p-type region is known as the *emitter junction.* It is also customary to refer to the low-resistivity *n*-type region as the *emitter.* The same designation is applied to the contact and the lead from this contact. The intermediate *p*-type region is known as the *base.* The corresponding contact and lead are similarly identified. Finally, the reverse-biased junction is known as the *collector*

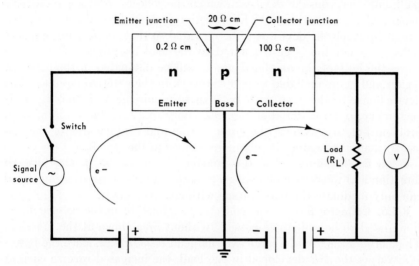

Fig. 8-2. An *n-p-n* transistor with signal source and external load added.

junction and, by analogy with the emitter, the entire high-resistivity *n*-type region, contact, and lead are known as the *collector.*

8-3. TRANSISTOR DESIGN REQUIREMENTS

Having described the power amplification of which a transistor is capable, the circuits which make this amplification possible, and the nomenclature applied to the various circuit elements, we may now enumerate the necessary and sufficient conditions under which a semiconductor structure may be made to function as a power amplifier. (These parameters are complete when applied to a junction transistor, but, as we shall see later, several additional conditions must be imposed in the case of a point-contact device.) It should be understood that the properties in the following list may be varied over a considerable range without destroying transistor action, which makes it possible to apply these principles to a large variety of special-purpose devices.

8-3.1. Emitter Resistivity. The emitter shall have a resistivity considerably smaller than that of the base.

8-3.2. Base Resistivity. The base region shall have a resistivity high enough that the carriers injected by the emitter do not recombine with majority carriers of the opposite type in the base region before reaching the collector junction.

8-3.3. Collector Resistivity. The collector region shall be, in general, of high resistivity to ensure that the carriers gathered by the collector junction will not recombine before reaching the collector contact. (We shall, however, consider devices in which the collector is of low resistivity in special configurations.) This resistivity should be matched to the circuit in which the device is to be operated so that the resistance of the collector is not too large compared to the load resistance. If it is too large, the bulk of the amplified power will be dissipated in the collector region and be unavailable to perform work in the external circuit. The device must be designed with both these requirements in mind. It is customary to use material of high resistivity to avoid the recombination problem but to maximize the cross section and contact area of the collector in order to reduce its ohmic resistance in the circuit.

8-3.4. Emitter Bias. The d-c emitter bias must be in the forward direction and must be high enough that the impressed alternating signal can only modulate the bias voltage without reversing it.

8-3.5. Collector Bias. The collector bias must be in the reverse direction and shall be as high as possible without breakdown at the junction. This permits the use of a large external load resistor and, since the power is given by the I^2R developed in the load, the increased reverse current will result in a maximum power output.

8-3.6. Base Width. The base width shall be as small as practical, the lower limit being interference from one junction with the other and the ability to make an electrical contact. These considerations should be obvious, since recombination in the base region must be minimized.

8-3.7. Base Lifetime. The material composing the base region shall have as high a lifetime as possible. This will further minimize recombination.

8-3.8. Open-circuit Collector Current. When the emitter circuit is open, the collector current shall be as small as possible. This will prevent excessive heating of the device, contribute to sensitivity, and facilitate the design of the external circuit.

8-3.9. Contacts. All contacts must be ohmic and nonrectifying for efficient operation. This subject will be treated in detail in a later chapter.

8-3.10. For Any Transistor. Since the nine conditions listed above apply to a power-amplifying transistor, it will be of interest to enumerate the four basic necessities for all transistor action, regardless of the application. They are:

1. The conductivity of the emitter must be greater than the base conductivity (cf. 8-3.1).
2. The emitter must be forward-biased (cf. 8-3.4).

3. The collector must be reverse-biased (cf. 8-3.5).
4. The base thickness must be minimized (cf. 8-3.6).

Note that nothing is said of the impurity concentration in the collector. It is in general lower than that in either the emitter or base, but transistor action does not basically depend on this.

Although we chose as our model an n-p-n transistor, other configurations involving two or more junctions are equally possible and sometimes desirable. The p-n-p configuration, for example, is widely used, and the same parameters apply to it. The designer has a wide latitude in both configurations and properties, which is evident from the large variety of transistors being produced in commercial quantities.

8-4. CURRENT RELATIONSHIPS

8-4.1. Junction Currents.

In our previous discussion we have considered only the flow of majority carriers, that is, electrons for an n-p-n structure and holes in the p-n-p case. We have also assumed that all emitted charge carriers arrive at the collector. Unfortunately, this is not entirely true. A small fraction of the emitted charge carriers (minority carriers in the base) disappear as the result of recombination which takes place in this region. Also, there is a finite hole current flow from the base connection to the emitter junction. We should, in addition, remember that in the flow of current through the collector and external circuit a portion must be carried by the minority carriers, or holes in this case. We shall now discuss these and other junction current components.

In order to assess the complete current relationships in our two-junction amplifying transistor, we shall reexamine the currents which flow at each junction under the conditions of bias that we have defined as necessary for amplification by the present device. (Later, we shall broaden our outlook to include certain special types of devices in which some of these assumptions, implicit for the amplifying transistor, are not required.)

We first go back to our original discussion of current flow at a p-n junction and remember that we identified four current components. Therefore, in a two-junction device we have to deal with eight current vectors, although we shall be able to ignore most of them by a consideration of their relative magnitudes.

We consider the emitter junction and remember that this is the forward-biased case of the unsymmetrical junction previously discussed. The principal mechanism of conduction is the injection of carriers into the region beyond the junction, where the injected carriers become minority carriers. Then, by diffusion and space-charge neutralization in the bulk material beyond the junction, these carriers produce a large forward current through the two regions adjacent to the junction and also through the external circuit. It was mentioned that 99 per cent of the forward current was carried by the injected carriers (majority carriers in the region of their origin and minority carriers once across the junction). In our present example, the emitter is an n-p junction and emitted electrons account for this high percentage of the current. The other 1 per cent is, of course, the corresponding reverse flow of holes. Note that these are both I_f currents that proceed by virtue of diffusion along a concentration gradient.

As before, we find the thermally generated drift currents I_g to be negligible in the case of the unsymmetrical forward-biased emitter junction. We need, therefore, to consider only the I_f components, both of which are significant in calculations of transistor performance. For the unsymmetrical reverse-biased collector junction, and especially at high values of this bias, we have seen that the I_f diffusion currents of majority carriers are negligible. Since, however, we are dealing with a reverse-biased junction, we must evaluate the reverse saturation current, composed of the I_g components. These correspond to the thermally generated holes and electrons, which are accelerated through the space-charge region and collected on opposite sides of the junction. Further, it is reasonable to expect, based on Fig. 6-8, that the hole current I_g corresponding to minority carriers from the high-purity region is greater than the opposing electron current from the impure toward the high-purity region owing to differences in collector efficiency. In other words, the flow of carriers due to drift current from the collector region to the base would predominate for the amplifying structure when the collector junction is more heavily doped in the base region.

The above mechanism states that the hole component of the reverse saturation current in an n-p-n transistor collector may be large in comparison to the electron component. Since transistor action depends on the injection of electrons and their subsequent collection at the reverse-biased collector, we must consider the possible effect on this mechanism of a reverse current composed largely of holes flowing from the n-type collector into the p-type base region.

By referring to Fig. 8-1, we find that a current of holes originating in the collector junction might easily find an alternate circuit path through the emitter, which is no longer considered to be open-circuited as in our previous example. This current may be larger or smaller than the I_s reverse current of 0.5 ma previously mentioned. Such a situation is suggested in Fig. 8-3, in which we have also shown the I_f component due to holes at the emitter. In the common lead to the base region, we have arbitrarily shown all possible currents, since we cannot now decide whether holes or electrons will predominate or in which direction they will flow. (If it is recalled that a flow of holes in one direction in a conductor cannot be distinguished from an equal flow of electrons in the opposite direction, the current flow in this lead might be represented either way. We shall, however, be interested in the relation of this current component to other components, and therefore we prefer to leave it unspecified at this point.) Note that the signal generator and load are omitted.

If we examine our model in terms of the earlier example, we recall that the emitter current was 100 ma (99 per cent of which is carried by electrons in the forward direction across the emitter junction). It was further postulated that the static reverse current I_s was 0.5 ma with the emitter circuit open and that 100 per cent of the electrons from the emitter reached the collector and were there collected. From these values we cannot tell how large I_s is under operating conditions or whether it is composed of electrons or holes. We know that the current flowing in the collector-base circuit as a result of the signal from the emitter must be (0.99)(100),

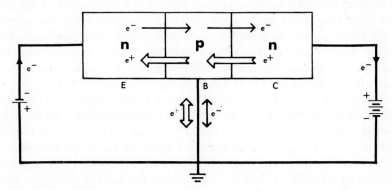

FIG. 8-3. Electron and hole currents in an n-p-n transistor circuit.

or 99 ma. This represents a change by a factor of almost 200 in the collector reverse current as a result of comparing the static value with the operating value (Sec. 8-2).

Let us now propose that, as a result of connecting the emitter lead, I_s no longer has the value of 0.5 ma but has some other value. Let us also postulate that this is because the hole component of I_s finds its way back through the circuit of the emitter. The ratio between the emitter signal and I_s may now be larger or smaller than 200. Theoretically, it can become infinite but cannot vanish.

A further complication might be mentioned. If the hole component of I_s were large and the holes were to recombine in the base region with electrons, the number of electrons might be reduced, thereby decreasing the collector output current and the apparent ratio of amplification.

Since we have no good means of measuring these internal currents and yet require a basis for interpreting transistor behavior, a set of parameters that allow us to analyze the internal conditions of an n-p-n structure has been devised. From our knowledge of junctions and minority carrier recombination mechanisms, we may identify three ratios each of which has a bearing on the response of the collector circuit to a signal from the emitter. Since we have touched on each of these in our previous discussion, they will be explicitly defined in these paragraphs and their further quantitative treatment will be left to the reader. For normal purposes they will appear henceforth as a combined parameter α, which will now be derived in terms of these ratios [1,2].

We have already mentioned that a forward-biased unsymmetrical emitter produces a forward current made up almost entirely of majority carriers in the low-resistivity region which become minority carriers upon crossing the barrier into the base region. We stated that about 99 per cent of this forward current is so comprised. Our first ratio is therefore the emitter minority carrier injecting efficiency, or η, defined as the fraction of the number of carriers supplied to the emitter from the external circuit which appear as minority carriers across the junction in the base region. Quantitatively, the ratio η is a function of relative impurity levels and the physical geometry and perfection of the junctions. Practically, it has values of 0.98 or better in most cases.

Next is the recombination of injected minority carriers in the base region. Since there are many carriers of opposite type, some of the injected carriers are annihilated before crossing the base region into the collector junction. Stated simply, this ratio, or transport efficiency, is a function of the minority carrier lifetime and the physical base width for ordinary frequencies. The ratio ζ defined as the ratio of minority carrier current arriving at the collector to minority carrier current introduced by the emitter, i_e, normally has values from 0.920 to 0.999 or higher, depending on a variety of factors many of which are strongly related to fabrication techniques.

Finally, there is a factor due to the junction at the collector. We have already discussed collector efficiency (Sec. 6-6) and avalanche breakdown (Sec. 5-8), both of which operate in determining the so-called "collector junction multiplication factor" M. Collector efficiency is an evaluation of the fraction of minority carriers arriving in the space-charge region which are swept up by the collector junction and added to the majority carriers already in the bulk collector region, thereby contributing to the appropriate current vector I_g. We have already seen that for an unsymmetrical junction the collector efficiency can be less than 100 per cent. However, for a transistor, because of the proximity of the emitter junction, the efficiency will be close to 100 per cent, depending on the relative impurity concentrations on either side of the collector junction. In any case, this effect is minor.

More important is the onset of incipient avalanche breakdown as a result of the normally high reverse bias on the collector junction. It can be shown that at reverse biases equal to about one-half of the PIV some avalanche breakdown is present; so that there is an increase in the conductivity. The increase represents a relatively small increment above the current carried by the collected minority carriers from the base region, but it can be significant in critical applications. This breakdown effect cannot be independently measured in the normal transistor structure. In silicon, for reverse biases of about one-half the PIV, M may be of the order of 1.05.

The ratios η, ζ, and M are usually lumped as a product and known as the α for the transistor. The parameter α is also defined as the current

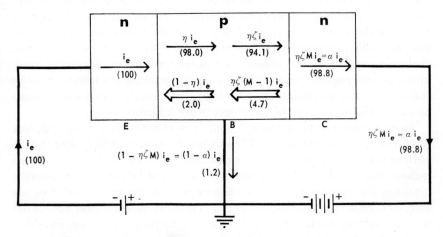

FIG. 8-4. Internal currents and current ratios for n-p-n transistor.

ratio I_C to I_E, where I_C is the collector current corresponding to an emitter current of I_E. The ratio α is normally less than but close to unity.

In terms of the above, using a value of 0.96 for ζ,

$$\alpha = \eta\zeta M \tag{8-1}$$
$$\alpha = (0.98)(0.96)(1.05) = 0.988 \tag{8-2}$$

which is a typical value.

Figure 8-4 shows the structure of Fig. 8-3 with the current arrows appropriately labeled in accordance with the definitions of η, ζ, M, and α. Note that we have elected to show the current flow in the base lead in terms of the input current i_e composed of electrons and known as the emitter current. Obviously, this could be represented by a hole flow with the arrow pointed toward the base region. Indeed, in the case of the conjugate, or p-n-p, structure it would seem more logical, from some points of view, to represent the current flow pattern as holes both within the semiconductor region and in the external circuit. If we recall, however, that there are no holes in metallic conductors, the convention of using only electron flow in external leads will be appreciated.

By applying the above ratios to the hole component of I_s from Fig. 8-4

$$I_s \text{ (holes)} = \eta\zeta(M - 1)i_e \qquad (8\text{-}3)$$

and, assuming that $i_e = 100$ ma, we have

$$(0.98)(0.96)(1.05 - 1.00)(100) = 4.7 \text{ ma}$$

and the hole current is found to be much larger than I_s (as previously measured with an open-circuited emitter). We still are unable to fix a value for the electron component of I_s with the transistor connected as shown. This is, however, unimportant, since we now have a means of evaluating both forward (electron) and reverse (hole) currents at each junction under operating conditions and of predicting the behavior of the external circuit, all from a knowledge of the separate ratios which make up α. (The magnitudes of the currents are indicated in Fig. 8-4.)

To determine the values of η, ζ, and M is experimentally difficult, but the method can be outlined. The emitter efficiency η will be less than 1 both at low emitter currents (owing to recombination in the space-charge region) and at high currents, owing to changes in the base resistivity resulting from conductivity modulation, causing the conductivity of the base to exceed that of the emitter. Between these extremes, and in the region of interest for transistors of the simple amplifying variety, the value of η reaches a maximum given by [3]:

$$\eta = \left(1 + \frac{L_b \sigma_b}{L_e \sigma_e}\right)^{-1} \qquad (8\text{-}4)$$

where L_b and L_e are the diffusion lengths of minority carriers in the base and emitter, respectively. In Eq. (8-4) we may express the diffusion length L as

$$L = (D\tau)^{1/2} \qquad (8\text{-}5)$$

where τ is the lifetime of the carrier whose diffusion length is L and D is the diffusion constant, in square centimeters per second, for the minority carriers in the region of interest. Equation (8-4) then becomes

$$\eta = \left[1 + \frac{(D_b\tau_b)^{1/2}\sigma_b}{(D_e\tau_e)^{1/2}\sigma_e}\right]^{-1} \qquad (8\text{-}6)$$

a result we shall return to later.

The base transport efficiency ζ is given by [3]

$$\zeta = \text{sech}\, \frac{W_b}{(D\tau)^{1/2}} \tag{8-7}$$

or

$$\cong 1 - \frac{W_b^2}{2D\tau} \tag{8-8}$$

where W_b is the effective geometric base width in centimeters, D is the diffusion constant (diffusivity) as before, and τ is the lifetime of minority carriers in the base region. In the normal case, where $W_b^2 < D\tau$, Eq. (8-8) is applicable and much simpler.

The quantity M for a step (alloy or fused) junction is given by [4,5]

$$M = \frac{1}{1 - (V/BV_{CB})^m} \tag{8-9}$$

where V is the applied reverse bias, BV_{CB} is the normal breakdown voltage of the collector-base junction (also called PIV), and m is an empirical quantity which depends on the resistivity and conductivity type of the material on the high-resistivity side of the junction. For silicon, m ranges from values of 1 to 4 when the high-resistivity side is n-type and from 1 to 2 when it is p-type. It is therefore possible to obtain reasonable estimates for M if the normal BV_{CB} value is known and the junction is of the abrupt type.

In Eqs. (8-6) and (8-7) all the factors D, σ, and τ are functions of the injection level and, therefore, of the emitter current. Lifetime τ is also dependent on many other factors affecting the base region, such as crystal perfection, surface effects, and impurities. Direct measurement of τ is very difficult under actual use conditions. It is therefore necessary to approach the problem indirectly. If W_b is small, the value of ζ will be very near to 1. If the injection level i_e across the junction is varied until i_c reaches a maximum, η can be approximated for at least one range of collector bias values.

Values for M, η, and ζ having been obtained (perhaps by using a special physical structure in which W_b is very small), it is possible to measure α for the transistor as a whole and check the calculations for this set of conditions only. For graded junctions and other values of bias and injection level, the problem is handled by a series of approximations. Other indirect techniques have been worked out, but they are beyond the scope of this discussion.

In summary, we have reduced the number of junction current components from the rigorous total of eight to the more workable and practical total of three. These are (1) the total emitter current I_E, (2) the total collector current (reversed for the collector junction) I_C, and (3) a third current component, identified as I_B, representing the net difference between the flow of holes from the base region through the emitter junction to the emitter contact and the hole component of I_s from the collector to the base.

From the foregoing discussion, we assign the following values to the n-p-n structure under consideration:

$$I_E = 100.0 \text{ ma} \qquad (8\text{-}10)$$
$$I_C = \alpha I_E = 98.8 \text{ ma} \qquad (8\text{-}11)$$
$$I_B = (1 - \alpha)I_E = 1.2 \text{ ma} \qquad (8\text{-}12)$$

Although I_B is small by comparison with I_E and I_C in this particular case, it can become larger and very significant under other circumstances and can represent a situation of poor efficiency. Note that the α for the above case is near unity and has the same value as before:

$$\alpha = \frac{I_C}{I_E} = \frac{98.8}{100.0} = 0.988 \qquad (8\text{-}13)$$

8-4.2. Input-Output Currents. We may now examine in more detail the current-voltage relationships in the collector circuit which correspond to a given current flow in the emitter circuit. This elementary discussion will provide us with a quantitative picture of the overall performance of the amplifying transistor in terms of only the input current and the current-voltage characteristics of the output. It will serve as an example of the power amplification properties of the transistor (as opposed to the voltage-amplifying vacuum tube) and as a point of departure for the more detailed circuit analysis to follow. Based on these understandings, it will then be possible to describe other multijunction structures and to derive their most important properties. Accordingly, we choose the structure of Fig. 8-2, which is typical of a sizable class of power amplifiers.

If no voltage is applied between emitter and base, the collector current will be that of a diode under a given reverse bias. With a forward bias on the emitter, most of the carriers injected into the base region will reach the collector and the collector current will be almost equal to the emitter current. If the collector voltage is varied, the collector current will change only slightly because the reverse-biased collector junction is already collecting all the carriers which reach it. These considerations lead to collector current-voltage characteristics of the form shown in Fig. 8-5 for small-signal devices, where the curves show the effect of increasing the emitter bias and therefore I_E from zero to 5 ma.

A similar plot for a power transistor is shown in Fig. 8-6, but for a different mode of connection and with different values of the regulating base current I_B. We shall examine this mode of connection more carefully in a subsequent chapter. For the present, the striking similarity to the

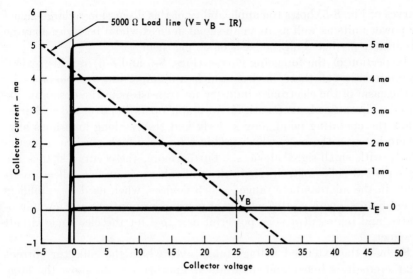

Fig. 8-5. Collector characteristics of small-signal n-p-n transistor (grounded-base).

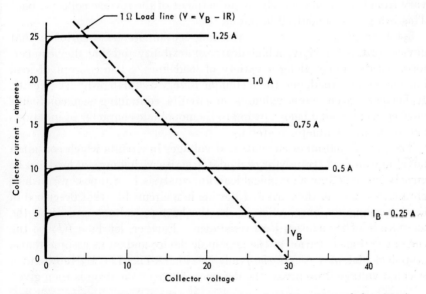

Fig. 8-6. Collector characteristics of power n-p-n transistor (grounded-emitter).

curves of Fig. 8-5 shows the applicability of this discussion to large-signal or power units as well as to small-signal devices whose behavior may be accurately predicted on the basis of simple theory.

Inspection of the foregoing curves (Figs. 8-5 and 8-6) reveals several characteristics of these devices which are important reasons for the enthusiasm of the electronics industry for transistors in the design of circuits. First, the characteristic shows stable operation over a broad area. Thus the operating point moves freely and stably along load lines anywhere between the axes, which provides for very high efficiency, particularly with small-signal devices. Furthermore, these curve shapes are preserved at collector voltages in the millivolt range and at collector currents in the microampere range. Such devices, when used as amplifiers or oscillators, operate satisfactorily at power consumptions of a few microwatts, and represent a very powerful new tool for the electronic circuit designer.

Second, the extended voltage range over which the collector current characteristic is linear and invariant is another way to show the large power gains and constant output current characteristic of transistors. Inspection of Fig. 8-5 shows, for an emitter current of 1 ma (and an input voltage, not shown, of perhaps 0.5 volt) a stable response in the collector from 0 to 45 volts. In other words, for an input of 0.5 mw the output can vary from 0 to 45 mw merely by adjustment of the reverse collector bias. This is a gain of about 90 in power.

8-4.3. Load Lines. In the preceding section it was mentioned that devices of this type have a high degree of flexibility and that they can perform with stability along a variety of load lines. The concept of load line analysis is broad, and the technique is very useful in ordinary circuits. It becomes even more valuable in circuits containing semiconductor devices, as discussed below, owing to the nonohmic character of junctions and bulk semiconductor materials.

The determination of currents and voltages in circuits which contain a nonohmic device* (transistor or diode) in series with a normal load can be greatly simplified by a graphical load line analysis.† Suppose we wish to obtain the value of the current I flowing in a circuit like that described in the preceding sentence. Let V_B be the battery voltage and R be the magnitude of the chosen load resistance. Further, let $V = f(I)$ be the current-voltage equation of the nonohmic device and let us assume that a graphical plot of $V = f(I)$ is available (Fig. 8-7). Let V_A represent a selected voltage drop across the device. This voltage drop is then given

---

* A nonohmic device is one which does not possess a linear current-voltage characteristic; that is, it does not obey Ohm's law.

† Adapted with permission from J. N. Shive, courtesy of D. Van Nostrand Company, Inc. [6].

analytically by the equation

$$V_A = V_B - IR \qquad (8\text{-}14)$$

The voltage drop at this same point is also given by

$$V_A = f(I) \qquad (8\text{-}15)$$

and V_A is obtained by the simultaneous solution of these two equations.

This solution can be obtained graphically as shown in Fig. 8-7 by drawing a plot of the device characteristic $V = f(I)$ and superimposing on it a straight-line plot of $V = V_B - IR$, where R is the selected load resistance.

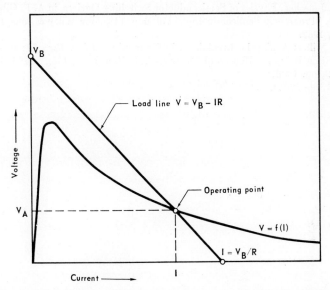

FIG. 8-7. Load line analysis. (*After J. N. Shive, courtesy D. Van Nostrand Company, Inc.* [6].)

This straight line, called the load line, has the intercept V_B on the voltage axis, the intercept V_B/R on the current axis, and a negative slope of R. The point at which it intercepts the device characteristic curve gives an operating point corresponding to the voltage drop V_A for the device in question and defines the current flowing through the device at this time as well.

When it is desired to d-c-bias a device to some chosen operating point, suitable values of battery voltage and load resistance are selected so that the resulting load line intercepts the plotted device characteristic at the desired operating point.

In conclusion to this section, it should be mentioned that the foregoing examples represent only a few typical cases for junction transistors and have been included as illustrations to help us understand the operation of two-junction devices. In the next chapters we shall discuss the use of two-junction devices in a number of applications, and we shall consider their current-voltage characteristics under several different common modes of operation.

REFERENCES

1. J. N. Shive: "Properties, Physics and Design of Semiconductor Devices," chap. 21, D. Van Nostrand Company, Inc., Princeton, N.J., 1959.
2. A. B. Phillips: "Transistor Engineering," pp. 157–162, McGraw-Hill Book Company, New York, 1962.
3. A. K. Jonscher: Physics of Semiconductor Switching Devices, in A. F. Gibson et al. (eds.), "Progress in Semiconductors," vol. 6, p. 148, John Wiley & Sons, Inc., New York, 1962.
4. S. L. Miller: Avalanche Breakdown in Germanium, *Phys. Rev.*, **99**: 1234 (1955); Ionization Rates for Holes and Electrons in Silicon, *Phys. Rev.*, **105**: 1246 (1957).
5. Ref. 2, pp. 133ff.
6. Ref. 1, pp. 52–54.

9: Transistor Amplifier Circuits

9-1. INTRODUCTION

In the preceding chapter we discussed a single type of transistor in a common, but by no means universal, circuit. Our purpose was to understand the physical mechanisms involved in the application of an n-p-n transistor (with a base lead common to both the emitter and collector circuits) to problems of current and power amplification. By thus limiting our range of interest we were able to achieve a basic understanding of semiconductor junction properties in a two-junction configuration. We now wish to broaden this approach and to consider other connection arrangements, circuit calculations, electrical performance calculations, and some of the limitations of such a simplified approach. We shall, in addition, learn enough transistor electronic theory to understand other devices of specialized types which will not be discussed in detail here, as well as new discoveries in the field.

We shall consider small-signal behavior of transistors in each of the three possible arrangements in a circuit, and we shall then learn how to calculate d-c and low-frequency a-c characteristics of the device in terms of such parameters as input impedance, output impedance, and amplification factors by means of the "black-box" equivalent. A qualitative discussion of performance under other conditions of frequency, etc., will also be offered, with comments on phase changes for each of the various connection modes.

It is obviously impossible to cover even a small area in this enormous and complex field completely. However, having obtained a clear understanding of the fundamentals and an appreciation of the controlling factors involved, the reader should be able to follow through a major part of the current literature without too much difficulty.

9-2. CONNECTION ARRANGEMENTS

We have thus far spoken of an arrangement wherein the base lead is common to both the emitter and collector circuits. This is known as the common-base or grounded-base configuration. The other possibilities obviously will involve the common- (or grounded-) emitter or collector configurations. Indeed, if the reader refers to Fig. 8-6, he will note that the characteristics of the power transistor were those of the grounded-emitter configuration rather than the common-base one. This choice was deliberately made to show the very similar behavior to that of a small-signal transistor. Curves for the common-base connection operating as a power transistor could have been shown, but since they deviate somewhat from a linear response (primarily because of base geometry effects) the parallel between Figs. 8-5 and 8-6 would not have been so striking.

In the paragraphs to follow, we shall illustrate the three* possible modes of connection, showing bias batteries (although we shall later see that more elegant bias sources are more desirable) and electron current flows through the two loops formed by each connection pattern. As in Fig. 8-2, we shall show the input generator and the output, or load, resistor. (We elect to show the components of the input signal source as two separate elements e_G, a zero-resistance voltage generator, and R_G, the internal resistance of any practical device if used as a voltage source. The reasons will become apparent as the discussion proceeds. Basically, this presentation permits us to "turn off" the signal source e_G or remove it to a new position in the circuit and, at the same time, leave R_G in place.) Finally, we shall present some overall considerations and a table of typical circuit parameters.

9-2.1. Grounded Base. This configuration is shown in Fig. 9-1. As previously discussed, the emitter is forward-biased and the collector is reverse-biased, which provides electron current flow as shown and as discussed earlier.

9-2.2. Grounded Emitter. In this configuration the input signal and bias battery for the emitter have merely been moved to the base lead. However, the effect of e_G has reversed itself, since in this circuit, if the signal from e_G to the base becomes more negative, less current will flow in the emitter circuit, and the collector current will reflect this drop. As a consequence, the signal seen across the output terminals is reduced, making the upper terminal more positive. In this case, therefore, there has been a phase reversal. Note that the current arrows still flow in the same basic direction, as must be true for a transistor structure according to our definition of its electric functioning. All that has happened is that

* There are obviously three other possible arrangements, since the transistor is a three-lead device. However, only those to be discussed here result in transistor action. The reader can easily verify this by recalling that the input must always include the emitter junction, the output must always include the collector junction, and, obviously, no single lead can serve as both input and output. Under these restrictions, any other modes of application will prove to be variations of the three to be discussed herein.

a variation of the input signal in the negative direction results in a corresponding variation of the output signal in the positive direction. In a-c terminology this is a phase reversal. Its d-c equivalent is a negative signal appearing as a positive output.

One might, by reference to Fig. 9-2, guess that if e_G becomes more negative, the influx of electrons into the base region will act to increase the current at the collector junction, just as if the electrons had originated by injection across the emitter junction. This is not true for the following reason. The relative number of electrons in the base region due to this signal is very small compared with the numbers which would be injected

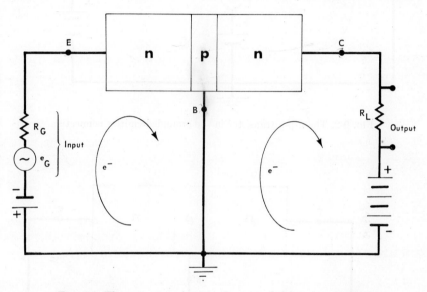

FIG. 9-1. The n-p-n transistor in the grounded-base connection.

by the emitter for a comparable voltage signal. This is due to the higher resistivity of the base region, which results in a relatively large reverse emf for a small current of electrons arriving at the base lead. This current therefore produces a reverse potential at the emitter junction which greatly reduces the injection.

Other aspects of the performance of a transistor in the grounded-emitter configuration will be apparent later when it is compared in detail with the grounded-base type. Some of these inherent properties account for the popularity of the grounded-emitter connection.

9-2.3. Grounded Collector. The last, and least common, connection for a transistor is the common-collector, as illustrated in Fig. 9-3. In this rather odd configuration the input remains in the base, while the output is now from the emitter. In this case, if the signal becomes negative,

132 *Semiconductor Junctions and Devices*

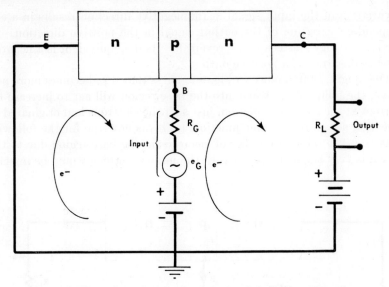

Fig. 9-2. The *n-p-n* transistor in the grounded-emitter connection.

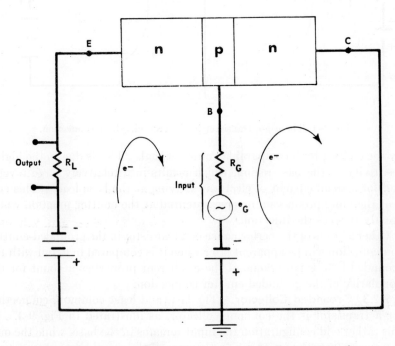

Fig. 9-3. The *n-p-n* transistor in the grounded-collector connection.

Transistor Amplifier Circuits   133

there is a corresponding negative output and therefore no phase reversal. Other distinctions of this type of configuration will be mentioned in later sections.

9-3. GENERAL COMPARISONS

In addition to the question of phase reversal, there are several other comparisons which may be made at this stage. Some of them will be stated qualitatively and later justified by actual circuit calculations.

9-3.1. Bias. In all cases shown above, the collector and emitter loops have separate bias sources. Careful inspection of Figs. 9-1 to 9-3 will show that only in the case of the first configuration (grounded-base) are two power supplies or bias sources needed. This results from the opposing polarities of the biases in this case. In both of the other cases it is possible to make a single bias source serve. This point is illustrated in Fig. 9-4, which also serves to introduce the standard symbolism for a transistor. [Note that the emitter is always drawn with an arrow in the direction of hole flow. This simple device tells us at once that the emitter is injecting holes or electrons (by reversal of the arrow) and therefore whether we are dealing with a p-n-p or an n-p-n structure.] In Fig. 9-4 we have shown the n-p-n transistor connected as common-base, common-emitter, and common-collector with the bias sources separate as in our earlier examples and again with the bias sources combined where possible. (In the case of the grounded-collector, an additional resistor R_A has been added. In series with R_G, it forms a voltage divider for proper bias of the base.) Also shown is a p-n-p structure.

9-3.2. Gain. In this discussion we shall anticipate the numerical results of later sections and speak qualitatively of the current, voltage, and power gains to be expected from the three configurations. We may do this from our previous knowledge of the resistivities of typical two-junction structures and a consideration of external circuit elements as they are normally rated.

To estimate the gain of a transistor, we must consider both current and voltage separately and, again by itself, the power gain. In the grounded-base connection we know that the emitter current cannot be amplified but that the voltage gain is very large. Therefore, the power gain will also be high by virtue of the large value of R_L, the load resistor. As we shall see, no other connection shows a startlingly higher voltage gain, although one greatly excels in overall power gain.

For the grounded-emitter case we have already mentioned that a very small change of signal current (applied to the base region in this case) produces a disproportionately large change in emitter current and, therefore, in the collector current. We therefore have a very considerable current amplification (of the order of 25-fold). Since this current now

134 Semiconductor Junctions and Devices

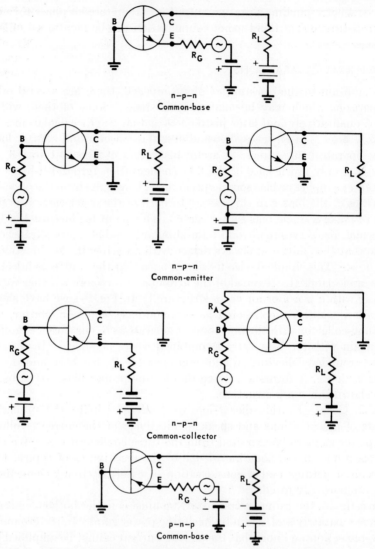

Fig. 9-4. Biasing connections for different configurations.

flows through R_L, we might expect a proportionate increase in voltage gain over the corresponding value for the grounded-base case. However, inspection of the circuit diagram in Fig. 9-2 shows that this current is actually flowing through R_G and its bias battery also. In practice, therefore, the voltage gain is in the range of several hundred to one, which is of the same order of magnitude as the grounded-base situation.

Since the current gain is high and the voltage gain is also high, it is not surprising that the grounded-emitter configuration has the highest power gain of all three and, in addition, a very respectable high-voltage output. Finally, we have the grounded-collector connection. By analogy with the grounded-emitter one, we expect high current gain, since the signal is again applied to the base lead. Indeed, for some applications, the current gain for this arrangement is even higher than that of the previous example. The voltage gain, however, is entirely different. In this configuration the current appearing at the collector terminal may divide between the base lead R_G and the emitter circuit containing R_L. As a result, and owing to bulk resistance, only a small part of the current flows through R_L and the voltage gain is essentially 1 or less. Consequently, the power gain is found to be the poorest, usually less than 1.

It should be noted that the calculation of power gain for any nonohmic device is substantially more complicated than determining the product of current and voltage gains, particularly since the latter are usually reported as maxima determined under optimized conditions and without regard for the other parameter. We shall later develop exact expressions for these gain ratios and their interdependence.

9-3.3. Input and Output Resistances. These properties of a device are of great significance in circuit design. This is particularly true because semiconductor devices depart widely from vacuum tubes in these areas and considerable ingenuity is required to devise compatible circuits to replace the older vacuum-tube versions.

Starting with the grounded-base connection, it is no surprise that the input resistance (low frequency) is very low. We know that signals of very small voltage can profoundly affect the emitter and, therefore, the collector current. Similarly, owing to the high resistivity of the collector region under reverse bias and the large load resistance in series, we are not surprised that the output impedance is high, indeed the highest of all three. We have, in fact, in this one method of connection the lowest input resistance and the highest output resistance.

Coming now to the grounded-emitter connection, we find that the signal enters the base region instead of the emitter. We know this region to be of higher resistivity (about 50 times) but also to be very thin, so that there is little distance to travel before reaching the emitter. It is therefore not surprising that these two effects balance off to give an input resistance about ten times larger than in the grounded-base situation but still, from a practical point of view, quite low.

For the grounded emitter the output resistance is less than that of the grounded base because, even if R_G is made very large (open circuit) as it usually is for this measurement, we still have a shunt path back to the emitter, which has very low resistivity and which therefore reduces the output resistance as seen at the terminals bracketing R_L. This effect is not overwhelming, but it reduces the very large output resistance of the grounded-base connection by a factor of 5 or more. This is still reasonably high, but it is not in the same class with many vacuum-tube circuit elements.

Finally, we come to the grounded-collector connection, and we find that the input resistance is quite respectably high, though still an order of magnitude lower than nontransistor circuits require in many cases. Reference to Fig. 9-3 shows us at once that we are now dealing with a current path made up of the base resistance plus two very high resistances, R_L and the collector region resistance, in parallel. Since both the latter are in the megohm range, it is not surprising that this configuration has the highest input resistance of the three, being about 1,000 times higher than that of the grounded-emitter case and comparing favorably with vacuum tubes.

TABLE 9-1. PROPERTIES OF TRANSISTOR AMPLIFIER CONNECTIONS*

Property	Grounded base	Grounded emitter	Grounded collector
Current gain	Low	Highest	High
Voltage gain	High	Highest	Low
Power gain	High	Highest	Low
Input resistance	Lowest	Low	High
Output resistance	Highest	High	Low

* After D. M. Warschauer, by permission [1].

On the other hand, the output resistance for the grounded-collector system is very low, about 1,000 times less than its nearest neighbor, the grounded-emitter system. The reason for this is clear from the position of the output in the emitter lead. The situation is almost the same (if R_G is very large) as the measurement of the input resistance for the grounded-base connection, except that in the latter case there was no external resistance in the base lead and R_L was in the collector lead. We now have an open circuit in the base lead, but no external resistance in the collector lead. Remembering the effect of reverse bias at the collector junction, it is not too surprising that the effective output resistance is very low.

9-3.4. Summary. The above data are summarized in Table 9-1. From all that has been said, the most important properties associated with the individual circuit hookups may be stated as follows, omitting, for the present, the question of high-frequency performance. The common-base configuration has the advantage of very high output resistance, good voltage and power gains, and no phase reversal. It resembles a step-up transformer, and it has low input resistance. It also has the disadvantage of very low or negative current gain, and it requires two separate bias power sources. Because of the voltage and power gain, this configuration is frequently seen, especially if a match from low to high resistance (impedance) is needed in connection with power amplification.

The grounded-emitter connection has the big advantage of highest current gain, voltage gain as high or higher than any other, and therefore the largest power gain. Another advantage is the ability to use a single bias supply. It suffers from low input resistance (intermediate between that of the grounded-base and grounded-collector types, but on the low side), has a somewhat lower output resistance than that of the grounded-base type, and always produces phase inversion. However, as will be seen in later sections, it is possible to adjust the transistor structure and the external circuit parameters so that high input impedances may be achieved, but with some sacrifice in gain.

The net result of these pros and cons is that the grounded-emitter configuration, which is the only one showing both highest voltage and highest power gain, also has a very high degree of flexibility and is consequently the most frequently used for almost all applications, particularly amplifiers.

In the grounded-collector circuit we find a high current gain and very high input impedance but very poor voltage and power gains as well as low output resistance. This configuration therefore resembles a step-down transformer analogous to the cathode-follower action of a vacuum tube. In addition, the range over which the output (from the emitter) may show voltage variation is very limited, which further limits application. In practice, therefore, this configuration is most widely used to match a high-resistance input to a low-resistance output. As further advantages it should be recalled that this circuit shows no phase inversion and requires only a single bias supply.

9-3.5. Typical Parameters. Before concluding this chapter, we present a list of the parameters needed for quantitative understanding of the general picture discussed above. Because it is usually easier to appreciate the significance of algebraic reasoning if numerical examples are given, we shall use the data of Table 9-2 in our future discussion. The values given are average values for a wide variety of possible conditions and circuit configurations. Nevertheless, they are valid within a factor of 2 for the overwhelming majority of situations, and they are excellent practical working values for small-signal, low-frequency junction transistors.

In accepted notation, lowercase letters refer in general to internal values of the transistor when speaking of resistances and capital letters refer to

138 Semiconductor Junctions and Devices

external circuit elements which may be independently varied. Therefore, r_e, r_b, r_c, and r_m are respectively the resistances of the emitter, base, collector, and a "mutual" resistance about which we shall learn shortly. R_G and R_L are the resistances of the voltage (signal) generator and the load. Both Greek letters α and β refer to current amplification; β will be defined later. Finally, e_G is the voltage output of the signal generator

TABLE 9-2. TYPICAL JUNCTION
TRANSISTOR PARAMETERS*

Parameter	Value†
r_e	25 ohms
r_b	500 ohms
r_c	1.0 megohm
r_m	0.96 megohm
R_G	500–5,000 ohms‡
R_L	0.1 megohm‡
α	0.98§
β	25¶
e_G	0.001–0.1 volt

* After A. Coblenz and H. L. Owens, by permission [2].
† Average values considering both n-p-n and p-n-p junction transistors.
‡ Often matched; with values of 0.1 megohm each.
§ Grounded base only.
¶ Grounded emitter only.

considered to be independent of the internal generator resistance R_G. (The use of a lowercase symbol for e_G is a matter of established practice, even though the signal generator is an external circuit element.)

REFERENCES

1. D. M. Warschauer: "Semiconductors and Transistors," p. 186, McGraw-Hill Book Company, New York, 1959.
2. A. Coblenz and H. L. Owens: "Transistors: Theory and Applications," p. 100, McGraw-Hill Book Company, New York, 1955.

10: Black-box Equivalent—Grounded Base*

We have now seen the transistor in three different circuit configurations, and we have deduced its approximate behavior with respect to a number of variables having direct and important bearing on the application of such a device in an actual circuit. We have not, however, been able to assign quantitative values to our variables. We now wish to undertake the mathematical solution of the transistor problem and to derive exact expressions for quantities such as gain and input impedance.

Since we know that the transistor is a nonohmic circuit element and must be treated as an unknown in terms of ordinary d-c circuit behavior, we shall approach the problem by means of the black-box or equivalent-circuit technique. The effect of this approach is to avoid physical equations whose solution is cumbersome, if not impossible, and to let us obtain information by the solution of ordinary algebraic expressions based on Ohm's law.

10-1. METHOD

To apply this technique, we adopt the viewpoint that the transistor is a sealed, mysterious mechanism with a pair of input leads and a pair of output leads. We shall make conventional measurements by means of these terminals and deduce therefrom the properties of the circuit elements in the sealed black box. We shall then try to construct a circuit which has the properties of the black box. Our first task, then, is to fully characterize the transistor in the black box by means of measurements at the external terminals [2]. From such measurements we shall then attempt to select a circuit which, if it were in the black box, would give rise to the same set of externally measured values. Any such circuit may be known as an equivalent circuit, and there will, in general, be a wide variety of

---
* Adapted with permission from the book by A. Coblenz and H. L. Owens [1]. Some parts are copied verbatim and are hereby acknowledged.

140 *Semiconductor Junctions and Devices*

possibilities. The second part of our task is to select the simplest among them, remembering that there need be no correspondence, in a physical sense, between the chosen equivalent circuit and the actual hardware within the black box.

It is beyond the scope of this book to consider all possible equivalent circuits in quantitative detail. Therefore, we shall present one of the simpler picturizations of the black-box equivalent of a two-junction transistor and show how the results derived from measurements of the black-box characteristics can be related to the three ways in which a transistor may be incorporated into an electronic circuit. (In this discussion we choose to refer to an n-p-n transistor; the behavior of the p-n-p transistor is known to be completely analogous.)

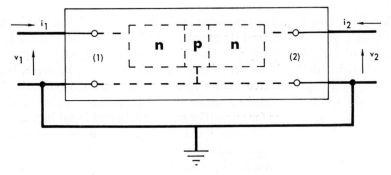

FIG. 10-1. The four-terminal black-box equivalent of a transistor (grounded base shown).

In its most general form the black box consists of an electronic element with four leads—two input leads and two output leads. In the specific case of the transistor, which is a three-lead device, it is obvious that one input and one output lead are connected in common to one of the three transistor regions. Nevertheless, the black-box concept can be used to characterize the transistor regardless of the manner in which the leads are attached. Four basic measurements can be made, and these will be found to be sufficient to characterize the four-terminal network. It is assumed that we have a current generator with zero internal resistance and a voltmeter which consumes no current. It is also necessary that we have an ammeter with no internal resistance. While these requirements cannot be fully realized, a proper selection of modern instruments will provide circuitry in which they are essentially complied with.

Figure 10-1 shows a transistor encased in a black box. The pairs of terminals labeled 1 and 2 are the input and output terminals, respectively. The lower terminals are assumed to have a common connection as shown.

We now wish to characterize the electronic element in the box by means of measurements of electrical quantities appearing at the terminals. Since, when dealing with any electrical device, we immediately think of current-voltage relationships (which in turn can be converted into resistances), we shall assume that the transistor can be characterized in terms of d-c measurements and that information derived from such measurements will permit us to make intelligent use of the transistor in our proposed circuit. (Later in our discussion we shall return to this subject and show that simple d-c measurements are not sufficient in every instance.)

The four measurements that are necessary and sufficient to characterize any four-pole d-c network are made as follows:

1. With a known current flowing through the input terminals, measure the input voltage drop.
2. With a known current flowing through the input terminals, measure the output voltage drop.
3. With a known current flowing through the output terminals, measure the output voltage drop.
4. With a known current flowing through the output terminals, measure the input voltage drop.

In cases 1 and 3, when the input and output circuits, respectively, are being measured, the opposite circuits must be open-circuited. The reader will undoubtedly be able to suggest other combinations of measurements which could be made, but experimentation will prove that no additional fundamental data on the system will be derived by other measurements.

Before proceeding with this circuit analysis, a word on the symbols which are conventionally used is needed. The resistance obtained by measuring any voltage and current of the four-pole network is denoted by two numbers in a subscript; the first number refers to voltage, the second to current. For instance, r_{12} refers to the resistance obtained from the ratio of the voltage across terminals 1 (or in circuit 1) and the current in circuit 2. In other words, the test, or signal, voltage e_G is applied to circuit 2; and the current in circuit 1, being only that drawn by the voltmeter, is essentially zero. Similarly, r_{21} is the resistance obtained from the ratio of the voltage in circuit 2 to the current in circuit 1. These are important conventions for transistor work. The other two ratios are, obviously, r_{11} and r_{22} (Fig. 10-2). Following these simple conventions, let us now write equations for the above tests which characterize the four-terminal box.

5. Apply the signal e_G to circuit 1 (which corresponds to the emitter). Then read i_1 and v_1 to obtain

$$\frac{v_1}{i_1} = r_{11} \qquad (10\text{-}1)$$

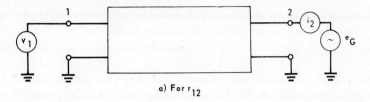

a) For r_{12}

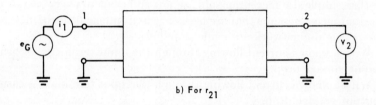

b) For r_{21}

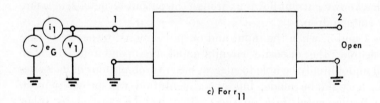

c) For r_{11}

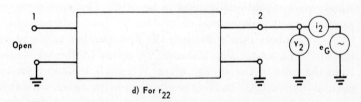

d) For r_{22}

Fig. 10-2. Circuit connections for measurement of resistance parameters.

6. Apply the signal e_G to circuit 1 as above, but read i_1 and v_2 to obtain

$$\frac{v_2}{i_1} = r_{21} \tag{10-2}$$

7. Apply the signal e_G to circuit 2 (which corresponds to the collector). Then read i_2 and v_2 to obtain

$$\frac{v_2}{i_2} = r_{22} \tag{10-3}$$

8. Apply the signal e_G to circuit 2 as above, but read i_2 and v_1 to obtain

$$\frac{v_1}{i_2} = r_{12} \tag{10-4}$$

In keeping with our earlier pattern, we have assumed, in identifying the above resistance ratio measurements, that we are dealing with a grounded-base configuration. For that particular case only, the following values of the resistance, or r parameters, would be typical:

$$r_{11} = 525 \text{ ohms} \tag{10-5}$$
$$r_{21} = 0.96 \text{ megohm} \tag{10-6}$$
$$r_{22} = 1 \text{ megohm} \tag{10-7}$$
$$r_{12} = 500 \text{ ohms} \tag{10-8}$$

Although we have been able to identify r_{11} and r_{22} with the emitter and collector circuits, respectively, by assuming a grounded-base configuration, a more general terminology is used; it has the further advantage of describing r_{12} and r_{21}. In logical fashion, therefore, r_{11} is known as the input resistance, since it is entirely determined from the input terminals with the output open-circuited. Similarly, r_{22} is known as the output resistance. (These quantities for actual transistors were discussed earlier as functions of the configuration.)

Parameters r_{12} and r_{21} are known as *transfer resistances*, with r_{12} being distinguished as the reverse transfer resistance because the signal is in the output with voltage read in the input. Conversely, r_{21} is known as the forward transfer resistance. Although these latter parameters are also d-c resistances, it is not readily possible to relate them to specific circuit paths through the transistor structure and thereby give them physical significance at this time.

If we now carry the discussion of the resistance parameters a step further, we may remove the restriction that the input and output be open-circuited (or connected through a voltmeter of infinite resistance) and consider the overall effect of applying a signal across circuit 1 and then determining the values of v_1 and v_2. If a voltage v_1 appears across the input terminals, there will usually be a voltage v_2 across the output terminals. Also, currents i_1 and i_2 will flow. Then, from Eq. (10-1) we know the contribution to v_1 caused by i_1, and from Eq. (10-4) we can find the contribution due to i_2. Therefore,

$$v_1 = r_{11}i_1 + r_{12}i_2 \tag{10-9}$$
and
$$v_2 = r_{21}i_1 + r_{22}i_2 \tag{10-10}$$

These equations relate the quantities v_1 and v_2 to i_1 and i_2 under any and all circumstances if the parameters r_{11}, r_{12}, r_{21}, and r_{22} are known and are constant. We have now accomplished our first objective, and we have

144 Semiconductor Junctions and Devices

completely characterized the behavior of the black box under all conditions of current and voltage, provided the frequency is low or the measurements are confined to direct current.

Continuing with our equivalent-circuit analysis, we recall that a black-box equivalent circuit is defined as a network of basic components which will give the same external test indications as the unknown device in the box gives. One of the simplest of the possibilities in the present case would be the T arrangement of resistors shown in Fig. 10-3 for the grounded-base example. In this arrangement, if we applied a signal between terminal 1 and ground and read the current and voltage in circuit 1 (or the emitter circuit), we could arrange matters to get the correct value of r_{11} of Eq. (10-1) and perhaps the other resistances as well.

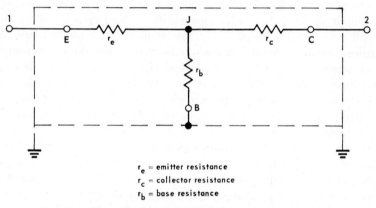

r_e = emitter resistance
r_c = collector resistance
r_b = base resistance

Fig. 10-3. Simple resistor T equivalent circuit for transistor.

However, if we pursue our tests on this equivalent T circuit, we shall find an important discrepancy. Consider tests 6 and 8. In test 6 the signal was applied to circuit 1, the current was recorded, and the voltage from terminal 2 to ground was measured. Since we are using a very high impedance voltmeter, we may consider that it draws no current and there is therefore no current flowing in circuit 2. The voltage drop across r_c can then be ignored, and v_2 measures the voltage drop from J of Fig. 10-3 to ground or across r_b. Since the test conditions are those of test 6, we have determined r_{21} as set forth in Eq. (10-2) and have found this to be r_b.

If we now perform test 8, we find that we are again measuring the voltage drop across r_b, but under conditions defined by Eq. (10-4) as yielding r_{12}. Therefore,

$$r_{12} = r_b = r_{21} \tag{10-11}$$

But from the values in Eqs. (10-6) and (10-8),

$$r_{12} \neq r_{21} \tag{10-12}$$

and they both cannot satisfy Eq. (10-11) as our simple analysis of an equivalent T circuit requires. We must look further for a more satisfactory equivalent circuit.

To fulfill our purpose, we must introduce a very important point regarding equivalent circuits or networks in general. We make use of the concept of an active network or circuit, as contrasted with a passive network. If a network is passive, it contains no generators or sources of voltage or current. Therefore, a signal passing through it can only lose energy as heat or otherwise attenuate. There can be no amplification because the circuit does not contribute to the amplitude of the signal. Therefore, in a passive network the r_{21} or forward transfer resistance is always equal to the r_{12} or reverse transfer (feedback) resistance. Since in our example we have found that r_{12} and r_{21} are very far from being equal, we correctly

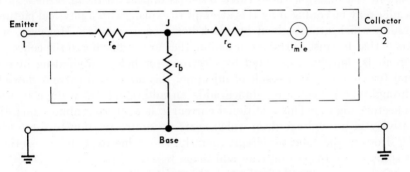

FIG. 10-4. Active equivalent network for transistor.

guess that the simple passive-resistance T equivalent circuit is not adequate for our transistor.

An active network, on the other hand, has a power source or generator which is capable of amplifying the input signal. A familiar example is the vacuum tube, whose active behavior is well known and is represented on its equivalent-circuit diagram by a generator of variable voltage output. As a result of this generator, the forward and reverse transfer resistances are very definitely unequal.

Since we must add a generator or voltage source of some kind to the equivalent network to show that we are dealing with an active circuit which will permit the forward and reverse transfer resistances to be unequal, we do so as follows. Classically, this generator is indicated by $r_m i_e$ in the equivalent circuit of Fig. 10-4. The resistor r_m is called the "mutual resistance" as mentioned earlier, and it will shortly be fully explained. It is also necessary to mention that i_e is the current through the emitter resistance r_e and is not necessarily the loop current in circuit 1.

The true emitter current i_e, which flows through the emitter resistance

r_e, is also assumed to flow through a part of the collector circuit (namely, r_m) in such a way as to produce a voltage $r_m i_e$. This is an important concept in interpreting the equivalent-circuit representation of the transistor. It should also be noted that whereas the term $r_m i_e$ represents a voltage drop in series with the collector circuit, caused by the flow of the emitter current i_e through r_m, the emitter current does not flow through r_c or r_b, even though these are part of the collector circuit. The reader will recognize that this concept involves electrical behavior which cannot be described by classical circuit diagrams. However, it has been found useful and accurate to describe transistor action in terms of familiar circuit elements.

Likewise, i_c, the collector current which flows through r_c and r_b in the collector circuit, does not flow through the element r_m, in spite of the fact that in the diagram these three resistors appear to be in series. The reader may be confused by this attempt to explain with a physical picture results which are obtained by black-box measurements. However, we know that because of transistor action, the current flowing in the collector circuit is sharply augmented by electrons (or holes) originating in the emitter circuit. As a result of injection, this additional current flowing through the collector resistance adds an additional IR voltage in the collector circuit. This additional current i_e is converted almost quantitatively into a voltage drop which bears no simple relation to the other IR drops in the collector circuit (namely, those due to r_b and r_c), so that it is necessary to postulate an additional resistance r_m, which is effective only when current i_e flows through the emitter circuit.

If we now reexamine the resistance parameters in the light of the proposed active network, we may easily derive the following from Eqs. (10-1) to (10-4) and Fig. 10-4:

$$r_{11} = r_b + r_e \qquad (10\text{-}13)$$
$$r_{12} = r_b \qquad (10\text{-}14)$$
$$r_{21} = r_b + r_m \qquad (10\text{-}15)$$
$$r_{22} = r_b + r_c \qquad (10\text{-}16)$$

Again we note that these equations are valid for the grounded-base connection only. There are similar relationships (as will be seen below) for the grounded-emitter and grounded-collector configurations, and either one, if chosen for our present analysis, would lead to the same active network but with numerically different values. In all cases (as shown for our example above) the inequality of the forward and reverse transfer resistances r_{12} and r_{21} emerges clearly.

To complete our analysis of the equivalent network and to relate the results to the transistor require only the application of Kirchhoff's laws to closed-loop circuit elements and the solution of the resulting equations.

If our choice of equivalent circuit is one of the numerous possible correct ones, we shall obtain consistent answers and our proposed circuit will yield calculated values of r_{11}, r_{12}, r_{21}, and r_{22} which will be equal to those previously measured experimentally. As we shall see, the circuit of Fig. 10-4 is one which satisfies our requirements for d-c or low-frequency applications.

Aside from the limitation on frequency, we have omitted one other consideration in setting up our black-box equivalent circuit, namely, bias. Our circuit has no external bias sources, and no provision for them is made. Yet we know that, just as in the case of the vacuum tube, the transistor is a biased device employing voltages at the emitter and collector of the appropriate sign and magnitude to bring the device into the desired operating condition. We have stated, for example, that two of the requirements for transistor action are that the emitter be forward-biased and the collector reverse-biased. How, then, can our equivalent circuit represent a valid analysis if the bias voltages are omitted? We know, for example, that the a-c signal used to measure the parameters r_{11}, r_{12}, r_{21}, and r_{22} is of small magnitude compared with the bias and that, if the bias voltages are omitted, we will be measuring the performance only under conditions of current and voltage near the origin of the characteristic where these devices are almost all sharply nonlinear. We may then wonder whether the parameters we have measured are of any real value.

The resolution of this apparent difficulty illustrates two important concepts involved in this type of equivalent-circuit analysis. First, we have been careful to construct our equivalent circuit from components such as resistors and zero-resistance voltage generators, none of which will undergo any change as the result of external bias. Therefore, we may perform our measurements (being careful that this does not upset the bias source or produce errors in the measurement due to the bias voltages) under any desired condition of external bias. Once we do this and determine the resistance parameters, the solution of the equations will give us valid and general results for this situation.

One such set of external bias conditions would be precisely those described above, that is, zero bias. It is not a very attractive condition for an amplifier, but in switching applications the nonlinearity, as will be seen in later chapters, is of considerable utility. Another set of conditions would be the normal forward and reverse voltages on emitter and collector defined as being necessary for transistor action. If our bias sources and measurement techniques are without effect on the internal elements of the equivalent circuit and do not themselves interact in any way, each set of measurements is meaningful for that particular set of conditions. It should also be pointed out that transistors are very nonlinear, nonohmic devices and that no means is available to make a single set of measurements under any single set of bias conditions serve to characterize the device under other conditions.

The answer to the initial point at issue is, therefore, twofold. First, the measurements made in the absence of bias have value, since many devices have usefulness in such ranges. Second, if we wish to broaden our approach, we have only to meet the criteria already stated with regard to measuring instrumentation and bias sources to obtain meaningful data on the device behavior under any arbitrary set of bias conditions.

We have already defined our measuring instruments in such a way that there is no interference with the elements of the equivalent circuit. We have chosen signal sources with zero internal resistance, a voltmeter of infinite resistance, and an ammeter which consumes no current. Such instruments can also be used in the presence of bias voltages without

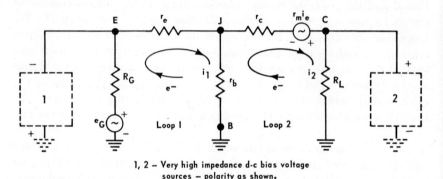

1, 2 — Very high impedance d-c bias voltage sources — polarity as shown.

Fig. 10-5. Active equivalent network for grounded-base transistor (n-p-n case shown).

causing any disturbance. If we now choose bias voltage sources with the same care, we can (theoretically, at least) avoid stray interactions between such sources and the equivalent circuit and between the biasing supply and the measuring instruments.

Suitable bias sources would be supplies of infinite resistance or impedance, so that they represent a voltage source with an effective open-circuit resistance. In this way we avoid the introduction of new resistance terms and the general solution for the case under consideration is valid. In practice, this is usually attempted by the use of high-voltage batteries with high resistances in series or constant-current sources. A schematic circuit is shown in Fig. 10-5, which also shows the other essentials of our equivalent circuit, R_L, e_G, $r_m i_e$, R_G, and the current loops according to the polarity of the normal bias voltage sources. These sources have their polarity as shown in keeping with conventional amplifying practice for an n-p-n transistor. Also, in Fig. 10-5, the voltage sources e_G and $r_m i_e$ are oriented with their polarities as shown for all circuit calculations, despite the fact that they are opposed to the electron-current flow arrows. (For a *p-n-p* unit this would not be true.) These conventions are more fully discussed below.

The scheme discussed above assumes certain qualities in the bias sources and measuring devices which are not always easy to realize. We shall come to appreciate some of these problems more fully when we consider frequency as a variable. For the present, however, we may assume that the required conditions have been met for d-c and low-frequency measurements and that any errors are therefore negligible.

With the bias sources no longer a complication we may proceed with the circuit analysis. Basically, Kirchhoff's law states that the net sum of all voltages (due to generators or IR drops) around any closed circuit path must be algebraically zero. If we then sum all these voltages around all closed loops in the network, we arrive at a series of simultaneous equations which may be solved for the resistance and current values.

Since we have two current loops in our equivalent circuit, we shall have two such equations, one for loop 1 (current i_1) and another for loop 2 (current i_2). In writing these closed-loop equations, we shall adopt a set of conventions found applicable in transistor work. As will become apparent, equally valid results are obtainable with opposite rules. It is important to use a uniform system so that the end results are self-consistent. The rules to be followed herein are:

1. Sources e_G and $r_m i_e$ are always so placed that their polarities oppose the electron loop currents i_1 and i_2, respectively.

2. Current direction arrows are taken as the flow of electrons (where known) as determined by the type of device (*n-p-n* or *p-n-p*) under consideration and the normal requirements of forward and reverse bias at the emitter and collector, respectively, to obtain transistor action. (If there is doubt about the direction of a current flow, an arbitrary arrow may be drawn, and the sign of the resulting current vector will determine the correctness of the choice.)

3. Voltage drops will be assigned plus signs, and generators or voltage sources will have positive signs if in the direction of the current arrow or negative signs if opposed.

4. If two loop currents flow through the same element, they add algebraically with respect to direction.

5. Current i_e, when flowing through the mutual resistance r_m, is taken to be the same as i_1 when both i_e and i_1 flow through r_e toward point J. If they are reversed, then $i_e = -i_1$.

The above assumptions are consistent with much recent literature. The reader will find, however, that there are many apparent inconsistencies in published work when different assumptions are made or when the polarities are changed during the evolution of a set of calculations. A common variation is to reverse the current arrows and to use the older

150 *Semiconductor Junctions and Devices*

concept of current flow from positive to negative. Such apparent anomalies are easily spotted, and, so long as a system is consistently followed, the end results must be the same.

We may now proceed to apply the above techniques to the analysis of the grounded-base *n-p-n* transistor whose equivalent circuit is represented in Fig. 10-5.

10-2. RESULTS—GROUNDED BASE [3,4]

10-2.1. Current Formulas. The first step in the analysis of an equivalent network is to obtain an expression for the currents i_1 and i_2 in terms of the current sources and resistances in each current loop by writing for each loop an expression of the fact that the sum of the voltage rises and voltage drops in a closed loop is zero. Beginning at the ground point of loop 1 and proceeding clockwise in the direction of the arrow, we may consider the voltage drops and rises as follows:

Since e_G is a generator or voltage rise and since the current flows through it from minus to plus, e_G will be assigned a minus sign by rule 3. Across R_G and r_e are two simple voltage drops which are given a positive sign. Since i_1 and i_2 flow through r_b in opposite directions, the voltage drops due to i_1 and i_2 are algebraically additive, and we represent their sum as $i_1 - i_2$. Based on this analysis, we may now write as follows:

$$i_1(R_G + r_b + r_e) - i_2 r_b - e_G = 0 \qquad (10\text{-}17)$$

In loop 2, starting at the base connection and proceeding around the loop in the direction of the current arrow, we have first the drop at r_b. We now represent the algebraic sum of the currents as $-(i_1 - i_2)$, since the effect in loop 2 must be the opposite of that in loop 1, and obtain the positive voltage drop $(i_2 - i_1)r_b$. Next we have $r_c i_2$, and then, since the current flow through $r_m i_e$ is from minus to plus and $i_e = i_1$, the next term becomes $-r_m i_1$. Finally, we have $R_L i_2$. We may now write the expression

$$-i_1(r_b + r_m) + i_2(R_L + r_b + r_c) = 0 \qquad (10\text{-}18)$$

Equations (10-17) and (10-18) completely define or describe the circuit of Fig. 10-5 in so far as low-frequency a-c voltages and currents are concerned and will provide information regarding resistances, voltage and current gains, power gains, and the stability of the circuit.

The solution of the linear simultaneous equations (10-17) and (10-18) for the currents reveals the following:

$$i_1 = \frac{e_G(R_L + r_b + r_c)}{(R_G + r_b + r_e)(R_L + r_b + r_c) - r_b(r_b + r_m)} \qquad (10\text{-}19)$$

$$i_2 = \frac{e_G(r_b + r_m)}{(R_G + r_b + r_e)(R_L + r_b + r_c) - r_b(r_b + r_m)} \qquad (10\text{-}20)$$

These expressions give currents i_1 and i_2 in terms of resistance parameters which, for a given transistor and equivalent circuit, are known. Henceforth in our discussion we may treat i_1 and i_2 as known quantities and express the circuit characteristics desired in terms of the current as well as the resistance parameters.

Having now obtained values for the input and output currents, we are ready to use these currents with the other known parameters of the circuit to find the input resistance, the output resistance, the voltage gain, the current gain, and the power gain.

10-2.2. Current Gain. We first consider the current gain of the equivalent circuit (grounded-base) and reintroduce the concept of the parameter α, as defined by Eq. (8-13), to be

$$\alpha = \frac{I_C}{I_E}$$

or the ratio of the collector current to the emitter current. Values for α were found to be less than, but very close to, unity. This expression was derived from a consideration of junction currents only, and it did not take account of any external resistances, such as R_G or R_L.

We have now obtained expressions for i_1 and i_2 in terms of resistance parameters. Obviously, for the grounded-base configuration,

$$\alpha = \frac{I_C}{I_E} \sim \frac{i_2}{i_1} \tag{10-21}$$

and, by substitution from Eqs. (10-19) and (10-20),

$$\alpha \sim \frac{r_b + r_m}{R_L + r_b + r_c} \tag{10-22}$$

If we now remove the external resistance R_L, we have α, which is sometimes referred to as the short-circuit current gain, or

$$\alpha = \frac{r_b + r_m}{r_b + r_c} \tag{10-23}$$

By substitution from Eqs. (10-15) and (10-16)

$$\alpha = \frac{r_{21}}{r_{22}} = \frac{I_C}{I_E} \tag{10-24}$$

This relationship will be found useful in succeeding sections.

Actually, in practice, it is not feasible to make R_L equal to zero. The same effect is achieved by setting $r_c \gg R_L$ or making the load resistor small compared with the internal resistance of the reverse-biased collector junction. Since this junction presents a very high resistance when reverse-biased (of the order of 1 to 3 megohms), it is easy to satisfy the requirements of $R_L \ll r_c$ and still have a very respectable load resistance (up to several hundred thousand ohms) for power amplification. As a result, although it is called the short-circuit current gain, α is not greatly changed in most typical circuit applications where, as shown above, $r_c \gg R_L$.

10-2.3. Input Resistance. In order to obtain the input resistance for the grounded-base connection, let us refer once again to Fig. 10-5 and to the circuit conditions seen by the generator e_G when it is connected from emitter to ground. The effect of the transistor and the rest of the circuit might be replaced by a single resistance. This equivalent resistance is, of course, the input resistance, or r_{11} from our previous measurements. Figure 10-6 indicates how R_i, the input resistance, replaces the effect of the rest of the equivalent circuit while current $i_G = i_1$ remains the same. If we now take e_G and divide by i_1, the result is the total

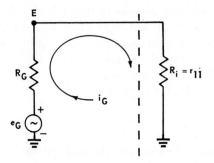

Fig. 10-6. Equivalent-circuit input-resistance measurement.

resistance in the circuit of Fig. 10-6. We now subtract R_G, the internal resistance of the generator, and the result is the input resistance R_i. That is,

$$R_i = r_{11} = \frac{e_G}{i_1} - R_G \qquad (10\text{-}25)$$

Since i_1 is defined by Eq. (10-19), it is possible to obtain the input resistance as a function of the circuit parameters. By making the substitution, we obtain

$$R_i = r_{11} = r_b + r_e - \frac{r_b(r_b + r_m)}{R_L + r_b + r_c} \qquad (10\text{-}26)$$

If R_L in the last term becomes very large (a condition analogous to the open-circuit restriction imposed initially when r_{11} was being measured), Eq. (10-26) reduces to

$$R_i = r_{11} = r_b + r_e \qquad (10\text{-}27)$$

which is exactly the relationship previously derived for the grounded-base configuration, Eq. (10-13).

Equation (10-26) expresses the input resistance of a grounded-base transistor in terms of all the circuit parameters. By studying this expression, and noting the negative sign before the third term on the right-hand side, we observe an interesting fact. In general, the circuit parameters of a transistor or its equivalent network are positive numbers. Therefore, a combination of parameters which will make R_i negative may occur in Eq. (10-26). This circumstance leads to the application of transistors in electronic switching arrangements in which the negative input resistance is intentionally utilized. When it occurs but is not intended, instability and parasitic oscillations occur. For more detailed information on these switching phenomena the reader is referred to Chap. 14.

By using the typical values given in Table 9-2, the reader can quickly verify that the input resistance for a junction transistor will be low as previously mentioned, about 90 ohms. Therefore, the transistor in the grounded-base connection is a low-input-impedance device. (In vacuum-tube practice the input resistance at ordinary frequencies may be of the order of a megohm or more.)

10-2.4. Output Resistance. To obtain the output resistance, which is the resistance determined by the ratio of a voltage applied in the output circuit to the resulting current, we employ a scheme which the reader will encounter in many other applications also. In Fig. 10-7a observe how the generator e_G has been placed in series with R_L (but with its polarity reversed to oppose the electron loop current) and the generator resistance has been retained where it was before, in the input circuit. (Figure 10-7b is a simplified version of this circuit, by analogy to Fig. 10-6.) We are now in a position to determine what value of resistance the generator e_G, if placed in the circuit as shown, would encounter. Note that this theoretical generator is considered to have zero internal resistance. Therefore, the resistance seen by the generator e_G, connected as in Fig. 10-7, less R_L, will be the output resistance. This may be written, as before,

$$R_o = r_{22} = \frac{e_G}{i_G} - R_L \qquad (10\text{-}28)$$

Now, i_G is not equal to i_2, since the signal generator e_G has been put in loop 2. Otherwise, the loop-current equations will be identical to Eqs. (10-17) and (10-18). They will now be as follows, using i_1' and i_2' to denote the change in location of e_G,

$$i_1'(R_G + r_b + r_e) - i_2' r_b = 0 \qquad (10\text{-}29)$$
$$-i_1'(r_b + r_m) + i_2'(R_L + r_b + r_c) - e_G = 0 \qquad (10\text{-}30)$$

154 Semiconductor Junctions and Devices

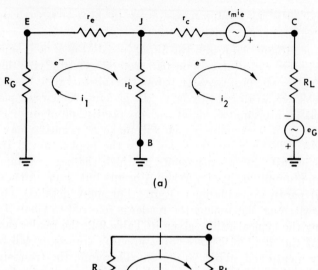

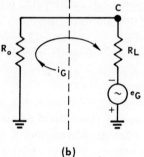

FIG. 10-7. Equivalent-circuit output-resistance measurement. (*After A. Coblenz and H. L. Owens, by permission* [5].)

We need to solve these two equations for i_2' only as can be seen from the equivalent circuit, since that portion of the circuit beyond the collector terminal may be replaced by a single resistance R_o. When we do this, we find that i_2' is given by

$$i_2' = \frac{e_G(R_G + r_b + r_e)}{(R_G + r_b + r_e)(R_L + r_b + r_c) - r_b(r_b + r_m)} \tag{10-31}$$

By using the value of i_2' from Eq. (10-31) in Eq. (10-28) and remembering that i_G equals i_2', we have

$$R_o = r_{22} = r_b + r_c - \frac{r_b(r_b + r_m)}{R_G + r_b + r_e} \tag{10-32}$$

If we now impose the condition that R_G be very large (or open-circuited), as when determining r_{22}, we have from Eq. (10-32)

$$R_o = r_{22} = r_b + r_c \tag{10-33}$$

which is identical with Eq. (10-16).

By using Table 9-2 to obtain typical values, it can be verified that the output resistance of the grounded-base transistor is approximately 0.6 megohm. Further, the negative sign in Eq. (10-32) suggests, as for the case of the input resistance, that a combination of parameters for which R_o is negative may be found. A negative output resistance will lead to instability or parasitic oscillations, just as a negative input resistance will. The reader will observe from Eq. (10-32) that very low values of input generator resistance R_G generally tend toward instability in transistor operation.

10-2.5. Voltage Gain. In general, voltage gain is defined for any device as the ratio of output voltage to input voltage. Since we know there is no phase reversal, both voltages will have the same sign at any given moment. [This is further indicated by the fact that i_1, i_2, and e_G are positive in Eqs. (10-19) and (10-20), despite the negative value assigned to e_G when setting up the loop-current equations in accordance with Kirchhoff's laws.] Since the effective voltage output is that available across the load resistance, we may write for the voltage gain

$$VG = \frac{i_2 R_L}{e_G} \qquad (10\text{-}34)$$

Knowing i_2 from Eq. (10-20), we substitute in Eq. (10-34) and obtain an expression for the voltage gain as follows:

$$VG = \frac{(r_b + r_m)R_L}{(R_G + r_b + r_e)(R_L + r_b + r_c) - r_b(r_b + r_m)} \qquad (10\text{-}35)$$

By substituting numbers for the parameters from Table 9-2, it is easily verified that the voltage gain for a junction transistor is of the order of 100 or 200.

It is useful for many applications to know the maximum theoretical voltage gain for a transistor. To obtain it, we idealize the circuit arrangement by assuming a perfect generator of zero internal resistance and an ideal load of infinite resistance. (If the internal generator resistance is zero, no voltage drop can occur across it; and if the load resistance is infinite, all the available output voltage will be developed across it.) We now substitute Eqs. (10-13) to (10-16) in Eq. (10-35) and obtain

$$VG = \frac{r_{21} R_L}{(R_G + r_{11})(R_L + r_{22}) - r_{12} r_{21}} \qquad (10\text{-}36)$$

and, by dividing the numerator and denominator by R_L, we then obtain

$$VG = \frac{r_{21}}{(R_G + r_{11})(1 + r_{22}/R_L) - (r_{12} r_{21}/R_L)} \qquad (10\text{-}37)$$

156   *Semiconductor Junctions and Devices*

If we now set R_G equal to zero and allow R_L to approach infinity, we obtain the maximum theoretical voltage gain:

$$VG_m = \frac{r_{21}}{(0 + r_{11})(1 + 0) - 0} = \frac{r_{21}}{r_{11}} \qquad (10\text{-}38)$$

We now multiply by r_{22}/r_{22} and get

$$VG_m = \frac{r_{21}r_{22}}{r_{22}r_{11}} \qquad (10\text{-}39)$$

From Eq. (10-24) we see that r_{21}/r_{22} is α, and upon making this substitution in Eq. (10-39) we have the interesting relation that*

$$VG_m = \alpha \frac{r_{22}}{r_{11}} \qquad (10\text{-}40)$$

In a previous analysis it has been mentioned that α, defined by Eq. (10-24), is also the ratio of collector current to emitter current. Equation (10-40) clearly shows that the maximum possible voltage gain from a transistor is the product of the current gain and the ratio of the collector circuit (output) resistance to the emitter circuit (input) resistance. Thus,

$$VG_m = \frac{i_c r_{22}}{i_e r_{11}} \qquad (10\text{-}41)$$

For the junction-type transistor, using the values of Table 9-2, the maximum theoretical voltage gain is several thousand, compared with the ordinary gain of 100 or 200 found earlier. This attractive possibility is utilized whenever possible by adjustment of circuit parameters in amplifying applications. Note, however, that if R_L is made large (near to r_c) the value of α, or current gain, will fall off, as previously discussed.

10-2.6. Power Gain. We have just developed equations which express the input and output resistances and the current and voltage gains for junction transistors in the grounded-base circuit. The power gain of an electrical device is obviously an important concept to the design engineer, and equations to express this parameter will now be developed.

In Fig. 10-8 we show a simple circuit consisting of a generator e_G (of internal resistance R_G) which is supplying a load whose resistance is R_L. We must determine under what conditions a given generator will achieve the maximum power output (the power developed across R_L). We recall that power may be expressed as I^2R or V^2/R. In Fig. 10-8 the

---

* Note that $r_{22} = r_c + r_b$ and r_b is negligible when compared with r_c. Consequently Eq. (10-40) is equivalent to $VG_m = \alpha r_c/r_{11}$, a result which will be of interest when the grounded-emitter case is considered.

current flowing through the circuit will be

$$i_G = \frac{e_G}{R_G + R_L} \quad (10\text{-}42)$$

We may also express the current flowing through the right-hand leg of Fig. 10-8 as follows:

$$i_G = \frac{V}{R_L} \quad (10\text{-}43)$$

We may now combine these equations to eliminate i_G:

$$V = \frac{e_G R_L}{R_G + R_L} \quad (10\text{-}44)$$

and since $V^2/R_L = P_o$, the power across R_L,

$$P_o = \frac{e_G^2 R_L^2}{(R_G + R_L)^2 R_L} = \frac{e_G^2 R_L}{(R_G + R_L)^2} \quad (10\text{-}45)$$

If we now differentiate P_o with respect to R_L and equate the resulting

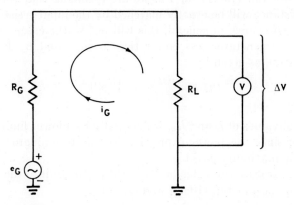

FIG. 10-8. Schematic equivalent circuit for power gain measurement.

differential to zero, we can determine when the power P_o will be at a maximum (P_m):

$$\frac{dP_o}{dR_L} = \frac{e_G^2(R_G + R_L - 2R_L)}{(R_G + R_L)^3} \quad (10\text{-}46)$$

The above differential equation will have a value of zero when R_G equals R_L, and P_o will be a maximum (P_m). (It is common, when using vacuum tubes, to match the plate load resistance to the internal plate resistance in order to obtain maximum power output. This result is therefore another example of the principle that, to obtain maximum power across a given load, the internal resistance of the generator must be matched to the load resistance.) If we make R_G equal to R_L in Eq. (10-45), the maximum power we can draw from the generator e_G will be given by

$$P_m = \frac{e_G^2}{4R_G} \qquad (10\text{-}47)$$

Note that this is expressed entirely in terms of the generator properties for a load whose resistance R_L equals that of the generator R_G. We have optimized the circuit to take best advantage of the input, considering the load only as a single resistor. This is a general principle and is not restricted to transistor theory. Returning to the problem of the transistor, however, the power output is given by $i_2^2 R_L$ where, from Eq. (10-20),

$$i_2 = \frac{e_G(r_b + r_m)}{(R_G + r_b + r_e)(R_L + r_b + r_c) - r_b(r_b + r_m)}$$

and the power gain of the transistor is the ratio of $i_2^2 R_L$, the output, to P_m, the maximum power which may be drawn from the generator. However, if we use Eq. (10-47) for P_m, we are assuming that the generator internal resistance will be exactly matched by the input resistance of the transistor. Of course, in general, this will not be the case, and we shall draw from the generator less power than is indicated by Eq. (10-47). The power gain, as given by

$$PG = \frac{i_2^2 R_L}{e_G^2/4R_G} = \frac{4R_L R_G i_2^2}{e_G^2} \qquad (10\text{-}48)$$

for any given values of i_2 and R_L, will actually be a lower limit or a minimum value, since the power expression in the denominator (the input power) is the maximum possible.

We now substitute the value of i_2 from Eq. (10-20) into the expression for the power gain in Eq. (10-48) and obtain

$$PG = \frac{4R_L R_G (r_b + r_m)^2}{[(R_G + r_b + r_e)(R_L + r_b + r_c) - r_b(r_b + r_m)]^2} \qquad (10\text{-}49)$$

This expression can be simplified somewhat if we note in Table 9-2 that r_b can be neglected compared to r_c or r_m:

$$PG = \frac{4R_L R_G r_m^2}{[(R_G + r_b + r_e)(R_L + r_c) - r_b r_m]^2} \qquad (10\text{-}50)$$

Equation (10-50) is an expression for the operating power gain of a grounded-base transistor in terms of its fundamental parameters. It allows the calculation of the actual power gains in circumstances where the transistor cannot be represented by a single load resistor R_L identical with R_G. If the maximum gain is to be approached, as stated in Eq. (10-47), we must so adjust the parameters as to make the entire transistor match R_G and thereby draw the maximum from the generator. Even then, however, the actual output power seen across R_L will be less than the maximum due to other resistances, notably r_m and r_c, which are part of the circuit but from which we cannot draw power.

Using the values of the parameters as given in Table 9-2, a typical operating power gain is over 400, representing approximately 25 db. (The decibel is obtained by multiplying the logarithm to the base 10 by a factor of 10.)

10-2.7. Additional Applications of Alpha in the Grounded-base Configuration. In preceding discussions of the two-junction transistor using the grounded-base configuration as a model we have had occasion to discuss α and the significance of this ratio. We have shown how it may be derived from the resistance parameters and have shown its typical value for junction transistors in Table 9-2 for the grounded-base configuration.

Before leaving our discussion of the grounded-base configuration, we wish to present several additional equations and approximations in which α replaces other resistance values for the elements in the transistor. These relationships are more convenient to use than the more complicated forms originally derived.

If we desire to know the power gain as a function of the α of the circuit, we divide numerator and denominator of Eq. (10-49) by $(r_c + r_b)^2$:

$$PG = \frac{4R_L R_G \alpha^2}{\{(R_G + r_b + r_e)[1 + R_L/(r_b + r_c)] - \alpha r_b\}^2} \quad (10\text{-}51)$$

The power gain may also be expressed as a function of the VG and α:

$$PG = \frac{4R_G \alpha (VG)}{(R_G + r_b + r_e)[1 + R_L/(r_b + r_c)] - \alpha r_b} \quad (10\text{-}52)$$

by using the following expression for VG [Eq. (10-35)] written in terms of α:

$$VG = \frac{\alpha R_L}{(R_G + r_b + r_e)[1 + R_L/(r_b + r_c)] - \alpha r_b} \quad (10\text{-}53)$$

The above equations demonstrate the reason for the inclusion of α as one of the important parameters in the evaluation and comparison of transistors. Also, the fact that α equals i_c/i_e shows the further usefulness of α as a comparison number for current gain.

Finally, if we again make the assumption that $R_L \ll r_c$, the following brief approximations are obtained. First, as seen before,

$$\text{Current gain} = \alpha \tag{10-54}$$

then, from the preceding group of equations,

$$\text{Voltage gain} = \frac{\alpha R_L}{R_G + r_e + r_b(1 - \alpha)} \tag{10-55}$$

$$\text{Operating power gain} = \frac{4 R_L R_G \alpha^2}{[R_G + r_e + r_b(1 - \alpha)]^2} \tag{10-56}$$

REFERENCES

1. A. Coblenz and H. L. Owens: "Transistors: Theory and Applications," chaps. 8–10, McGraw-Hill Book Company, New York, 1955.
2. E. A. Guillemin: "Communication Networks," John Wiley & Sons, Inc., New York, 1935.
3. R. M. Ryder and R. Kircher: Some Circuit Aspects of the Transistor, *Bell System Tech. J.*, **28**: 367 (1949).
4. H. W. Bode: "Network Analysis and Feedback Amplifier Design," D. Van Nostrand Company, Inc., Princeton, N.J., 1945.
5. Ref. 1, pp. 103, 104.

11: Black-box Equivalent—Grounded Emitter and Grounded Collector*

11-1. RESULTS—GROUNDED EMITTER

So far, discussions have been restricted to the grounded-base connection for transistors. However, the theory of black-box parameters and equivalent circuits developed earlier applies equally well to other configurations, and we shall continue our discussion with the equivalent circuit for the grounded-emitter configuration.

Since we have treated the grounded-base configuration in considerable detail and since the other connections are analogous, we shall be brief and concentrate chiefly on the significant differences which will affect our comparison. We shall begin by presenting the equivalent active network, without bias, for the grounded-emitter case. From this we shall derive directly the values of the black-box parameters r_{11}, r_{12}, r_{21}, and r_{22}. We shall then discuss the significance of α in this configuration, after which we shall proceed with circuit analysis as before, leading to a set of analogous expressions for the gain, resistance, and other factors.

11-1.1. Resistance Parameters. Figure 11-1 shows an equivalent active circuit diagram for the grounded-emitter connection that is analogous to the grounded-base configuration in the preceding section. Note that the polarities of the signal generator e_G and mutual resistance $i_e r_m$ in current loop 2 are based on the rules used in the grounded-base equivalent network. Once again, in accord with our basic technique, arbitrary current directions proceed according to bias conditions for the n-p-n transistor in this configuration. Likewise, we must remember that when the emitter current i_e flows through the emitter resistance r_e toward J, it is arbitrarily assigned a positive value and the term $r_m i_e$ is likewise positive. In Fig. 11-1 it is obvious that the current flowing through r_e (the emitter resistor) is the sum of the two loop currents i_1 and i_2 and that

* Adapted with permission from the book by A. Coblenz and H. L. Owens [1]. Some parts are copied verbatim and are hereby acknowledged.

they are flowing toward point J. Therefore, following our basic rule for network evaluation, we must consider that i_1 plus i_2 equals i_e. Finally, we note that the current in loop 2 flows through $r_m i_e$ from minus to plus, so that it becomes a negative quantity. As before, e_G, a voltage rise, is negative.

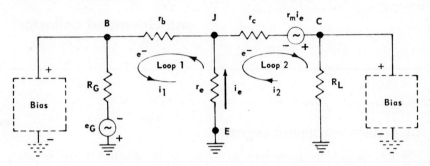

FIG. 11-1. Active equivalent network for grounded-emitter transistor (n-p-n case shown).

Before writing the current equations, we refer back to the general relationships which characterize the four-pole network. Repeating Eqs. (10-9) and (10-10),

$$v_1 = r_{11}i_1 + r_{12}i_2 \qquad v_2 = r_{21}i_1 + r_{22}i_2$$

By inspection of Fig. 11-1 we may also write, omitting e_G, R_G, and R_L,

$$v_1 = (r_b + r_e)i_1 + r_e i_2 \tag{11-1}$$
$$v_2 = (r_e - r_m)i_1 + (r_c + r_e - r_m)i_2 \tag{11-2}$$

From these equations for the black box, we may now identify the r parameters for the grounded-emitter connection

$$r_{11} = r_b + r_e \tag{11-3}$$
$$r_{12} = r_e \tag{11-4}$$
$$r_{21} = r_e - r_m \tag{11-5}$$
$$r_{22} = r_c + r_e - r_m \tag{11-6}$$

It is to be noted that the terms equivalent to the resistance parameters ($r_{11}, r_{12}, r_{21}, r_{22}$) are different, as would be expected, from the black-box analysis employing a different arrangement of the leads from the transistor element. We shall summarize the differences in these and other characteristics for the various circuit arrangements in a later chapter.

11-1.2. Current Gain. We have spoken of the significance of α in the case of the grounded-base configuration when the load resistance R_L is zero. In the grounded-emitter case we wish to make a similar derivation in terms of output and input current. We therefore write the Kirchhoff equations for the current loops:

$$(R_G + r_b + r_e)i_1 + r_e i_2 - e_G = 0 \tag{11-7}$$
$$(r_e - r_m)i_1 + (R_L + r_c + r_e - r_m)i_2 = 0 \tag{11-8}$$

Solving these for i_1 and i_2 yields

$$i_1 = \frac{e_G(R_L + r_c + r_e - r_m)}{(R_G + r_b + r_e)(R_L + r_c + r_e - r_m) + r_e(r_m - r_e)} \tag{11-9}$$

$$i_2 = \frac{e_G(r_m - r_e)}{(R_G + r_b + r_e)(R_L + r_c + r_e - r_m) + r_e(r_m - r_e)} \tag{11-10}$$

Then the ratio of output to input current is

$$\frac{i_2}{i_1} = \frac{r_m - r_e}{R_L + r_c + r_e - r_m} \tag{11-11}$$

If we now set R_L at zero, for the short-circuit case,

$$\frac{i_2}{i_1} = \frac{r_m - r_e}{r_c + r_e - r_m} = \beta \tag{11-12}$$

This is the current gain expression for the common-emitter connection, commonly referred to as β. We recall from an earlier discussion that α, pertaining to the common-base connection, is written [Eq. (10-23)]

$$\alpha = \frac{r_b + r_m}{r_b + r_c}$$

and by reference to Table 9-2 we find that r_b and r_e are very small compared to r_c and r_m. Consequently, these ratios may be expressed without appreciable error as follows:

$$\beta = \frac{r_m}{r_c - r_m} \tag{11-13}$$

$$\alpha = \frac{r_m}{r_c} \tag{11-14}$$

If both numerator and denominator of Eq. (11-13) are divided by r_c, it is seen that

$$\beta = \frac{\alpha}{1 - \alpha} \tag{11-15}$$

This equation gives the relation between the short-circuit current gain of the grounded-emitter circuit and that of the grounded-base circuit for the same transistor. Since the approximations made in its derivation are good in practice, the relationship may be regarded as exact. As pointed out earlier, the current gain of a grounded-base circuit is approximately unity, while, as we shall see below, that applicable to a grounded-emitter circuit can be much higher. For example, assume α equals 0.96 for some transistor.

$$\beta = \frac{0.96}{1 - 0.96} = 24 \tag{11-16}$$

The current gain when this transistor is connected in a grounded-emitter circuit is 25 times higher than when it is connected in a grounded-base circuit.

Before leaving the subject of current gain, it is well to point out that, although both α and β are measures of the ratio of output to input current, or i_2/i_1, and are interrelated as shown in Eq. (11-15), there is a fundamental difference which may be overlooked. In the case of α, i_2 is a collector current, or i_c, while i_1 is an emitter current, or i_e. The parameter α might then be written

$$\alpha_{ce} = \frac{i_2}{i_1} = \frac{i_c}{i_e} \tag{11-17}$$

where α_{ce} refers to the currents comprising the ratio, the first letter of the subscript being output current and the second being input.

In the grounded-emitter case, $i_2 = i_c$, but $i_1 = i_b$, or base current. Therefore, we may write

$$\beta = \beta_{cb} = \frac{\alpha_{ce}}{1 - \alpha_{ce}} = \frac{i_2}{i_1} \tag{11-18}$$

To illustrate the difficulties which may arise, particularly in the older literature, the following identity was common:

$$\alpha_{cb} = (\beta) = \frac{\alpha_{ce}}{1 - \alpha_{ce}} \tag{11-19}$$

In more recent work α is reserved for the grounded-base configuration and β for the grounded-emitter.

11-1.3. Input Resistance. To compute the input resistance R_{ie} for the grounded-emitter connection, we shall use precisely the same approach as we used in the corresponding calculation for the grounded-base connection. Thus,

$$R_{ie} = \frac{e_G}{i_1} - R_G \tag{11-20}$$

On substituting for i_1, solving, and eliminating r_e because it is negligible compared to r_m, we have

$$R_{ie} = r_e + r_b + \frac{r_e r_m}{R_L + r_c - r_m} \tag{11-21}$$

(Note the convention adopted here: If no additional subscript is used, for example, R_i, the value refers to the grounded-base connection. R_{ie}

refers to the grounded-emitter connection, and similarly, R_{ic} will refer to the grounded-collector connection.)

From Eq. (11-21) it can be seen that the input resistance will in general be greater than r_e plus r_b by the quantity $r_e r_m/(R_L + r_c - r_m)$. Therefore, since α is less than unity, and since α equals r_m/r_c, it is clear that $r_c - r_m$ is always positive. Hence the above fraction is always positive and R_{ie} is always greater than $r_e + r_b$. Using the typical values from Table 9-2, we find that R_{ie} is almost 1,000 ohms for junction transistors in the grounded-emitter connection. In general, the input resistance of these units will not exceed a few thousand ohms. The reader may verify, from Eq. (11-21), that if R_L is made very small and the α approaches unity (say 0.997), high input resistances may be achieved. However, such α's are exceedingly uncommon; and if R_L is made small, the possible voltage and power gains are sharply reduced.

11-1.4. Output Resistance. By proceeding as before, we obtain the expression for the output resistance R_{oe}:

$$R_{oe} = \frac{e_G}{i_2' - R_L} \tag{11-22}$$

As before, we move e_G to the output circuit and write new loop-current equations according to Kirchhoff's law. These become

$$(R_G + r_b + r_e)i_1' + r_e i_2' = 0 \tag{11-23}$$

and
$$(r_e - r_m)i_1' + (R_L + r_c + r_e - r_m)i_2' - e_G = 0 \tag{11-24}$$

By solving for i_2', we obtain

$$i_2' = \frac{(R_G + r_b + r_e)e_G}{(R_G + r_b + r_e)(R_L + r_c + r_e - r_m) + r_e(r_m - r_e)} \tag{11-25}$$

By substituting in Eq. (11-22), rearranging, and collecting terms, we obtain

$$R_{oe} = r_c + r_e - r_m + \frac{r_e r_m - r_e^2}{R_G + r_b + r_e} \tag{11-26}$$

A typical value for the output resistance of a transistor in the grounded-emitter connection would be about 60,000 ohms. Also, because r_c is always greater than r_m for these junction transistors, the output resistance is always positive and the grounded-emitter connection is unconditionally stable.

An inspection of Eq. (11-26) will show that, unlike the hypothetical high R_i value discussed in Sec. 11-1.3, high values of resistance are not theoretically possible for R_o. Therefore, this connection is usually found with both input and output impedances in the intermediate range of 5,000 to 75,000 ohms.

166 *Semiconductor Junctions and Devices*

The reader should satisfy himself that the input resistance R_{ie} and output resistance R_{oe} are identical with the parameters r_{11} and r_{22} for the grounded-emitter connection. Make the same assumptions concerning the values of R_G and R_L that correspond to open-circuit conditions as were previously made for the grounded-base case.

11-1.5. Voltage Gain. When we considered the grounded-base connection, Eq. (10-34) gave us an expression for the voltage gain. In a similar fashion the voltage gain for the grounded-emitter connection (VG_e) is given by

$$VG_e = -\frac{i_2 R_L}{e_G} \tag{11-27}$$

where the minus sign reflects phase reversal as indicated by the opposed current arrows in Fig. 11-1.

By substituting for i_2 and rearranging, we have

$$VG_e = \frac{-(r_m - r_e)R_L}{(R_G + r_b + r_e)(R_L + r_c + r_e - r_m) + r_e(r_m - r_e)} \tag{11-28}$$

A typical voltage gain for the grounded-emitter connection is about -600, with the voltages across the load and input 180° out of phase [2].

In our discussion of voltage gain for the grounded-base connection, Eqs. (10-36) to (10-40) were derived to express the maximum voltage gain possible. These equations expressed the theoretical maximum voltage gain in terms of α and the circuit resistance parameters r_{22} and r_{11}. The maximum voltage gain for the grounded-base connection was determined to be $\alpha(r_{22}/r_{11})$.

A similar derivation in the case of the grounded-emitter connection gives the following [cf. Eq. (10-40)]:

$$VG_e \text{ (max)} = -\frac{(\alpha - r_e/r_c)r_c}{r_{11}} \tag{11-29}$$

The maximum theoretical voltage gain for the grounded-emitter connection is somewhat less than that for the grounded-base connection. However, since for most transistors r_e/r_c is negligible compared to α, there is, in practice, no appreciable difference between these two theoretical gains.

11-1.6. Power Gain. The expression for the power gain in this configuration is identical with the expression used [Eq. (10-48)] for the grounded-base configuration. However, since the equation includes i_2, the final form for the expression once i_2 has been substituted will be considerably different. The complete expression is as follows:

$$PG_e = \frac{4R_L R_G(r_m - r_e)^2}{[(R_G + r_b + r_e)(R_L + r_c + r_e - r_m) + r_e(r_m - r_e)]^2} \tag{11-30}$$

Using typical values for the parameters involved, the operating power gain is about 7,000 for transistors in this configuration. The grounded-emitter connection provides the maximum power gain using the typical values of the parameters given in Table 9-2. There are other combinations of circuit design which can result in even higher power gain, but this subject will not be expanded further in this chapter.

11-2. RESULTS—GROUNDED COLLECTOR

The third principal method of connection of transistors is in the grounded-collector or common-collector connection. An equivalent-circuit diagram is given in Fig. 11-2.

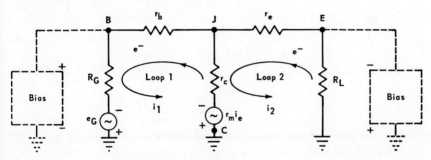

FIG. 11-2. Active equivalent network for grounded-collector transistor (n-p-n case shown).

11-2.1. Resistance Parameters. In writing the circuit equations, bearing in mind the rules previously laid down, the following must be considered for the grounded-collector circuit: The current through r_e equals i_e, which flows toward junction J and is, in this case, identical with i_2. Also, the polarity of $r_m i_e$, consistently with the usage for the grounded-emitter and grounded-base connections, is chosen to oppose electron current arrow i_2. Finally, for the grounded-collector connection, i_c equals current through r_c, or i_1 minus i_2. Exactly as in the other configurations, the current-determining equations are

$$(R_G + r_b + r_c)i_1 - (r_c - r_m)i_2 - e_G = 0 \qquad (11\text{-}31)$$
$$-r_c i_1 + (R_L + r_c + r_e - r_m)i_2 = 0 \qquad (11\text{-}32)$$

We shall now consider the voltages which appear at the input terminals and output terminals of the black box. In this configuration it is again assumed that e_G, R_G, and R_L are all external elements of the circuit (outside the black box) and are therefore effectively equal to zero. Consequently, the black-box equations for this circuit are

$$v_1 = (r_b + r_c)i_1 + (r_c - r_m)i_2 \qquad (11\text{-}33)$$
$$v_2 = r_c i_1 + (r_c + r_e - r_m)i_2 \qquad (11\text{-}34)$$

168 *Semiconductor Junctions and Devices*

As before, from these equations we obtain

$$r_{11} = r_b + r_c \tag{11-35}$$
$$r_{12} = r_c - r_m \tag{11-36}$$
$$r_{21} = r_c \tag{11-37}$$
$$r_{22} = r_c + r_e - r_m \tag{11-38}$$

11-2.2. Current Gain. In the grounded-collector configuration it is again possible to set up an analytical expression for α in terms of the resistance parameters r_{11}, r_{12}, r_{21}, and r_{22}. To demonstrate this, we shall solve the loop equations (11-31) and (11-32) for i_1 and i_2:

$$i_1 = \frac{e_G(R_L + r_c + r_e - r_m)}{(R_G + r_b + r_c)(R_L + r_c + r_e - r_m) - r_c(r_c - r_m)} \tag{11-39}$$

and

$$i_2 = \frac{e_G r_c}{(R_G + r_b + r_c)(R_L + r_c + r_e - r_m) - r_c(r_c - r_m)} \tag{11-40}$$

The theoretical expression for α becomes [from Eqs. (11-39) and (11-40)]:

$$\alpha_{eb} = \frac{i_2}{i_1} = \frac{r_c}{R_L + r_c + r_e - r_m} \tag{11-41}$$

If, as in the past, we assume that R_L has been set at zero by short-circuiting, Eq. (11-41) becomes

$$\alpha_{eb} = \frac{r_{21}}{r_{22}} = \frac{1}{0.04} = 25 \tag{11-42}$$

by substitution of typical values from Table 9-2. However, if R_L is assumed to have its typical value of 100,000 ohms, Eq. (11-41) becomes

$$\alpha_{eb} = \frac{1}{0.14} = 7.1 \tag{11-43}$$

This strong dependence on R_L is typical of the grounded-collector and grounded-emitter configurations. In expressing current gain it is therefore necessary to adopt the convention which leads to the result of Eq. (11-42).

Obviously, the circuit designer employing either the grounded-collector or grounded-emitter configuration must carefully select his load resistance if proper performance is to be expected. As a matter of interest, let us consider the current gains for the other configurations previously discussed. In both of these examples the load resistance R_L was also assumed to be zero. If we now compare the value of α for R_L equal to zero and R_L equal to 100,000 ohms, we find that in the case of the grounded-base configuration the α changes only from 0.96 to 0.87.

However, in the case of the grounded-emitter configuration, the β values are 24 and 6.9 for R_L equal to zero and 100,000 ohms, respectively, almost the same as in the grounded-collector case. Therefore some care must be exercised in circuit design employing these configurations, whereas the grounded-base configuration is much less sensitive to changes in R_L.

11-2.3. Input Resistance. The input resistance of the grounded-collector circuit is

$$R_{ic} = \frac{e_G}{i_1} - R_G \qquad (11\text{-}44)$$

Substitution for i_1 from Eq. (11-39) yields

$$R_{ic} = r_b + r_c - \frac{r_c(r_c - r_m)}{R_L + r_c + r_e - r_m} \qquad (11\text{-}45)$$

When typical values (from Table 9-2) are inserted in the above equation, we find that the typical input resistance for the grounded-collector connection is 0.7 megohm.

11-2.4. Output Resistance. The output resistance of the grounded-collector circuit is

$$R_{oc} = \frac{e_G}{i_2''} - R_L \qquad (11\text{-}46)$$

where i_2'' is the loop 2 current when the signal generator e_G is placed in this loop. Without rewriting the Kirchhoff equations, which are analogous to previous cases, we may simply state that the loop 2 current i_2'' is

$$i_2'' = \frac{e_G(R_G + r_b + r_c)}{(R_G + r_b + r_c)(R_L + r_c + r_e - r_m) - r_c(r_c - r_m)} \qquad (11\text{-}47)$$

If we now substitute this in Eq. (11-46) and rearrange terms, we obtain

$$R_{oc} = r_e + r_c - r_m - \frac{r_c(r_c - r_m)}{R_G + r_b + r_c} \qquad (11\text{-}48)$$

By using values from Table 9-2 in the above equation, we find that the typical output resistance for the grounded-collector configuration is 65 ohms. The grounded-collector connection has a very low output resistance.

11-2.5. Voltage Gain. Continuing our analysis of the grounded-collector connection, the voltage gain is given by

$$VG_c = \frac{i_2 R_L}{e_G} \qquad (11\text{-}49)$$

If we now substitute in the above equation the value for i_2 from Eq. (11-40), we have

$$VG_c = \frac{r_c R_L}{(R_G + r_b + r_c)(R_L + r_c + r_e - r_m) - r_c(r_c - r_m)} \qquad (11\text{-}50)$$

Typical values for voltage gain are slightly less than unity, but positive. If we attempt to find a maximum value $VG_c(\max)$ by setting R_G equal to

170 Semiconductor Junctions and Devices

zero and R_L equal to infinity, as before, we find that the maximum approaches a positive value of unity but cannot exceed it so long as r_c and r_m are both much larger than r_e or r_b. These conditions are always satisfied in practice.

11-2.6. Power Gain. With a high input resistance and low output resistance, the grounded-collector connection, having a very low voltage gain, is truly analogous to the cathode follower of vacuum-tube practice. The current gain of 25 further improves the analogy. With this analogy in mind, we now develop the equations for power gain:

$$PG_c = \frac{4R_L R_G i_2^2}{e_G^2} \quad (11\text{-}51)$$

On substituting for i_2 we have

$$PG_c = \frac{4R_L R_G r_c^2}{[(R_G + r_b + r_c)(R_L + r_c + r_e - r_m) - r_c(r_c - r_m)]^2} \quad (11\text{-}52)$$

When values from Table 9-2 are substituted in Eq. (11-52), we find that the power gain in the grounded-collector connection is about 0.02.

The power gain may be made somewhat higher by suitable choice of R_L and R_G, but not much. In practice, the grounded-collector connection is used principally for impedance matching or current gain. It is clearly no good for voltage or power amplification.

REFERENCES

1. A. Coblenz and H. L. Owens: "Transistors: Theory and Applications," chaps. 8–10, McGraw-Hill Book Company, New York, 1955.
2. R. M. Ryder and R. Kircher: Some Circuit Aspects of the Transistor, *Bell System Tech. J.*, **28**: 367 (1949).

12: Transistor Performance and Frequency Effects

12-1. INTRODUCTION

Having completed detailed equivalent-circuit analyses of the three basic connections for the simple transistor amplifier, we shall next examine the practical consequences. In addition to applications in simple amplifying functions, transistors and similar structures are useful in a variety of other circuits, a few of which will be described in succeeding chapters. Space permits only a very limited selection, and yet the greater part of the remaining text will be concerned with utilitarian extensions of theory, including special junction devices, contacts, reliability, and a short orientation in the point-contact active-element family. In preparation for this change in emphasis, there are a few topics for discussion in this and the next chapter which will augment our present understanding and provide the nomenclature of commercial practice.

This chapter will be devoted to a summary of the performance of transistors as we have so far discussed them and a look at an extension into a new area where frequency of the input signal becomes a major consideration. In Chap. 13 we shall become familiar with commercial terms, symbols, and specifications.

Based on the detailed calculations of the last two chapters, it is now possible to discern overall relationships which can be of considerable practical value. From such an understanding we may derive useful information about any application regardless of the connection mode, simply from a knowledge of the basic current gain parameter α and the external load resistance R_L. Following this analysis (Sec. 12-2), and to conclude the discussion of the simple transistor amplifier in terms of the resistance parameters, the results of the preceding sections are summarized with complete typical values of the r_{jk} resistance parameters, the "operating" characteristics such as current, voltage, and power gains, as well as input and output resistance. Such typical values for the three

172 *Semiconductor Junctions and Devices*

connection modes completely describe the d-c or low-frequency behavior of a simple two-junction device under small-signal or equilibrium operating conditions (Sec. 12-3).

Finally, we must also consider the effect of higher frequency. Our entire discussion thus far has been oriented toward low-frequency (audio) a-c signals. In this sense, our approach has paralleled the history of the solid-state device, since performance was limited to low frequencies during the early development of the junction transistor. (Point-contact units were exceptions to this general rule; they will be discussed in a separate chapter.) We shall therefore conclude our basic treatment of simple junction transistors by examining the problems incidental to the development of units to operate in the megacycle and higher frequency range (Secs. 12-4 to 12-6). Although a variety of new problems arise, the basic technique of circuit analysis and parameter measurement will be found to be entirely analogous to the d-c studies just described. The added instrumentation problems and the somewhat more complex form of the current-voltage relationships where frequency becomes an independent variable add somewhat to the effort required, but the principles are basically unchanged.

12-2. CURRENT PATTERNS

By reviewing the results of the preceding equivalent-circuit network analysis and by recalling a basic principle of transistor action, it is possible to state by inspection the current through any part of a transistor structure, regardless of its connection. Further, by relating this mechanism to other quantities, we may obtain a rapid estimate of such variables as current gain for the circuit as a whole and input resistance of the transistor. This treatment depends on values of the current ratio α and the external load resistance R_L [1].

12-2.1. The Ratio α. In Chaps. 10 and 11 we discussed α as a function of load resistance R_L for each of the connection modes. To do this, we substituted the complete current equations in the ratios of output to input currents i_2/i_1 (which, by definition, are expressions for α and the other current gain parameters) and obtained numerical values showing the effect of increasing the load resistance from zero to 100,000 ohms. We also defined the new parameter β, applicable to the grounded-emitter case, and related it to the normal α (or α_{ce}) for the grounded-base connection. For the grounded-collector, we defined a new α, α_{eb}, again representing the current ratio i_2/i_1. However, these parameters for current ratios were derived by means of a somewhat lengthy calculation, and it was not necessarily apparent that they have a common basis inherent in the operation of any transistor structure. This relationship not only clarifies the current ratio parameters but also materially assists in understanding the wide variations which result from the three methods of connecting the structure to the external circuit.

To illustrate these principles, we shall redraw the equivalent network diagrams in terms of electron flow. We shall adjust the polarity of e_G where necessary and shall complete the circuit through the ground leads. We shall then consider the current vectors in the resulting network for the n-p-n case, remembering the basis of transistor action, that is, when a unit change of current occurs in the emitter (through r_e) a current change of $\alpha_{ce}(= i_2/i_1)$ times unity must occur in the collector circuit through r_c. By combining this basic requirement with the common principle that the sum of all currents at any junction in a network must be zero (with currents flowing toward the junction taken as positive and those flowing away from it as negative) we may represent the situation by the line drawings in Fig. 12-1. In this representation the heavy arrows near the point J in the network correspond to the electron current vectors unity, α_{ce}, and $1 - \alpha_{ce}$ occurring in any transistor structure and flowing through the emitter r_e, the collector r_c, and the base circuit r_b, respectively. Note that in all cases the current vectors represent the split at point J into the components α and $1 - \alpha$ and that the loop-current arrows i_1 and i_2 are drawn for the vectors $1 - \alpha$ and α, respectively.

12-2.2. Current Gain—R_L Shorted. We may now reconsider the current gain ratios for the case where R_L has been set equal to zero, the short-circuited condition. It will be recalled that this was implicit in our original definition of α for the grounded-base case. (As has been seen, the influence of variations in R_L can be strongly felt in some of the transistor parameter values, and it has become customary to use the short-circuit case as a standard basis for comparison.)

For the grounded-base connection we see at once from Fig. 12-1 that the following relationships hold; they are a restatement of our earlier definition of α:

$$i_2 = \text{current through } r_c = \alpha_{ce} \tag{12-1}$$
$$i_1 = \text{current through } r_e = 1 \tag{12-2}$$
$$\frac{i_2}{i_1} = \frac{i_c}{i_e} = \frac{\alpha_{ce}}{1} = \alpha_{ce} \quad \text{with } R_L = 0 \tag{12-3}$$

For the grounded-emitter configuration we have the following, by the same reasoning:

$$i_2 = \text{current through } r_c = \alpha_{ce} \tag{12-4}$$
$$i_1 = \text{current through } r_b = 1 - \alpha_{ce} \tag{12-5}$$
$$\frac{i_2}{i_1} = \frac{i_c}{i_b} = \alpha_{cb} = \frac{\alpha_{ce}}{1 - \alpha_{ce}} = \beta \tag{12-6}$$

again for the case that $R_L = 0$.

174 *Semiconductor Junctions and Devices*

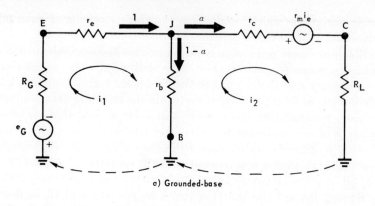

a) Grounded-base

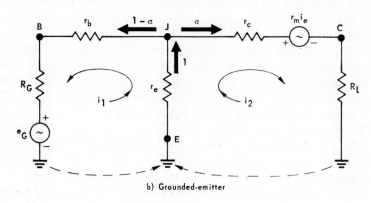

b) Grounded-emitter

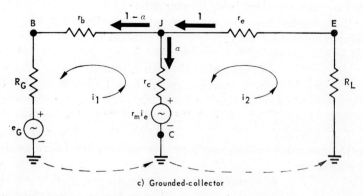

c) Grounded-collector

Fig. 12-1. Equivalent network diagram showing current vectors. (*After A. Coblenz and H. L. Owens, by permission* [1].)

Finally, for the grounded-collector case, again with R_L equal to zero, we may write by analogy

$$i_2 = \text{current through } r_e = 1 \quad (12\text{-}7)$$
$$i_1 = \text{current through } r_b = 1 - \alpha_{ce} \quad (12\text{-}8)$$
$$\frac{i_2}{i_1} = \frac{i_e}{i_b} = \frac{1}{1 - \alpha_{ce}} = \alpha_{eb} = \beta_c \text{ (or } \beta_{eb}) \quad (12\text{-}9)$$

Note that the parameter β_c (or β_{eb}) from Eq. (12-9) is of different form than that of Eq. (12-6), and it is not commonly used. Ordinarily, the parameter β refers to the grounded-emitter case, and current gain for the grounded-collector configuration (which is relatively rarely used) is clearly identified (see Chap. 13) to distinguish it from the common parameters α and β, most frequently seen without subscripts.

Analysis of the equivalent networks in terms of current vectors has now emphasized the universal applicability of α (or α_{ce}) and permitted us to derive all three current gain expressions in terms of this single inherent characteristic of all transistors for the case of R_L equal to zero. Note that if α is chosen to have a value of about 0.96 (α_{ce} for the case $R_L = 0$) the values for β and α_{eb} will be the same as those previously calculated.

12-2.3. Input Resistance versus R_L. We may derive a further bit of information from these circuit diagrams when R_L equals zero and the output circuit contains only the network elements. The definition of α_{ce} in terms of a short-circuited output is complied with. Also, in this amplifier application we have already stated that α_{ce} is normally close to unity. (There are other circumstances, to be dealt with in subsequent chapters, when the value of α can vary over wide limits, in which case the present arguments do not necessarily hold true.)

If we now examine the network diagrams of Fig. 12-1, we see that in the cases of grounded-emitter and grounded-collector connections the current in the loop containing e_G is $1 - \alpha$. Therefore, in both cases, we expect i_1 to become smaller the nearer α approaches unity. In other words, the input signal generator sees a very high resistance in the balance of the circuit and we may say that the input resistances are high, as previously determined from quantitative considerations.

For the grounded-base situation the loop containing e_G always has a unit current vector flowing through it, independently of the value of α. Therefore, the input resistance would not be expected to be high, if R_L equals zero, under any circumstances. Note that this is the conclusion previously arrived at by more complicated reasoning. The above information, confined to the special case of R_L equals zero, is not particularly useful if the transistor is to be used in connection with a functioning external circuit. We shall now reexamine the above cases as R_L varies, usually in the range of 10^3 to 10^6 ohms.

For the grounded-base connection it is intuitively clear that, no matter how large R_L becomes, the current division at point J would not permit i_1 to become small for normal values of r_e. Even if α were made smaller, the balance of the current would still flow through the base resistance and i_1 would not be greatly affected. A final look may be provided by reexamining the loop current for i_1, Eq. (10-19):

$$i_1 = \frac{e_G(R_L + r_b + r_c)}{(R_G + r_b + r_e)(R_L + r_b + r_c) - r_b(r_b + r_m)}$$

By inspection, there is no way i_1 can be made to approach zero by variations in R_L. Indeed, even if we also permit realistic variations in R_G, the same conclusion is reached. The grounded-base transistor cannot be made to exhibit a high input resistance.

In the case of the grounded-emitter there are, regardless of the value of α, two current paths for the input signal, which must simultaneously present a high resistance to make the current value i_1, corresponding to $1 - \alpha$, small. Therefore, if both

$$e_G \approx (i_2 + i_1)r_e \tag{12-10}$$
and
$$-e_G \approx i_2(R_L + r_c) - (i_2 + i_1)r_m \tag{12-11}$$

then i_1 becomes very small and the drop across $R_G + r_b$ is negligible.

If we now require that i_1 approach zero (recognizing that i_2 is also approaching zero), we may derive from Eqs. (12-10) and (12-11)

$$-i_2 r_e = i_2(R_L + r_c - r_m) \tag{12-12}$$
or
$$R_L + r_c + r_e - r_m = 0 \tag{12-13}$$

[which is in the numerator of the loop-current equation previously derived, Eq. (11-9), and the denominator of the expression for input resistance, Eq. (11-21)]. Obviously, if R_L is so chosen that Eq. (12-13) is satisfied, i_1 approaches 0 and R_{ie} approaches ∞. [Note that for the particular transistor described in Table 9-2, Eq. (12-13) cannot be satisfied.]

By similar reasoning, we may examine the grounded-collector situation and show that if both

$$e_G \approx i_2(r_e + R_L) \tag{12-14}$$
and
$$e_G \approx i_2(r_m - r_c) \tag{12-15}$$
then
$$R_L + r_c + r_e - r_m = 0 \tag{12-16}$$

This is the expression that occurs in Eqs. (11-39) and (11-45) for the grounded-collector loop current i_1 and input resistance R_{ic}, respectively. Obviously, the same value of R_L satisfies this relationship.

We have now obtained confirmation of our earlier conclusions for current gain and input resistance by a simple examination of the current flow diagrams of the basic transistor connected in three different configurations. This could be extended, but the qualitative approach must eventually break down and require the more rigorous solutions obtained earlier.

However, the approach employed in this chapter, which relates the behavior of a transistor in all configurations back to the fundamental relationship between i_c and i_e in terms of α, will be of great help in understanding complex configurations and devices which may be encountered outside this discussion, and it suggests that considerable information may be derived from such a qualitative analysis and a knowledge of typical parameter values, like those of Table 9-2, for any new situation.

12-3. TYPICAL LOW–FREQUENCY PERFORMANCE

We now summarize the numerical data obtained thus far for the three connections. It will be seen that these values, using the data of Table 9-2, are merely numerical results for the parameters r_{11}, r_{12}, r_{21}, and r_{22} and typical values of the characteristics which were qualitatively set forth in Table 9-1. The practical usefulness of the three modes of connection again becomes evident from the numerical examples.

12-3.1. Grounded-base Resistance Parameters. [Previously given as Eqs. (10-5) to (10-8)]

$$r_{11} = r_e + r_b = 525 \text{ ohms} \qquad (12\text{-}17)$$
$$r_{12} = r_b = 500 \text{ ohms} \qquad (12\text{-}18)$$
$$r_{21} = r_b + r_m = 0.96 \text{ megohm} \qquad (12\text{-}19)$$
$$r_{22} = r_b + r_c = 1 \text{ megohm} \qquad (12\text{-}20)$$

12-3.2. Grounded-emitter Resistance Parameters

$$r_{11} = r_b + r_e = 525 \text{ ohms} \qquad (12\text{-}21)$$
$$r_{12} = r_e = 25 \text{ ohms} \qquad (12\text{-}22)$$
$$r_{21} = r_e - r_m = -0.96 \text{ megohm} \qquad (12\text{-}23)$$
$$r_{22} = r_c + r_e - r_m = 0.04 \text{ megohm} \qquad (12\text{-}24)$$

12-3.3. Grounded-collector Resistance Parameters

$$r_{11} = r_b + r_c = 1 \text{ megohm} \qquad (12\text{-}25)$$
$$r_{12} = r_c - r_m = 0.04 \text{ megohm} \qquad (12\text{-}26)$$
$$r_{21} = r_c = 1.0 \text{ megohm} \qquad (12\text{-}27)$$
$$r_{22} = r_c + r_e - r_m = 0.04 \text{ megohm} \qquad (12\text{-}28)$$

Finally, we include Table 12-1, which gives numerical values for a typical *n-p-n* or *p-n-p* transistor in the three connection modes. Figures

178 Semiconductor Junctions and Devices

TABLE 12-1. CONNECTION MODES AND TRANSISTOR PARAMETERS*

Parameter	Grounded base	Grounded emitter	Grounded collector
Current gain.........	$\alpha = 0.96\text{--}0.99$	$\beta = 35$	$\alpha_{eb} = \beta_{eb} = 25$
Input resistance......	$R_i = 90$ ohms	$R_{ie} = 700$ ohms	$R_{ic} = 0.7$ megohm
Output resistance.....	$R_o = 0.6$ megohm	$R_{oe} = 63{,}000$ ohms	$R_{oc} = 70$ ohms
Voltage gain.........	$VG = 150$	$VG_e = -600$	$VG_c = <1$
Operating power gain.	$PG = 400$ (25 db)	$PG_e = 7{,}000$ (38 db)	$PG_c = <0.05$

* After A. Coblenz and H. L. Owens, by permission [2].

are given for current gain, input resistance, output resistance, voltage gain (but not maximum voltage gain), and operating power gain. These are based on the parameters given in Table 9-2 and other values as discussed in preceding chapters.

12-4. HIGH-FREQUENCY PROBLEMS

In the preceding discussion we used the resistance, or low-frequency a-c, parameters r_{11}, r_{12}, r_{21}, and r_{22} as the basis for the characterization of junction transistors and the calculation of their behavior in useful circuits. Both this information and the method by which it was obtained are of fundamental value in understanding transistor electronics. In addition, the calculations were relatively straightforward, permitting the focus to be primarily on the philosophy underlying the basic approach.

However, as the use of transistors at frequencies beyond the audio range became a reality, and as fabrication techniques grew more specialized in an effort to maximize performance characteristics favorable to a particular application, new parameters were found to be necessary and a more universal set of derived characteristics, based on the new parameters, was evolved. Basically, the trend has been toward a set of electrical quantities which can be uniformly applied to characterize the thousands of multijunction devices in order to encompass variables such as frequency, bias, method of connection, temperature, and geometry. As uses for transistors have multiplied, starting with the basic amplifier described herein, to include oscillators, switches, computer elements, logic circuits, regulators, and many others, it has become desirable to select numerical specifications on such a basis that the designer can evaluate the potentialities and predict the performance of each device for any of a wide variety of applications.

Obviously, such a field is always expanding, and new developments are sometimes difficult to fit into the existing terminology. It is not intended that this discussion include detailed information on such a rapidly changing situation, and there will be no attempt to probe into the more esoteric applications of transistors or to deal with the specialized developments of terminology, test methods, etc. However, there exists a broad gulf between the approach used in this discussion thus far and the majority of contemporary data and performance characteristics as applied in a standardized fashion to most of the commercial devices now being sold. Since the development of this present system represents a logical outgrowth of the original approach, with only two new variables, frequency and bias, and since the new terminology is really a logical generalization of effects we have already noted, it is desirable to trace these evolutions and define the new quantities in terms of familiar parameters.

12-5. RESISTANCE PARAMETER INADEQUACIES

In addition to the low-frequency limitation, there are other areas in which the simple resistance parameters have proved to be unsatisfactory. These include basic difficulties in experimental measurement and lack of an adequate procedure to ensure the presence of standard, reproducible bias when the measurements are made. This last difficulty may not be entirely resolved within the framework of the measurement itself and may sometimes require an additional parameter. However, the more recent techniques are relatively well standardized and give consistent results in a wide variety of practical situations.

Leaving the matter of frequency response until later, we shall first examine the measurement technique. It will be recalled that, in determining r_{11}, r_{12}, r_{21}, and r_{22}, one pair of terminals was required to be open-circuited. This is, in practice, difficult to achieve. True, if a voltage is to be determined, a voltmeter of very high impedance (>100 megohms) may be employed and, compared to values of the parameters, may be large enough to be regarded as an open circuit. Unfortunately, however, there is the bias supply still to be reckoned with, and although this can theoretically be made to have as high an impedance as necessary, it becomes somewhat unattractive and cumbersome. Finally, even if these external impedances are high in relation to the internal circuit being measured, there are always the internal capacitances, both junction and distributed.

In actual devices, these capacitances may add up to produce a stray current through the transistor itself, independent of the "open" circuit external to it, of such magnitude as to cause serious error even at relatively low frequencies (of the order of 270 cps). At higher frequencies the problem becomes worse, and this inherent property of the transistor structure plus the relatively cumbersome biasing arrangements led to a search for other parameters which might be more readily determined experimentally and still supply the necessary information.

12-6. HYBRID PARAMETERS

The first attempt to avoid the pitfalls of the resistance parameters was to measure the reciprocal of resistance, or conductance. This very neatly substituted a short circuit for the open circuit and avoided the problems mentioned above. Further, by measuring the conductance parameters with a-c signals, it was easy to maintain a d-c bias on a "shorted" pair of terminals, that is, to create an a-c short circuit (by means of a capacitor) while effectively leaving it open for d-c bias voltage.

It is not necessary to go into the exact formulation of conductance parameters, since they are the inverse of the resistance situations. However, a little thought will make it clear that the conductance parameters g_{11}, g_{12}, g_{21}, and g_{22} are not simply the reciprocals of the resistance parameters. It is easy to show that

$$g_{11} = \frac{r_{22}}{r_{11}r_{22} - r_{12}r_{21}} \tag{12-29}$$

$$g_{12} = \frac{-r_{12}}{r_{11}r_{22} - r_{12}r_{21}} \tag{12-30}$$

$$g_{21} = \frac{-r_{21}}{r_{11}r_{22} - r_{12}r_{21}} \tag{12-31}$$

$$g_{22} = \frac{r_{11}}{r_{11}r_{22} - r_{12}r_{21}} \tag{12-32}$$

Corresponding circuit analyses can be worked out by using these parameters. A slightly modified equivalent circuit, known as the *pi form*, has been found to be most adaptable, but it will not be further elaborated here.

Unfortunately, in practice there is still an experimental difficulty. In the r-parameter approach there was trouble achieving a true open circuit at the high-resistance (r_{22}) end of the network, and there is a similar problem in creating a true short circuit at the low-resistance (r_{11}) end of the network for measurement of conductance. This arises from the fact that, associated with any capacitance, there is a certain reactance, or a-c resistance. This quantity is inversely proportional to the capacitance, so that to make it truly small, in comparison to the r_{11} value of the network, enormous capacitances are required across the input terminals to effect a true a-c short circuit while still leaving the circuit open for d-c biasing. Significant errors of the same type as seen with the r parameters are otherwise observed.

In the light of the above troubles with the r and g parameters,* it is a

---

* Before proceeding to the final step in this discussion, it should be mentioned that in many texts and other publications, the r parameters are known as Z parameters and the g parameters as Y parameters.

logical step to find a new set in which it is never proposed to have the output open-circuited or the input short-circuited. Also, when dealing with a high-resistance circuit such as a collector, where large voltage changes produce small current changes, it is easier to let the voltage be the independent variable and the current be dependent on it. Conversely, for the emitter, of low resistance, we choose current for the independent and voltage for the dependent variable.

These conditions result in the following definitions:

h_{11} = input resistance (output shorted) (12-33)
h_{12} = reverse voltage amplification (input open) (12-34)
h_{21} = forward current amplification (output shorted) (12-35)
h_{22} = output conductance (input open) (12-36)

The measuring circuits are shown in Fig. 12-2. Note that h_{11} has the units of resistance, h_{22} the units of conductance, while h_{12} and h_{21}, being ratios, are pure numbers. Note also that the previous problems of bias have disappeared by virtue of eliminating the open-circuit-biased collector (h_{22}) and the short-circuit-biased emitter (h_{11}). As required for these measurements, the bias provisions present no problems.

In terms of these parameters, the input-output equations take the form

$$v_1 = h_{11}i_1 + h_{12}v_2 \quad (12\text{-}37)$$
$$i_2 = h_{21}i_1 + h_{22}v_2 \quad (12\text{-}38)$$

Therefore, unlike the r or g parameters, the h parameters do not yield equations which are immediately amenable to circuit analysis. They may easily be converted, however, to either r or g parameters. For example, if Eqs. (12-37) and (12-38) are solved for v_1 in terms of i_1 and i_2 and the result is compared with Eq. (10-9), values for r_{11} and r_{12} are obtained directly:

$$r_{11} = h_{11} - \frac{h_{12}h_{21}}{h_{22}} \quad (12\text{-}39)$$

$$r_{12} = \frac{h_{12}}{h_{22}} \quad (12\text{-}40)$$

Repeating for v_2 [Eq. (10-10)], we obtain similar relations defining r_{21} and r_{22}:

$$r_{21} = -\frac{h_{21}}{h_{22}} \quad (12\text{-}41)$$

$$r_{22} = \frac{1}{h_{22}} \quad (12\text{-}42)$$

In the last case, the definition of h_{22} is identically the reciprocal of r_{22}.

182 *Semiconductor Junctions and Devices*

To obtain the value of α in the h-parameter system, we recall that $\alpha_{ce} = \alpha = r_{21}/r_{22}$ [Eq. (10-24)]. Then from Eqs. (12-41) and (12-42) we find $\alpha = -h_{21}$. The negative sign results from the definition of the h parameters and the form of the input-output equations (12-37) and

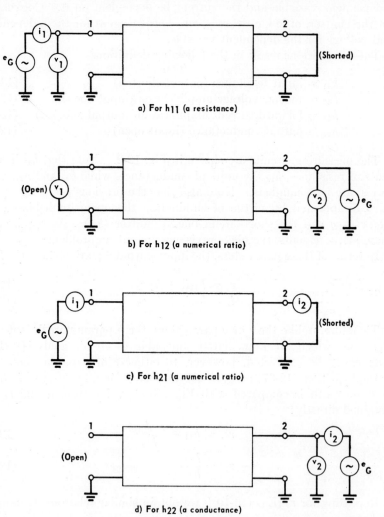

Fig. 12-2. Circuit connections for measurement of hybrid parameters.

(12-38). Note that the numerical value, sign, and meaning of α (which we define, in any parameter system, as the ratio I_C/I_E) is unchanged. Only the sign of the h-parameter equivalent for α has been reversed, an algebraic consequence with no physical significance.

There is another short-form notation for the h parameters which identifies them as follows:

$h_{11} = h_i$ input impedance with output shorted (12-43)
$h_{12} = h_r$ reverse voltage transfer ratio, input open (12-44)
$h_{21} = h_f$ forward current transfer ratio, output shorted (12-45)
$h_{22} = h_o$ output admittance or conductance, input open (12-46)

An additional letter is sometimes added to the above subscripts to indicate the transistor configuration. Thus, h_{ib} is input impedance with output shorted for the grounded-base mode and h_{re} is the reverse voltage transfer ratio with input open for the grounded-emitter mode. Similarly, h_{fb} is the forward current transfer ratio with output shorted for grounded-base ($= -\alpha$) and h_{fe} is the same parameter for grounded-emitter and is equal to β, with no sign reversal.

The final and most important property of the h parameters is that they give valid results for all frequencies, provided they are measured at the desired frequency. This is largely due to the elimination of spurious current paths at collector and emitter as a result of the short- or open-circuit requirements and the independence of bias response with respect to frequency changes. Therefore, the h parameters are usable over all frequencies and the loop equations (12-37) and (12-38) are valid, even though nonlinear with respect to frequency. However, unlike the other systems of parameters, if the values are obtained for the desired frequency range, they may then be converted to r or g parameters and these values can be used as described earlier to fully characterize the transistor.

Finally, as in the previous situations, the h parameters are most commonly measured for the grounded-base configuration, despite the relatively wider use of the grounded-emitter configuration in amplification, switching, and other applications.

REFERENCES

1. A. Coblenz and H. L. Owens: "Transistors: Theory and Applications," pp. 139, 140, McGraw-Hill Book Company, New York, 1955.
2. Ref. 1, p. 144.

13: Commercial Transistor Specifications

13-1. INTRODUCTION

With the introduction of the hybrid parameters and a qualitative look at high-frequency problems, the theory of the simple two-junction transistor has been covered in sufficient detail to permit an extension to commercial terminology. Only by understanding the transition from theory to practice can we appreciate the essential link between the disciplines of physical science and applied electronics. In the everyday language of the design and applications engineer we see the principle of the blackbox technique carried to a highly useful conclusion. This has produced a set of parameters and specifications which define each transistor in unambiguous terms familiar to the designer and user and which accurately forecast its performance in any proposed circuit. This chapter will accordingly be devoted to detailed definitions of these terms and interpretations based on fundamental properties already discussed.

In a brief survey of this type it is impossible to be all-inclusive in the encyclopedic sense. The intent is to point out only the most important guideposts and to encourage the reader to explore side roads of his own volition. There will, therefore, be some new ideas introduced in the course of this survey of commercial data sheets. For example, quantities such as rise time, storage time, and saturation values are first encountered in this chapter. We also anticipate the discussion of switching applications and note briefly that one of our examples is an excellent high-speed low-power switch. Most of these definitions will be amplified in subsequent chapters to relate the phenomena to physical principles, but no difficulty should be experienced in comprehending their significance at this stage.

13-2. SELECTED PARAMETER SYMBOLS [1]

The following tables contain definitions of selected parameter symbols used in commercial specification sheets. This group includes parameters such as the four-pole and derived or measured quantities, and it was selected to illustrate the following points: First, some of the symbols used do not relate directly to the h parameters. These ratios can be obtained, however, either from knowledge of the test procedures or from additional data from the device manufacturer. Second, the symbols are as unambiguous as possible with respect to connection mode of the transistor

TABLE 13-1. VOLTAGE SYMBOLS*

$BV_{CBO}\ (=V_{CBO})$	Collector-to-base d-c reverse breakdown voltage, emitter open-circuited (I_C must be specified)†
$BV_{CEO}\ (=V_{CEO})$	Collector-to-emitter d-c reverse breakdown voltage, base open-circuited; this may be a function of M, the junction multiplication factor, and the h_{fb} of the transistor (I_C must be specified)†
$BV_{CER}\ (=V_{CER})$	Same as BV_{CEO}, but with resistor R between base and emitter†
$BV_{CES}\ (=V_{CES})$	Same as BV_{CEO}, but base shorted to emitter†
$BV_{EBO}\ (=V_{EBO})$	Emitter-to-base d-c reverse breakdown voltage, collector open-circuited (I_E must be specified)†
$BV_R\ (=V_R)$	D-c breakdown voltage, reverse-biased diode
$V_{BE(SAT)}$	Saturation voltage measured at base terminal for grounded-emitter case†
$V_{CE(SAT)}$	Saturation voltage measured at collector terminal for grounded-emitter case†

* After General Electric Company, "Transistor Manual," by permission [1].
† Test conditions must be specified.

TABLE 13-2. CAPACITANCE AND FREQUENCY SYMBOLS*

C_c	Barrier capacitance at the collector junction
C_{eb}	Emitter-to-base capacitance, collector open (common-base configuration)†
C_{ob}	Collector-to-base output capacitance (common-base configuration)†
C_{oe}	Collector-to-emitter output capacitance (common-emitter configuration)†
f	Frequency at which measurement is made
$f_{hfb}\ (=f_{\alpha b})$	Short-circuit forward current transfer ratio cutoff frequency (common-base configuration)‡
$f_{MAX}\ (=f_{OSC})$	Maximum frequency of oscillation

* After General Electric Company, "Transistor Manual," by permission [1].
† Test conditions must be specified.
‡ Small-signal restriction applies.

and bias. Finally, most of the parameters are d-c values to avoid measurement problems and the necessity of supplying data at a variety of frequencies. Note that in the cases of symbols marked † the test conditions must be specified. It must also be remembered that the small-signal requirement is still imposed in some instances.

Table 13-1 gives the symbols used for various breakdown and saturation voltages in device terminology. The use of the notation V_{XYZ} for voltage, whether forward or reverse, is universal. For added clarity BV_{XYZ} is used to mean reverse breakdown or peak inverse voltage. Note also

186 Semiconductor Junctions and Devices

TABLE 13-3. HYBRID PARAMETER SYMBOLS*

$h_{fb}\ (=\ -\alpha)$	Forward current transfer ratio, output a-c shorted, common-base†
$h_{fe}\ (=\ \beta)$	Forward current transfer ratio, output a-c shorted, common-emitter†
$h_{fc}\ (=\ \beta_c)$	Forward current transfer ratio, output a-c shorted, common-collector†
$h_{ib'}\ h_{ie'}$ $h_{ic'}\ h_{ij}$	Input impedance, output a-c shorted, for common-base, -emitter, and -collector configurations and general case†
$h_{ob'}\ h_{oe'}$ $h_{oc'}\ h_{oj}$	Output impedance, input a-c open, for common-base, -emitter, and -collector configurations and general case†
h_{FE}	Static forward current transfer ratio I_C/I_E, common-emitter‡
$h_{rb'}\ h_{re'}$ $h_{rc'}\ h_{rj}$	Reverse voltage transfer ratio, input a-c open, for common-base, -emitter, and -collector configurations and general case†

* After General Electric Company, "Transistor Manual," by permission [1].
† Small-signal restriction applies.
‡ Test conditions must be specified.

TABLE 13-4. CURRENT SYMBOLS*

	I_B	Base current
	I_C	Collector current
$I_{CBO}\ (=\ I_{CO})$		D-c collector current, collector junction reverse-biased, emitter open†
	I_{CEO}	D-c collector current, collector junction reverse-biased, base open†
	I_{CER}	Same as I_{CEO}, but with resistor R between base and emitter†
	I_{CES}	Same as I_{CEO}, but with base shorted to emitter†
$I_{EBO}\ (=\ I_{EO})$		D-c emitter current, emitter junction reverse-biased, collector open†
	I_{ECS}	D-c emitter current, emitter junction reverse-biased, base shorted to collector†
	I_R	D-c reverse current, general case

* After General Electric Company, "Transistor Manual," by permission [1].
† Test conditions must be specified.

TABLE 13-5. MISCELLANEOUS SYMBOLS*

	P_o	Power output
	P_T	Average continuous total power dissipation
$R_{KJ(\text{SAT})}\ =\ R_{K(\text{SAT})}$		Saturation resistance or ratio of voltage to direct current measured at terminal K; mode of connection must be given or specified as $J\ =\ B,\ C,$ or E for common terminal†
	R_L	Load resistance
	T_A	Ambient operating temperature
	T_J	Junction temperature
	T_{STG}	Storage temperature
	t_r	Pulse rise time or time duration for an increase from 10 to 90% of pulse amplitude (associated with collector junction capacitance and emitter-base current flow)
	t_s	Pulse storage time or time duration for decrease of 10% of pulse amplitude (associated with base width, minority carrier mobility, and capacitance)
	t_f	Pulse fall time or time duration for decrease from 90 to 10% of pulse amplitude (reverse of rise time)

* After General Electric Company, "Transistor Manual," by permission [1].
† Test conditions must be specified.

that if BV_{XYZ} refers to a single junction, the reverse saturation current at which the junction is rated is denoted by I_X. Under other circumstances, I_X might also be the forward current.

Table 13-2 gives the capacitance and frequency symbols commonly used in device characterization. Note that the α cutoff frequency is defined as the frequency at which α has fallen to 0.707 of its d-c value.

Table 13-3 lists the symbols used for hybrid parameters commonly employed by device manufacturers.

In Table 13-4 are listed the various currents encountered in device terminology.

Table 13-5 lists the remaining miscellaneous symbols commonly used.

13-3. SPECIFICATION SHEET—2N332 TRANSISTOR [2]

The above symbols and the restrictions placed on their use are common in transistor designations now in use. As a concrete example, let us examine the following set of typical specifications for a relatively unsophisticated device widely used in simple applications such as audio amplification in the communication field. This is a versatile silicon transistor of the grown-junction type and is now available in a model which meets military specifications. It is a high-gain, high-temperature amplifier for the audio-frequency range.* We shall now review the tabulated specifications in detail and correlate the information with our definitions, parameters, etc.

13-3.1. Absolute Maximum Ratings. These are stresses beyond which the device may not be pushed without impairing service life or initial performance. Many devices of a given batch might exceed these ratings, but the manufacturer assumes no responsibility and makes no tests for higher ratings. In this case stresses are those of voltage, current, power, and temperature. Note that no conditions of test are given except temperature. This means that there are no conditions under which these values may be exceeded.

The value V_{CBO} could also be listed as BV_{CBO}, since it is a breakdown value referring to the reverse-biased breakdown for a single junction. For this class of devices the manufacturer has elected to use 50 μa as the reverse current corresponding to the breakdown of a single junction.

The collector and emitter currents are also absolute maxima, regardless of other conditions except that of ambient temperature. The values are given for 25°C, and the corresponding dissipation values set limits on the power both at 25°C and at higher temperatures. The temperature ratings are straightforward, except that estimation of actual service

---
* Audio frequencies are usually taken as 20 to 20,000 cps.

junction temperatures is difficult without more information from the manufacturer.

13-3.2. Electrical Characteristics. The d-c characteristics compare with those of a single junction. Note that the voltage, or BV_{CBO}, is the same as the maximum, but that conditions of measurement (temperature and conditions of the input circuit) are uniquely determined. Since

TABLE 13-6. SILICON TYPE 2N332 (n-p-n)*
Absolute Maximum Ratings (25°C unless Specified)

Voltage:
Collector to base, V_{CBO} .... 45 volts
Emitter to base, V_{EBO} .... 1 volt
Current:
Collector, I_C .... 25 ma
Emitter, I_E .... 25 ma
Transistor dissipation:
At 25°C .... 150 mw
At 100°C .... 100 mw
At 150°C .... 50 mw
Temperature:
Operating junction .... −65 to 175°C

Electrical Characteristics (T_J = 25°C unless Specified)

D-c characteristics (common-base):
Collector breakdown voltage (I_C = 50 μa), BV_{CBO} .... 45 volts
Collector cutoff current (V_{CB} = 30 volts), I_{CBO} .... 2 μa
Collector cutoff current (100°C, V_{CB} = 5 volts), I_{CBO} ...... 10 μa
Collector cutoff current (150°C, V_{CB} = 5 volts), I_{CBO} ...... 50 μa
Input impedance (V_{CB} = 5 volts, I_E = 1 ma), h_{ib} .... 55 ohms†
Output admittance (V_{CB} = 5 volts, I_E = 1 ma), h_{ob} ...... 0.5 μmho†
Feedback voltage ratio (V_{CB} = 5 volts, I_E = 1 ma), h_{rb} ... 195†
Current transfer ratio (V_{CB} = 5 volts, I_E = 1 ma), $h_{fb}(=-\alpha)$ ...... −0.925†
Saturation resistance (I_B = 2.2 ma, I_C = 5 ma) measured in common-emitter connection, $R_{C(SAT)}$ .... 70 ohms

High-frequency Characteristics

Frequency cutoff (V_{CB} = 5 volts, I_E = 1 ma), f_{ab} .... 6 mc
Output capacitance (1 mc, V_{CB} = 5 volts, I_E = 1 ma), C_{ob} ..... 10 pf

* After Texas Instruments Incorporated Bulletin DL-S 1035, by permission [2].
† Measured at 1 kc.

this is a breakdown voltage, no other parameters, such as bias, are applicable.

We omit further discussion of $R_{C(SAT)}$ at this time, since this parameter does not become important until we discuss switching applications.

In this set of specifications, where the frequency is presumed to be in or near the audio range, we are given values of the h parameters for a frequency of 1 kc at an operating point corresponding to V_{CB} = 5 volts and $I_E = 1$ ma. Values of the h parameters so obtained will be close to accurate at frequencies of less than 20 kc, or the audio range, thereby permitting us to make circuit calculations as previously described for the equivalent network.

At first glance, it would seem that our definitions of the h parameters have not been complied with, especially as to the requirement that the output be shorted and the input open. We remember, however, that the signal used to measure these values is 1 kc and that it is easy to a-c-short the output at this frequency and at the same time to d-c-bias it to

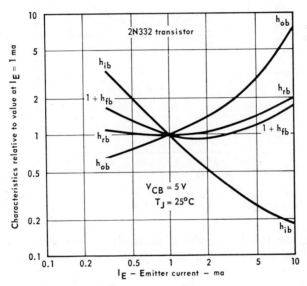

FIG. 13-1. Common-base characteristics versus emitter current. (*After Texas Instruments Incorporated, Bulletin DL-S 1035, by permission* [2].)

5 volts as specified for V_{CB}, the selected operating point. Similarly, we may open-circuit the input with respect to the 1-kc signal by means of a large impedance and at the same time supply a d-c bias voltage large enough to cause a current I_E of 1 ma to flow. In interpreting the data from the specification sheet, we assume that the h parameters were so measured.

Since we now have a set of h-parameter values for one operating point, namely, at $V_{CB} = 5$ volts and $I_E = 1$ ma, we may use the curves of Figs. 13-1 and 13-2 to adjust these values to different temperatures and emitter currents. In these plots, since they are for the common-base connection where h_{fb} approaches -1, it is easier to read the graphs if the function $1 + h_{fb}$ is plotted. Note that the ordinates are ratios, not absolute values. In other words, at an emitter current of 10 ma, $h_{rb} = 2 \times 195$ (the value at 1 ma), or 390.

Finally, to obtain values for other operating points, we refer to Figs. 13-3 and 13-4, which give us values of I_C for other combinations of V_{CB} and I_E, as well as corresponding data for the grounded-emitter case.

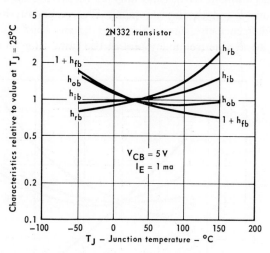

Fig. 13-2. Common-base characteristics versus junction temperature. (*After Texas Instruments Incorporated, Bulletin DL-S* 1035, *by permission* [2].)

Note that, for the latter, the h parameters must be converted from grounded-base values to grounded-emitter values as follows:

$$h_{ie} = \frac{h_{ib}}{1 + h_{fb}} \tag{13-1}$$

$$h_{re} = \frac{h_{ib}h_{ob}}{1 + h_{fb}} - h_{rb} \tag{13-2}$$

$$h_{fe} = \beta = -\frac{h_{fb}}{1 + h_{fb}} = \frac{\alpha}{1 - \alpha} \tag{13-3}$$

$$h_{oe} = \frac{h_{ob}}{1 + h_{fb}} \tag{13-4}$$

We are therefore in a position to calculate any small-signal circuit involving this device at frequencies in the audio range.

The specification sheet of Table 13-6 was chosen as an example of a wide-range, general-purpose device with no particular specialized function except high-temperature performance. The data were found to be sufficient for all practical purposes, and the h-parameter values were given completely, so that direct circuit calculations could be made.

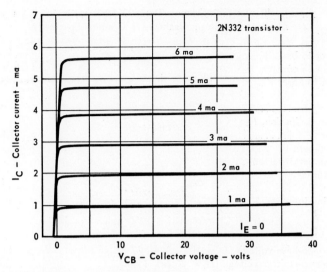

FIG. 13-3. Common-base output characteristics. (*After Texas Instruments Incorporated, Bulletin DL-S* 1035, *by permission* [2].)

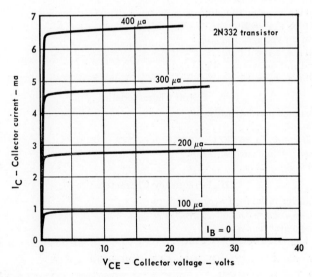

FIG. 13-4. Common-emitter output characteristics. (*After Texas Instruments Incorporated, Bulletin DL-S* 1035, *by permission* [2].)

TABLE 13-7. SILICON TYPE 2N2193 (n-p-n)*
Absolute Maximum Ratings (25°C unless Specified)

Voltage:
Collector to base, V_{CBO}.. 80 volts
Collector to emitter, V_{CEO}................................... 50 volts
Emitter to base, V_{EBO}.. 8 volts

Current:
Collector, I_C.. 1 amp

Transistor dissipation, P_T:
Free air 25°C†.. 0.8 watt
Case temperature 25°C‡.. 2.8 watts
Case temperature 100°C‡....................................... 1.6 watts

Temperature:
Storage, T_{STG}... −65 to 300°C
Operating junction, T_J....................................... −65 to 200°C

Electrical Characteristics (25°C unless Specified)

D-c characteristics:
Collector-to-base voltage ($I_C = 100$ μa), V_{CBO}............ 80 volts
Collector-to-emitter voltage ($I_C = 25$ μa), V_{CEO}§........ 50 volts
Emitter-to-base voltage ($I_E = 100$ μa), V_{EBO}............. 8 volts

Forward current transfer ratio, h_{FE}:
$I_C = 150$ ma, $V_{CE} = 10$ volts§........................... 120
$I_C = 10$ ma, $V_{CE} = 10$ volts............................. 30
$I_C = 1{,}000$ ma, $V_{CE} = 10$ volts§....................... 15
$I_C = 0.1$ ma, $V_{CE} = 10$ volts............................ 15
$I_C = 500$ ma, $V_{CE} = 10$ volts§........................... 20
$I_C = 10$ ma, $V_{CE} = 10$ volts, $T_A = -55°C$.............. 20

Base saturation voltage, $V_{BE(SAT)}$:
$I_C = 150$ ma, $I_B = 15$ ma................................... 1.3 volts

Collector saturation voltage, $V_{CE(SAT)}$:
$I_C = 150$ ma, $I_B = 15$ ma................................... 0.35 volt

Cutoff Characteristics

Collector leakage current, I_{CBO}:
$V_{CB} = 60$ volts... 10 mμa
$V_{CB} = 60$ volts, $T_A = 150°C$.............................. 25 μa

Emitter-base leakage current, I_{EBO}:
$V_{EB} = 5$ volts.. 50 mμa

High-frequency Characteristics

Current transfer ratio, h_{fe} ($= \beta$):
$I_C = 50$ ma, $V_{CE} = 10$ volts, $f = 20$ mc................ 2.5

Collector capacitance, C_{ob}:
$I_E = 0$, $V_{CB} = 10$ volts, $f = 1$ mc..................... 20 pf

Switching Characteristics

Rise time t_r.. 70 nsec
Storage time t_s.. 150 nsec
Fall time t_f.. 50 nsec

* After General Electric Company, "Transistor Manual," by permission [3].
† Derate 4.6 mw per °C increase above 25°C.
‡ Derate 16.0 mw per °C increase for case above 25°C.
§ Pulse width ≤300 μsec, duty cycle ≤2%.

There are, however, other types of devices for which different properties are more important and for which the data do not initially seem adequate.

13-4. SPECIFICATION SHEET—2N2193 TRANSISTOR [3]

We now select a highly specialized device designed for high-speed switching and high-frequency amplifier circuits. The device is produced by the "planar epitaxial" technique, which gives it great advantages in terms of leakage current, switching speed, current gain over a wide range, and high reverse breakdown voltage. Again, we review the specifications and compare them with our definitions and parameters.

13-4.1. Absolute Maximum Ratings. Once again these are really breakdown values and could have been symbolized as BV_{XYZ}. Note that $V_{CEO}(=BV_{CEO})$ is now a voltage across two junctions, where the first, or collector, junction is reverse-biased and the second is in a specified state of bias. In this case it is zero, since the base lead is open.

The maximum collector current, stated as 1 amp, is actually a pulse current, as is seen from the list of h_{FE} values. It corresponds to the highest value for which an h_{FE} is given and to the highest (pulse) current which can be tolerated owing to thermal ratings of the junction. In no case should this value be regularly reached in device service.

The values of transistor dissipation are actual life test thermal ratings. The free-air case is a pure convective cooling situation, while the case temperature values assume an infinite heat sink for cooling. By use of both values and their derating coefficients, intermediate situations may be estimated. The temperature ratings are self-explanatory. Once again determination of T_J presents problems.

13-4.2. Electrical Characteristics. Note that the manufacturer specifies a reverse current of 100 µa as the current corresponding to the breakdown of single junctions, as in $V_{CBO}(=BV_{CBO})$ and $V_{EBO}(=BV_{EBO})$ values. For breakdown across two junctions, the picture becomes more complex. For reasons beyond the scope of this discussion, the reverse breakdown voltage from collector to emitter will be lowest if the base lead is open. In other words, the voltage so obtained at a current of 25 ma is a conservative value, but it is of interest in some applications.

With respect to the h parameters, in this set of specifications we are given only values of h_{FE}, which has a maximum value of 120 for a pulsed signal of 150 ma and falls to a value of 15 for smaller continuous signals. In addition to h_{FE}, we know the corresponding output current, output voltage, and pulse conditions. We may also obtain a value for the input current, since h_{FE} is also β for d-c behavior. A reasonably good choice for the small-signal case might be represented by choosing I_C at about 10 ma for a β of about 20 to 25 and an input current I_B of about 0.4 to 0.5 ma. However, we still lack values for the input voltage and bias conditions.

The base and collector saturation voltages are of principal interest to designers of fast switching circuits, and they will not be discussed further at this time. Cutoff characteristics are measures of transistor leakage currents, which are also of importance in high-speed (high-frequency) switching applications. It is worth noting that these currents are 10^{-3} to 10^{-6} times the normal operating currents.

The high-frequency response is given in terms of the effective small-signal $h_{fe}(=\beta)$ at 20 mc, which has fallen to a value of 2.5 compared to its d-c value of about 25. Also given is the effective collector capacitance at 1 mc, another parameter of interest in high-frequency design. Finally, the switching times, all functions of geometry, capacitance, frequency response, lifetimes, and mobilities, are also shown for the benefit of the applications engineer who needs high-speed response.

13-4.3. h Parameters. In the specification sheet of Table 13-6 the h-parameter values were given under small-signal conditions and for the normal frequency range for which the transistor was designed. In the present case these values are not given explicitly. From the foregoing, if we assume that bias conditions are fulfilled, we can estimate values for h_{22} and h_{21} as follows:

$$h_{22} = h_{oe} = \frac{I_C}{V_{CEO}} \quad (13\text{-}5)$$

This is the ratio of the output current to the output voltage (conductance) with the input (base) circuit open. However, the V_{CEO} value is really a reverse breakdown (BV_{CEO}) and it was necessary to pulse the current I_C to avoid overheating. This is not, therefore, a small-signal value, and it could not be used with accuracy. Similarly, we cannot use the value of $V_{CE} = 10$ volts from the table, since this was not obtained with the input circuit open and therefore does not comply with the definition of h_{22}.

The other value is

$$h_{21} = h_{fe} = \frac{I_C}{I_B} \quad (13\text{-}6)$$

which we have estimated at 25 for this grounded-emitter case by using low-current values for I_C which correspond to a collector bias of 10 volts, the only value shown. This seems entirely reasonable, but it may or may not fulfill the requirements for small-signal-parameter values.

Obviously, more information is needed. In most cases, more complete data sheets are available from the manufacturer. In the present case, the additional data take the form of characteristic performance curves, like Fig. 13-5, which relate the collector voltage V_{CE} to collector current I_C for various values of input, or base, current I_B. This information frequently represents the extent of the published data.

The limitation in data arises from two considerations. First, devices such as the 2N2193 are rather specialized and have been designed for a relatively limited and difficult range of application. To potential users of such high-speed, high-frequency transistors, the circuit applications may be quite specific, and there is a considerable tendency to design the circuit to fit the device on the theory that this represents the best any device maker can do to meet a set of stringent requirements. Also, the operating range of the device may be limited, owing to its special nature, and it may be relatively costly, so that optimum use must be made of the performance capabilities.

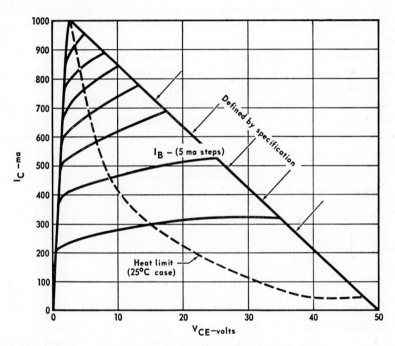

FIG. 13-5. Collector current-voltage curves for 2N2193 transistor. (*After General Electric Company,* "*Transistor Manual,*" *by permission* [4].)

The limited data normally supplied tell the designer enough about the operating conditions for such a device to permit him to incorporate it in his circuit and to obtain a high level of performance. The data might include values of the h parameters for all possible combinations of input, load, frequency, and temperature under which the transistor would give a reasonably good performance. However, such information would be of little interest, since other, less expensive units would serve equally well for a large percentage of the less exacting applications. Therefore, the need for the performance of the specialized device also narrows the range of interest, and the data supplied are usually found to be sufficient.

The above argument applies to a large segment of devices made today. Special high-speed switching transistors for computer applications, for example, are designed with a fixed set of operating parameters in mind so that they will be compatible with the numerous other replicating elements of the circuit. In this and many other areas, the designer needs to know primarily wherein the new device differs from previously available units, and he must usually make only minor changes to exploit its capabilities.

Reasoning of the opposite kind applies to the many-purpose unit, such as our high-temperature, high-gain audio-frequency amplifier, the 2N332. Here there are literally thousands of possibilities because of the economy and wide capability of the transistor. It is, in a sense, a standard component capable of useful application in almost any low-frequency, medium-power amplification job, and the data required to make intelligent use of it in these many ways are usually supplied.

In closing this section, it should be mentioned that the conventions and symbols given herein as examples are not universally followed (except for the basic definitions of the small-signal parameters) and that the amount and manner of presentation of data vary from the specification sheets of one manufacturer to another.

13-5. TRANSISTOR STRUCTURE TERMINOLOGY [5]

The discussion of the preceding chapters has been presented in terms of a nomenclature based on the physical configuration (n-p-n or p-n-p) and the functional purpose of the device (diode, transistor, etc.). It has been mentioned that usage has perpetuated some anomalies and that the same terminology may have different meanings in different contexts. Several examples, such as the p-n-p-n diode, have been encountered. Unfortunately, this is not the only such area in nomenclature of current semiconductor devices. We still face an array of descriptive phrases and abbreviations which refer to manufacturing techniques. In this section we shall define some of this terminology used to describe transistors (and in some cases diodes) available from commercial sources.

These terms refer only to the method of manufacture, and they do not in themselves state the fundamental properties of the transistor. However, each special manufacturing technique represents an answer (from at least one manufacturer) to problems of frequency response, switching speed, junction capacitance, recovery time, charge storage, etc. Therefore, a device which is produced by a process known to give good high-frequency response may be so identified by reference to the manufacturing details. It is still, of course, necessary to obtain more information to know just how good the high-frequency response actually is, but the structural terminology identifies the groups to be investigated.

Basically, all transistors and other junction structures are the result of one or more processing steps chosen from the general group comprised of grown junctions, epitaxial junctions, alloyed junctions, diffused junctions, and electroetched and/or electroplated junctions. All of these have been described earlier, in the discussion of junction formation. Variations of these include rate-grown, melt-back, microalloy (same as surface-barrier, except the forming temperature is high enough to produce alloying at the junctions), and a variety of combinations.

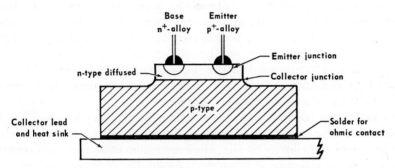

FIG. 13-6. A diffused-base mesa p-n-p transistor.

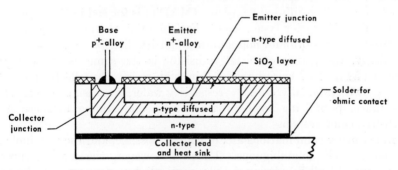

FIG. 13-7. A double-diffused planar n-p-n transistor.

Finally, there are two principal geometric terms: mesa and planar. A mesa transistor is one in which the metal around the emitter and base lead connections has been etched away, leaving the junctions exposed and with very small cross sections. A typical geometry is shown in Fig. 13-6. In some cases, two discrete mesas are formed. This construction provides high-frequency response, low capacitance, and high α.

A planar transistor is exactly what the term implies, as shown in Fig. 13-7. The most important feature is not so much the planar nature as the fact that the diffusion has produced junctions whose edges are protected at the upper surface by the impervious layer of SiO_2. This reduces leakage, improves PIV, and increases current gain at low-signal levels.

198 Semiconductor Junctions and Devices

From the above, it becomes reasonably simple to understand the terms in common use. As examples, we may cite the following. A diffused-base epitaxial mesa transistor is made by a combination of diffusion, alloying, and epitaxy. The epitaxial layer (almost always the high-resistivity collector region) is grown on a low-resistivity wafer of the same conductivity type. The base layer is then diffused to provide a nonuniform base impurity concentration (higher in impurities near the emitter) and a base-collector junction in the epitaxial layer. Finally, an emitter alloy junction is made to the base, and the entire structure is etched to the mesa geometry. Because of a thin base region of graded purity and mesa geometry, plus a very thin collector, these units are excellent for high-frequency, small-power operation.

Second, an alloy-diffused or post-alloy-diffused transistor (PADT) is made by using alloy dots for the emitter and base connections, except that the alloy dots contain both n- and p-type impurities. After forming the emitter and base contacts, the structure is diffused and a base region of opposite type to the emitter and collector is gradually spread around the points of the alloy. Finally, the base region embraces both contact points and the structure is complete. Again, the very small dimensions are useful for high-speed switching.

The microalloy-diffused transistor (MADT) is formed in the same way as the surface-barrier type, except that the original wafer (base) is first diffused to give optimum nonuniform impurity distribution.

Finally, the planar epitaxial transistor is the same as the diffused planar of Fig. 13-7 except that a thin, high-resistivity collector region is first grown epitaxially on a low-resistivity wafer, decreasing the collector resistance, permitting smaller base widths, and adding to speed and high-frequency response.

Many other combinations are, of course, possible, and the terminology is not entirely unambiguous. Usually, however, the manufacturer's literature will fill in missing details, particularly for any new technique aimed at the solution of a specific applications problem.

REFERENCES

1. J. F. Cleary (ed.): "Transistor Manual," 6th ed., pp. 38–43, General Electric Company, Syracuse, N.Y., 1962.
2. n-p-n Grown-junction Silicon Transistor, *Bull. DL-S* 1035, *Type* 2N332, Texas Instruments Incorporated, Dallas, Texas, 1959.
3. Ref. 1, pp. 28, 29.
4. Ref. 1, p. 37.
5. J. A. Walston and J. R. Miller (eds.): "Transistor Circuit Design," chap. 1, McGraw-Hill Book Company, New York, 1963.

14: Switching Applications of Transistors

14-1. INTRODUCTION

In the preceding chapters on junction transistors the discussion has chiefly concerned their function as amplifiers, although mention has occasionally been made of switching capabilities. In the present chapter the switching action of these devices will be explored more fully both because of the widespread use of certain classes of devices as ON-OFF switches and because certain basic physical phenomena, such as collector junction multiplication, contribute significantly to switching, in contrast to their relatively unimportant role in amplification.

Ideally, an electronic ON-OFF switch is a device displaying zero resistance when closed, infinite resistance when open, and having a convenient (nonmechanical) actuating or control circuit. Such an ideal switch has no moving parts, no contact problems, and no arcing, and it can be actuated by any of a variety of impulses from an electric circuit. In practice, the ideal conditions of zero and infinite resistance for closed and open conditions are never realized in electronic devices. Actual resistance ratios of the order of 10^4 to 10^5 are achieved in semiconductor applications; somewhat higher ratios can be achieved under special conditions. Vacuum-tube switches are capable of higher ratios than semiconductors, but at great sacrifice in space and simplicity.

Any triode amplifier vacuum tube is a switch to some extent, depending on the ON-OFF characteristics, and the same thing may be said for the transistor (and a number of other multijunction devices, as will be seen later). We shall therefore concern ourselves with semiconductor switching devices, starting with the standard amplifying transistor circuit as an example. This familiar junction configuration will then be reexamined after some of the simplifying conditions imposed during the initial treatment of the n-p-n transistor have been removed.

14-2. THE TRANSISTOR AMPLIFIER AS A SWITCH

In Fig. 14-1 a typical n-p-n transistor, connected in the common-emitter mode, is shown in two equivalent basic diagrams for switching applications. In this circuit two resistors, R_2 and R_3, are added to the basic circuit. R_3 is, in effect, the R_G of the bias battery in series with the switch in the base input. R_2 is a device to reduce leakage in the reverse direction (collector to emitter) when the base is open-circuited (switch off). R_2 merely shunts the emitter junction with a relatively low resistance, thereby putting most of the open-circuit drop across the reverse-biased collector junction. In these configurations, the current

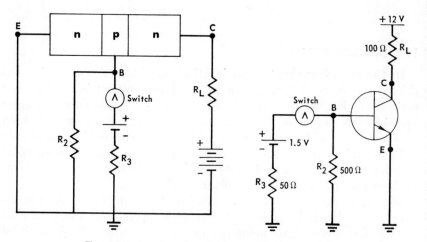

Fig. 14-1. An n-p-n transistor amplifier as a switch.

I_C through R_L is controlled by the voltage across the device V_{CE} and the input base signal I_B.

A typical set of characteristics for a common type of alloy-junction switching transistor is shown in Fig. 14-2. (For reference to an earlier discussion of such an n-p-n transistor when connected as a grounded-emitter amplifier, the corresponding set of complete current-voltage characteristics was presented as Fig. 8-6.) From an examination of this plot we can immediately recognize that for $I_B = 0$ the intersection of the load line with the characteristic, point A, gives the OFF position of the switch. Similarly, at some point B, along the steep part of the slope of the characteristic, there will be a position of high current and low voltage drop, or the ON position. For the unit under consideration, point A might correspond to a voltage of 25 volts and a current of 75 μa, giving a resistance value of about 33,000 ohms. Conversely, point B might correspond to a current of 200 ma and a voltage drop of 0.3 volt, or an effective resistance of about 1.5 ohms. We thus have an OFF to ON resistance ratio of about 2×10^4, a useful value for most switching operations. Note also that this has been achieved well within the allowed small-signal range of operation and at levels where steady current flow remains within the power limit for the device in question. This operation is simple, involving variation of the base current I_B and corresponding changes in the external circuit to adapt to the larger changes in V_{CE} and I_C.

In actual practice, however, a number of additional considerations enter the picture as a result of the variety and importance of semiconductor switch applications. It is not intended to present an exhaustive discussion of the many variables and combinations at the disposal of the

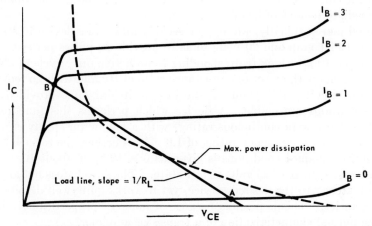

FIG. 14-2. Typical collector characteristic for n-p-n transistor as a common-emitter switch. (*After General Electric Company, "Transistor Manual," by permission* [1].)

designer. However, a brief reexamination of the transistor structure when used as a switch will help in understanding large-signal intermittent operation and the physical meaning of some parameters and terminology frequently used in characterizing such devices.

Among the objectives of the switch designer who is seeking improved switch performance would be the following: (1) to reduce the OFF current to a value as near to zero as possible, with a corresponding increase in voltage drop, (2) to decrease the forward resistance, or voltage drop, for a given current, (3) to increase the current which may be handled by a given device, in terms of size and cost, (4) to achieve the fastest possible turn-off and turn-on times, and (5) to achieve stability of the circuit in the presence of other variables, with the switch in the selected mode. These are, of course, interrelated in most cases and in a rather complex manner.

It will be remembered that we derived the quantitative picture of an amplifying transistor within certain limitations. These included preservation of small-signal characteristics (near-equilibrium, or theoretical, behavior), power levels well within the allowable heat dissipation of the device under continuous operation, and definition of the test conditions by adjusting bias values to provide linear, stable operation. In switching applications, however, we want maximum swings in signal from the highest to lowest possible current values so that the transistor is driven to the limit of its range. We may also define the duty cycle for a switch in terms of limited operation, so that large instantaneous thermal effects may become more important than the continuous thermal dissipation capability. Finally, the conditions of stability may well involve only the static condition of the switch, either on or off, without concern about intermediate regions of instability.

Stated more explicitly, we may be seeking an ON-OFF device to switch a current of 1 amp and 20 volts so as to provide 10 pulses per sec, each pulse to be 10 μsec long (corresponding to an active duty cycle of about 0.01 per cent) with an ON-OFF resistance ratio of 10^5 and with no requirement for circuit stability during the brief switching time provided the final conditions are stable. A device which performs this function is rated at 200 mw (a continuous rating) with a maximum steady current of 0.2 amp and a frequency rating of 1.0 mc. Further, the reverse leakage under standard conditions is given as 6×10^{-6}, or about 10^{-5} amp at a reverse voltage of 25 volts steady state. The switch does not appear to be matched to the desired service, yet it works quite well despite the discrepancies. This is possible because of inherent properties not given by the normal characterization of a transistor amplifier.

The proposed service takes the device out of the small-signal realm, since the current to be switched is five times greater than the maximum allowed value for continuous service, and the short duty cycle makes the steady-state power dissipation limit meaningless. What we really need to know is whether the short, high-current pulses will result in damage and whether the switch is in a stable condition in the ON and OFF positions. Finally, we do not know the ON-OFF resistance ratio, since the forward voltage drop is not given. However, the reverse leakage current corresponds to a resistance of 4×10^6 ohms, so that a forward resistance of as much as 40 ohms would be tolerable. Actual forward resistances rarely are higher than 20 ohms, so there should be no difficulty.

In practice, it is found that some of these questions (such as the instantaneous pulse damage problem) can be settled only by experiment. In other cases, however, information may be obtained from data available from the device maker, provided the large-signal, nonequilibrium mode of operation is reasonably well understood and the proposed application is covered by such information.

14-3. LARGE–SIGNAL d–c RESPONSE

Large-signal behavior does not, in general, lend itself to the simple, rigorous equivalent-circuit treatment used in previous calculations. In most cases, switching conditions are well outside the equilibrium range of linear response and each combination of conditions requires that a model be selected, certain assumptions be made, and a specific numerical solution be found. These solutions cannot be generalized and often

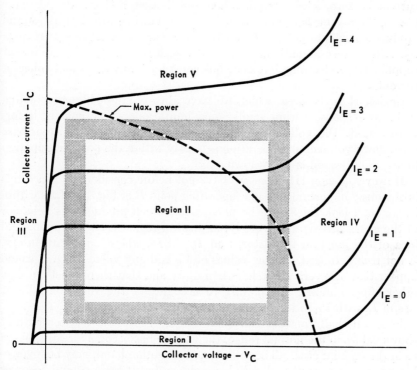

FIG. 14-3. Collector characteristics of grounded-base n-p-n transistor. (*After L. B. Valdes, by permission* [2].)

must be approximated. In many cases, there are discontinuities in the current-voltage curves. Exact calculations of these properties are beyond our scope, although the principles upon which any such solution must be based are no different from those used in the equilibrium cases previously considered.

As a first step in extending our appreciation of the basic transistor structure under this more general set of conditions, we shall go back to the n-p-n grounded-base configuration, which served as our original model. In Fig. 14-3 we see the familiar family of collector characteristics (compare Fig. 8-5) plotted for various arbitrary values of the emitter current I_E. (This is not the same device as that shown in Fig. 8-5, but the similarity is adequate.) We have shown the maximum permissible power dissipation, and we have further divided the area covered by the characteristic plots into five regions, which we may identify and describe as follows [3–6].

In region I we have a very low collector current, regardless of the collector voltage, and I_E is zero or very near to it. This region therefore corresponds to a reverse-biased collector, and, in order that the emitter current be zero, it also requires a reverse-biased emitter. If this were not so, the reverse collector current of holes, flowing into the base, would produce a small but positive forward emitter current (electrons) to remove the holes by recombination. Under reverse emitter bias, the base current supplies these electrons but does not contribute to the normal collector current.

Region II is the area, which we have already studied, corresponding to linear response within small-signal limitations. The emitter is forward-biased; the collector is reverse-biased below the collector-base breakdown voltage; and operation is mostly within the power dissipation limit. This region is well known.

However, region III is one we have not before considered. Since this is obviously important, because operating point B of Fig. 14-2 falls within it, we shall look briefly at the phenomena which produce this behavior of the characteristic. Obviously, in most of region III the fundamental idea of the junction transistor, that $I_C = \alpha I_E$, where α is nearly unity, is not true. Indeed, for low values of V_C and any value of I_E, it would appear that α is very much less than unity and that there must be some other effect entering the picture to account for the very steep linear rise of I_C with V_C in the low range.

First we shall consider the bias values. Since the emitter current shows values in the normal range, the emitter has become forward-biased as in the familiar case. The collector, on the other hand, may be reverse-biased at zero current and very low voltage, but it reverses and becomes forward-biased as soon as an appreciable current flows. The cause of this phenomenon is the load resistor R_L, which, for these very low collector bias values, develops a voltage drop $I_C R_L$ that opposes the bias voltage and results in reversal of the bias. Under these conditions, α is quite low compared to unity, owing to the recombination of emitted carriers, poor collector efficiency, and the tendency for most of the emitted carriers to short-circuit to the base, since the collector is no longer reverse-biased to receive them.

In addition, with the collector now forward-biased, the collector-to-base current reinforces the carrier current from the emitter, causing a steep rise in collector current and almost no rise in collector voltage.

Actually, this is a runaway condition, and it would result in complete instability if it were not for the effect known as *saturation*. Under injection from both junctions, the base region soon reaches a condition where no further increase in carrier current can occur, owing to space charge and ohmic effects. The currents are thus limited. With a further increase in V_C, the collector again becomes reverse-biased, α rises to a normal value near unity, and, without increase in I_E or I_C, the characteristic enters the stable pattern of region II.

The concept of saturation is an important one, since the operating point B lies on the saturation curve as near as possible to the point where it breaks over into region II. At this point, the structure presents the best value for ON conduction as a switch. The slope of the nearly vertical set of superimposed curves gives the value of the saturation resistance $r_{CB(SAT)}$. (More frequently, the grounded-emitter configuration is used, in which case it becomes $r_{CE(SAT)}$. Similarly, a value for $V_{CE(SAT)}$ will often be given for a given value of I_C.) The numerical value of the saturation resistance or voltage is of great importance for switching applications, since it not only describes the minimum ON resistance, as mentioned above, but is also a clue to the amount of charge in the base region which must be removed to desaturate the structure, a factor of importance in switching speed. Experimentally, it is measured with both junctions forward-biased at very low V_C values. It can be reduced by reverse-biasing the emitter (region I) or by open-circuiting the base region.

Finally, we may consider regions IV and V. Both are extensions beyond the normally allowed power dissipation curve. In region IV the collector voltage has been increased to incipient breakdown. A combination of increased multiplication at the junction and avalanche breakdown (indistinguishable under these conditions) causes the beginning of a runaway current characteristic leading to instability. Under some conditions, if the external circuit can limit the current, momentary excursions into region IV as part of a switching operation might be tolerated by the structure. All of the above remarks apply with even greater force to region V, since here there is almost no range of voltage beyond saturation within the normal power range. As in the case of region IV, if the structure would withstand short-term overloads, this region might be used for some switching applications.

In closing this section we turn to the avalanche transistor [7], a device which has much the same electrical properties as those we have discussed but which, by virtue of special physical design, avoids the runaway characteristic of region IV and is as a result a device having a switching characteristic which embodies a new concept. If, in the characteristics of Fig. 14-3, we were able to cause an increase in junction multiplication factor or avalanche effect without runaway breakdown, we would obtain a very different kind of behavior.

First, we recall that the parameter α is made up of the product of the emitter junction efficiency η, the transport efficiency or recombination rate in the base region ζ, and the multiplication factor M. The product of these, for an ordinary device in the small-signal range, was very close to but never quite equal to unity (Sec. 8-4.1). We now construct a device having a collector junction of the proper profile and impurity level so that M can become significantly greater than unity before the onset of runaway breakdown. Such a structure would be an alloyed transistor with step junctions and with the resistivity of the base greater

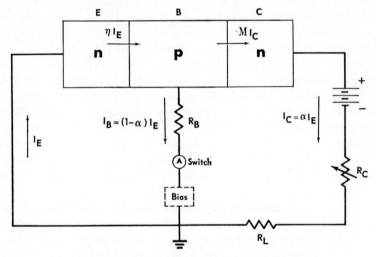

FIG. 14-4. Avalanche transistor switch circuit.

than that of both the collector and emitter (of the order of 0.1 to 5 ohm-cm). For the collector junction in such a configuration, Eq. (8-9) [8,9]

$$M = \frac{1}{1 - (V/BV_{CB})^m}$$

will hold (since the high base region resistivity tends to isolate the junctions) and M can become as high as 2 for a voltage corresponding to $0.7 BV_{CB}$. The region in which avalanche begins corresponds to reverse biases V (or V_{CB}) of 0.1 to 0.4 of BV_{CB}, depending on the resistivity and type of the base region. The increase is exponential, so relatively small increases in reverse bias produce large increases in the avalanche effect [7].

In the above structure, owing to the base resistance, emitter currents are low for low values of V_{CB}, η and α are <1, and most of the emitter current flows out through the base resistor R_B (Fig. 14-4). Therefore, as the collector bias increases, a typical single-junction reverse-breakdown curve results, with no appreciable increase in I_C (part A, Fig. 14-5). This approximates the situation of an open emitter. As the bias increases and the avalanche process begins (M and α are now both >1), the collector current increases (part B, Fig. 14-5) and the base becomes more conducting. The emitter current then increases, but the path direct to the collector is now of lower impedance than that through the base resistance. The emitter-collector path therefore takes over and η approaches unity. As the emitter efficiency increases owing to the increasing current, there is eventual "breakdown" from collector to emitter. This is actually only the effect of lowering the base region

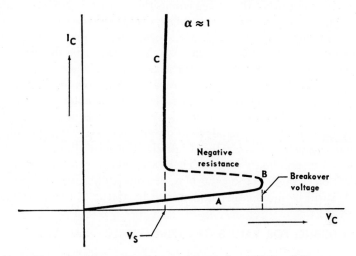

FIG. 14-5. The avalanche transistor and negative resistance.

resistance by increased charge accumulation, so that the structure enters the normal region of transistor behavior. Since the junctions are no longer independent, current continuity requirements limit α to a value of 1, and the product ηM must therefore approach this value. This produces a reduction of M, reflected as a drop in the voltage producing the reverse collector current. V_C therefore decreases until M is able to satisfy the relation $\eta M = \alpha \approx 1$, and the characteristic shifts back to a lower collector voltage for the same current (part C, Fig. 14-5).

This can also be described in terms of external currents. If the three current components I_E, αI_E, and $(1 - \alpha)I_E$ (Fig. 14-4) are considered, it is obvious that the base current $(1 - \alpha)I_E$ will reverse if α becomes greater than 1 because of increasing I_C, and there will be an increase in collector current with a decrease in voltage. This increase in current and decrease in resistance produce a negative slope, or negative-resistance region, as shown in Fig. 14-5. The avalanche transistor therefore passes quickly through an unstable region of negative resistance and stabilizes again at a higher current and lower voltage. Once again, the limiting effect which finally stabilizes the structure is saturation of the base region, causing increased resistance to injection from the emitter region. A minimum resistance for the structure as a whole is thus established. Note that the device is now locked into the high conducting mode unless the voltage drops below the sustaining voltage V_S.

The avalanche transistor may be operated by fixing a collector bias value V_C and varying I_C by means of the input base lead as shown in Fig. 14-4. A separate bias source for this purpose may or may not be desirable in the grounded-emitter connection. In this sense, the device is not unlike the conventional transistor, except for the controlled excursion into regions of $\alpha > 1$ and the negative-resistance characteristic. Another mode of operation is also possible, wherein the device becomes a two-terminal (emitter and collector in this case) element and is caused to switch from on to off through the region of negative resistance simply by varying R_C. In this case the base lead is not externally available, but is shorted to the emitter through the base resistor R_B. It should also be mentioned that these devices are not now commonly used, since the sustaining voltages are rather high and the saturation, or ON, resistances are correspondingly large. There are, however, other devices, also dependent on changes in the value of α, in which this difficulty is overcome and which will be more easily understood by virtue of this introduction.

14-4. SWITCHING FOR VALUES OF ALPHA GREATER THAN UNITY

In the preceding sections we have considered two versions of the triode transistor as a switch, and we have explored the large-signal d-c response of such units under circuit conditions previously excluded from the discussion of such devices in their normal, small-signal equilibrium mode of operation. We have shown how switching action occurs between the saturation region (or ON condition) and the reverse-biased emitter condition (or OFF condition). It was also pointed out that the saturation resistance limited the ON voltage drop to a minimum value below which it could not be reduced.

In this discussion of switching operation we found that the value of α, or current gain ratio, varied over a range from 0.1 (as a function of η) to more than 1. In the case of the avalanche transistor, a higher value of α was achieved by avalanche (or current multiplication) at the collector junction. The resulting switching action, involving a region of negative resistance, showed that a smaller sustaining voltage was sufficient to hold the unit in the high-conductance mode once avalanche had changed the current pattern. However, even this device did not result in a greatly improved ON resistance value. It was, in fact, only slightly more attractive (high current capability, negative-resistance region, and two-terminal operation) than an ordinary transistor within the normal range of the parameter α.

There are other devices, to be described in the next chapters, whose operation is also dependent in one way or another on values of α in excess of unity. In addition, many of these are capable of holding in the ON position with very small sustaining voltages (or currents). In order to make a systematic approach to these devices, we shall review the ways in which the sustaining signal (that is, V_S) may be reduced. These ideas will then be related to the avalanche transistor just discussed and will show the need for new configurations. We begin by considering the subparameters η, ζ, and M, whose product we know as α.

We have already seen that to increase M or to bring about the onset of avalanche multiplication (without runaway breakdown) requires a special structure and high collector reverse bias voltage. Then, if α depended only on M, this would increase α beyond the value of unity. However, if high values of M were required in the ON or holding position, we could not expect to achieve them at low voltages. There is, of course, the possibility of using highly doped material for the junction and thereby reducing the PIV, but this would not be compatible with the required high OFF resistance. It is therefore unlikely that improved switching devices can depend only on the parameter M.

The next variable, transport factor ζ, is primarily a function of base width (allowing for space-charge region variations), the diffusion constant, and lifetime. If we accept the constancy of D, the diffusion constant, which is a valid assumption for almost all circumstances, we find that the controlling variables are base width and lifetime. Both of these are subject to practical limits. A very narrow base region is desired, and it has been constantly improved on as the technology of fabrication has improved. There is, however, a working limit, and great enhancement of this property is not likely with device technology in its present form. A major breakthrough, such as thin-film techniques, might reduce this dimension by an order of magnitude, but other considerations such as doping level control might make it difficult to take full advantage of such an improvement.

Similarly, the lifetime of the base region is a complex compromise starting with the raw material from which the device is made and involving each individual step in an empirical fashion. Every effort to optimize this value is usually made, but the practical limit under current methods is not far away. We therefore find that values of ζ are now very close to unity, as previously shown, and that only a few per cent improvement may be realistically expected. For present purposes, we shall assume a value approaching unity, but no higher.

Finally, we have the emitter efficiency η, which we have already considered in the very low collector voltage range during our discussion of the saturation portion of the characteristic. In this range, it will be recalled, the emitter efficiency was low, owing to recombination at the emitter-base junction and diversion of a part of the emitter current through the base lead. We may eliminate this second effect by open-circuiting the base, but the improvement is not startling.

If we now define α' to be a function only of η (and therefore equal to η if both ζ and M are unity), with rising emitter current and low collector voltages, α' approaches unity, remains there for values corresponding to small-signal behavior, and then falls slowly. In this case, also, if we open-circuit the base, we again improve α'; but there is still, in general, a departure from the optimum value. (The region of normal α is that chosen as region II in Fig. 14-3. The higher current and voltage effects now under discussion are large-signal behavior.)

The mechanism for the decrease in α' with high current lies in the conductivity modulation of the base region. When injection levels are high, a space-charge region forms just beyond the junction and attracts majority carriers from the base contact (if not open-circuited) and from the collector junction to neutralize the nonequilibrium situation. This space charge becomes stationary in a state of dynamic equilibrium and opposes the emitter injection bias. If the base is also connected, a current flows through its resistance, further reducing the forward bias on the emitter and aggravating the problem. These situations are shown in Fig. 14-6. Here we have plotted the value of α' as a function of emitter current for a fixed collector bias. The collector voltage has been chosen in the low range, meaning that range from the saturation region out to the beginning of avalanche (or 0.1 to 0.4 BV_{CB} for the avalanche device). Here we see the result of the space-charge effect just described: a decrease in α' with increasing I_E. In Fig. 14-6 the low-current behavior is represented as a single curve, since the base connection is not noticeably effective, while the higher-current behavior is shown for the base lead open and connected.

We may now combine the effects due to η and M to show in a general way how the negative-resistance switching characteristic is generated. If we plot collector-emitter voltage versus collector current, we obtain the curve shown in Fig. 14-7 for a two-junction device (with high-resistivity base region) and a dependence of α' on emitter current as shown in Fig. 14-6. As the voltage V_{CE} increases (part A), almost all of it appears as reverse bias across the collector junction. The forward bias on the emitter is so low that there is virtually no emitter current. With a

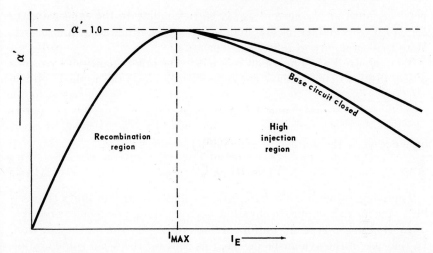

FIG. 14-6. Current gain α' plotted as a function of emitter current for low collector voltages. (*After A. K. Jonscher, by permission* [10].)

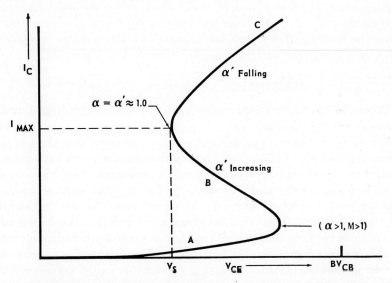

FIG. 14-7. Collector-emitter characteristic showing large-signal behavior as a function of α. (*After A. K. Jonscher, by permission* [10].)

further increase, avalanche multiplication occurs and produces an overall α greater than unity, owing to an increase in the value of M. At this point, α' is still quite small. As the collector current I_C increases, α' also increases (part B), owing to the current continuity requirement. The multiplication factor M simultaneously decreases, as a result of decreasing V_{CE}, until at the current I_{MAX}, corresponding to the same value in Fig. 14-6, the voltage is low and equal to V_S, while $\alpha' = \eta = \alpha \approx 1$. With increased current, α' falls off (part C).

To minimize V_S, we may now relate it to the other quantities by means of Eq. (8-9), written to show the dependence of α on the applied voltage,

$$\alpha(V) = \alpha' M(V) = \frac{\alpha'}{1 - (V/BV_{CB})^m} \tag{14-1}$$

or, since V_S is the voltage at which $\alpha(V)$ becomes unity,

$$V_S = BV_{CB} \sqrt[m]{1 - \alpha'} \tag{14-2}$$

Therefore, we need a low BV_{CB} and a value of α' as near unity as possible. We may also improve our situation by making m as small as possible or by choosing p-type silicon for the high-resistivity side of the junction. In the case of the avalanche device, this calls for the n-p-n configuration (Sec. 8-4.1). In the next chapter we shall consider other configurations in which α achieves values as high as 100, α' is held near unity, and low-voltage avalanche is provided, all under conditions which result in a small voltage drop in the ON position and varying degrees of power capability. Thus the most significant drawbacks of the two-junction device are overcome and many new switching applications are made possible.

14-5. SWITCH CHARACTERISTICS, LOAD LINES, AND STABILITY

Before going further into the consideration of other multijunction devices, and particularly switching configurations, we shall conclude this general discussion of switching properties by examining three typical switching modes. To illustrate these performance mechanisms, we shall look at yet another example of the negative-resistance switching behavior of a junction triode, namely, the large-signal input characteristic, or emitter current-voltage curve. For this purpose we assume that suitable low-bias conditions on the collector are fixed and that the base lead is common to both junction circuits. (Any of several other characteristics might have been chosen with the right combination of bias voltages and device physical properties, including the open-circuited base configuration. These wide choices are part of the reason for the seemingly endless circuits which employ semiconductor multijunction elements as electronic switches.)

Having chosen a characteristic, we shall show the various load line interactions and identify the monostable switch, the bistable switch, and the astable switch, also known as the multivibrator or oscillator. As a final example we shall apply the same criteria to the avalanche transistor previously discussed and discover it is a bistable switch.

The emitter characteristic for the device under consideration is shown in Fig. 14-8. In part I of the curve we have the behavior characteristic of region I, Fig. 14-3, with the emitter reverse-biased. The corresponding curve is shown as typical for a reverse-biased diode. Then, in part II, corresponding to region III, we have the normal saturation procedure, starting with the collector bias turned around so that it is forward-biased, and a runaway situation occurs as described earlier. Finally, with saturation, the collector again becomes reverse-biased and the unstable negative-resistance range ends, turning to the positive resistance value of part III, which corresponds to the break in the curve between regions

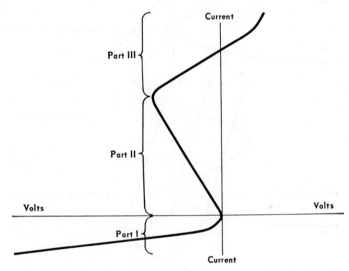

Fig. 14-8. Emitter characteristic for large-signal switch response.

II and III of Fig. 14-3. Note that α has just reached the value of unity in part III of Fig. 14-8.

We have now defined three sections: the cutoff zone (part I) wherein the emitter acts like a reverse-biased diode, the collector has a very high impedance, and the switch is therefore off; the negative-resistance zone (part II) which is unstable for both emitter and collector; and the saturation zone (part III) with a small positive emitter resistance, very low collector impedance, and the switch in the ON position.

We may now examine this emitter characteristic in terms of load lines and from this determine the switching mode to be expected. In Fig. 14-9 the first example shows a load line of slope and location such that it intersects the emitter characteristic at point A, with no disturbing input signal. In other words, the circuit is quietly operating under the conditions defined at point A. If we now apply a positive input signal, the load line will shift toward the right relative to the characteristic but retain its slope (if the other circuit elements are unchanged). The operating point goes then to point B, encounters instability, and slips instantaneously to point C or D, where the situation is again stable. If

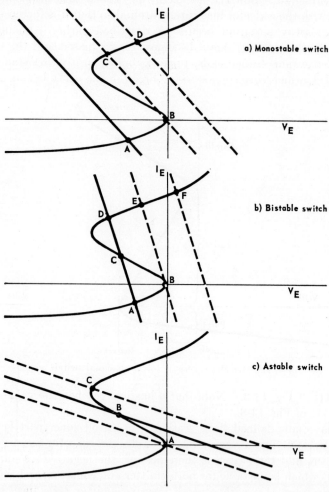

Fig. 14-9. Load lines and emitter characteristic versus switch stability. (*After D. M. Warschauer, by permission* [11].)

the signal is removed, the circuit will revert to point A. Thus the switch has been turned on by a signal, held there, and then reverted to off when the signal was removed. This is, by definition, a monostable switch.

In the middle figure, we proceed along the same lines, except that now the load line intersects the characteristic at points A and D. If we now drive to point B, the circuit will flip to point D and may be driven on to points E and F. Removal of the signal, however, merely returns the operating point to D, where it finds a stable, signal-free condition. If, while the circuit is at point D, a reverse polarity signal of sufficient size to overcome the slight hump to the left and below point D is applied, the circuit will return to point A. Note that it can never be stable at point C. This is known as a bistable circuit, since one pulse turned the switch on and a reverse pulse was required to turn it off again.

In the lower drawing we see that the quiescent intersection point B is in the unstable region and that any random change can swing the circuit about between points A and C. Although such a circuit could be

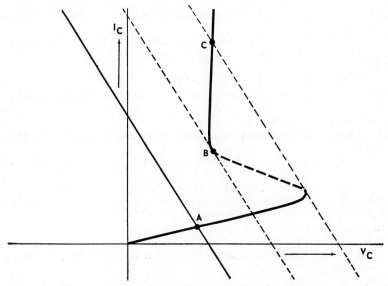

Fig. 14-10. Avalanche transistor switch and load lines.

stabilized by a fixed bias, its real value lies in the ability to follow oscillations of point B and to amplify them. This is known as an astable switch or multivibrator. Its chief use is in oscillators.

We now look once more at our avalanche transistor and its load line behavior. In Fig. 14-10 we show such a configuration. If we merely increase the collector voltage, the switch, stabilized at point A, will flip over, past point B to point C. It will remain there unless the voltage is reduced enough to bring it back to point B, whence it will flip down to a new value near point A, a typical bistable behavior. Such "self-contained" locking-type switches are similar to gas thyratron vacuum tubes. They find wide application as such, and they are not used as pulse-controlled devices except in a few special cases.

216   Semiconductor Junctions and Devices

As a final note, it is of interest to see the close analogy between the switching behavior of the emitter characteristic of Fig. 14-9, where the operation was entirely below or in the saturation region with α values ranging from much less than unity to a final value approaching this limit, and the similar action of Fig. 14-10, where the switch has been driven past the stable, small-signal region into the overload region, with α values in excess of 1, and then back to a stable state at an α of unity. The wide utilization and importance of this common, negative-resistance switching mode for semiconductor devices is one of their most distinctive attributes.

REFERENCES

1. J. F. Cleary (ed.): "Transistor Manual," 6th ed., p. 73, General Electric Company, Syracuse, N.Y., 1962.
2. L. B. Valdes: "The Physical Theory of Transistors," p. 297, McGraw-Hill Book Company, New York, 1961.
3. Ref. 2, pp. 334–340.
4. C. le Can, K. Hart, and C. de Ruyter: "The Junction Transistor as a Switching Device," pp. 115–139, Reinhold Publishing Corporation, New York, 1962.
5. D. M. Warschauer: "Semiconductors and Transistors," pp. 217–224, McGraw-Hill Book Company, New York, 1959.
6. A. E. Anderson: Transistors in Switching Circuits, *Bell System Tech. J.*, **31**: 1207 (1952).
7. S. L. Miller and J. J. Ebers: Alloyed Junction Avalanche Transistor, *Bell System Tech. J.*, **34**: 883 (1955).
8. S. L. Miller: Avalanche Breakdown in Germanium, *Phys. Rev.*, **99**: 1234 (1955); Ionization Rates for Holes and Electrons in Silicon, *Phys. Rev.*, **105**: 1246 (1957).
9. A. B. Phillips: "Transistor Engineering," pp. 133ff., McGraw-Hill Book Company, New York, 1962.
10. A. K. Jonscher: Physics of Semiconductor Switching Devices, in A. F. Gibson et al. (eds.), "Progress in Semiconductors," vol. 6, p. 149, John Wiley & Sons, Inc., New York, 1962.
11. Ref. 5, pp. 223, 224.

15: Multijunction Four-layer Devices

15-1. INTRODUCTION

Having considered the two-junction transistor in some detail, both as a power amplifier and switching triode, we may now make a brief survey of the most important commercial examples of other multijunction devices and their performance characteristics. Their operation will be qualitatively related to the junction phenomena with which we are familiar and will demonstrate a few of the remarkable variations of electrical performance which are possible within the framework of the basic transistor operation patterns. No such discussion could hope to be all-inclusive, since the variations and applications are almost unlimited. We shall therefore concern ourselves only with the detailed operation of active devices, most of which have three junctions and two or more leads.

These high-efficiency, high-speed switches, high-gain transistors, and versatile high-power rectifiers are among the most widely used devices now commercially available. They have in common the ability to achieve α values far in excess of unity by mechanisms which do not depend entirely on high reverse-bias voltages to produce avalanche breakdown, but which instead produce high emitter efficiencies, low-voltage avalanche, and a simultaneous reduction in "base" resistances. Also, as will be seen, these devices may be regarded as two transistors sharing certain common regions and leads, so that the overall α for the structure is effectively the product of two individual α's.

Qualitatively, the behavior of these structures is characterized by the presence of a third junction in place of the collector contact. Usually, the fourth region is of low purity, so that the new junction will become an emitter under conditions of forward bias and relatively high current. In the forward direction, as will be seen, the original emitter junction is forward-biased, as is the new or third junction. The normal collector junction is thus reverse-biased between two forward-biased junctions. The impurity levels are so chosen that this junction does not require high collector-emitter voltages for avalanche breakdown. After breakdown, both outside junctions become emitters, saturating the two base regions and making the middle junction also appear forward-biased. The corresponding voltage drop is low; the overall α value is greater than 1; and a very desirable forward characteristic is obtained. (In the reverse, two junctions are reverse-biased, and, since these are of normal configuration, the reverse characteristic is typical of a reverse-biased diode. This mode is not much used in actual switching applications, however.) It is also possible to alter the bias voltages on the base regions and thereby achieve the forward, or ON, condition at lower voltages and currents than if the bases are floating.

Very simply, then, these structures overcome the forward characteristic problems we have observed in the case of the avalanche transistor switch and provide for $\alpha > 1$ and high emitter efficiency by effectively wiping out collector junction behavior. Although rather costly compared to ordinary single-purpose devices of limited range, they have very broad operating characteristics, and they can perform many specialized functions for which no simple device is made. For exacting applications in switching circuits they have virtually no limitations except extreme speed. Some idea of the versatility of this basic active structure can be obtained from the brief descriptions of typical performances given below.

15-2. THE p-n HOOK TRANSISTOR [1,2]

This was the earliest four-region, three-junction device produced as a p-n-p-n structure. Since it was used as an amplifier, it was provided with three leads, leaving one of the four regions inaccessible. In its classic form, the p-n hook transistor was an effort to overcome the limited current gain characteristic of the common-base configuration of the junction transistor. (We have already noted that the common-emitter configuration produces very good current gains, and we shall be reminded of the fact by analogy to other p-n-p-n structures to be discussed later. For power gain, however, the grounded-base with $\alpha > 1$ would be superior.)

As originally produced, fairly early in semiconductor technology, the structure is shown by the diagrams of Fig. 15-1. The device could be made by conventional techniques, with the main requirement being that region p_3 be dimensionally thin. Also, in many cases, the pseudocollector n_4 was of lower resistivity than p_3, so that injection might occur across the p_3-n_4 junction.

Note particularly the bias polarities. If we consider only regions

Multijunction Four-layer Devices 219

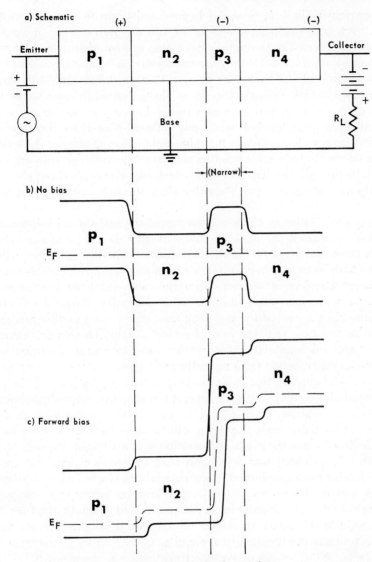

Fig. 15-1. The *p-n* hook transistor.

p_1-n_2-p_3, we have an ordinary transistor in the grounded-base configuration. We may then identify the emitter, base, and pseudocollector regions as p_1, n_2, and p_3, respectively. With the addition of the fourth region (n_4) we find references to the hook, which frequently encompasses part of region n_2 and all of p_3 and n_4. The junction at p_3-n_4 is forward-biased in order to provide the correct reverse bias at junction n_2-p_3.

In operation, the additional region acts to increase the output- to input-current ratio i_o/i_i, or α, far beyond unity, up to values of 100 or more. When the emitter injects holes into the normal base region n_2, they are transferred to the collector junction at n_2-p_3 and there collected. Once in region p_3, however, they are unable to reach the external collector circuit because of the barrier at p_3-n_4. This results in a positive space charge in p_3 which opposes the space charge normally present as the result of junction formation and thereby lowers the barrier at p_3-n_4. Note also that this junction is already forward-biased, so that electron injection from n_4 is possible. By this combination of effects, the space-charge barrier is reduced markedly and large electron currents are permitted to flow in the collector-base, or hook, circuit. These electron currents are, of course, much greater than the hole current producing them.

The analogy between the p-n hook transistor and the injection diode previously discussed (Sec. 6-7) is very striking. It will be recalled that, in that mechanism, the forward-biased diode resulted in carrier injection into the high-resistivity region, setting up a nonequilibrium space-charge condition. Carriers of opposite sign then entered from the external circuit to neutralize this situation. The similarity should be readily understood in the case of the p-n hook transistor. As a further analogy, we find in this case that the critical region p_3 must be thin so that the carriers are not annihilated before the nonequilibrium condition can cause a suitable reaction from the adjacent region.

From the above description, it is obvious that the function of the p-n hook could also be represented in general terms as two transistors sharing two common regions, as discussed in the opening paragraphs of this section. This more basic approach will be used to discuss other four-layer devices, where the mode of operation is less straightforward.

Finally, the p-n hook was never more than a research or special-purpose device, having been supplanted in commercial practice by other four-layer devices and/or common-emitter connections of ordinary transistors. Although somewhat limited in performance and hard to produce by early methods, the hook transistor represented a unique solution to a difficult problem in grounded-base circuits.

15-3. THE p-n-p-n TRANSISTOR SWITCH [2–4]

In this and the next two sections we shall look at three transistors having the same configuration as the hook transistor but different properties. These units are in part interchangeable, since their ranges of operation may overlap; but each has a unique design purpose, and typical units will be examined to emphasize the differences.

The p-n-p-n transistor switch (or the n-p-n-p complement) is a four-region, three-junction device with all regions accessible by means of exterior leads. This definition is somewhat more restrictive than that in common use some years ago, but it has been adopted by the General Electric Company and others to distinguish this type from the p-n-p-n controlled rectifier (with only three regions accessible). Originally, the controlled rectifier and p-n-p-n switch or transistor switch were not consistently differentiated by name, despite their applications and power ratings and the number of leads. We shall follow the current industrial example and identify the device on the basis of the number of leads, calling the four-lead (or "double-gate") configuration, referring to the

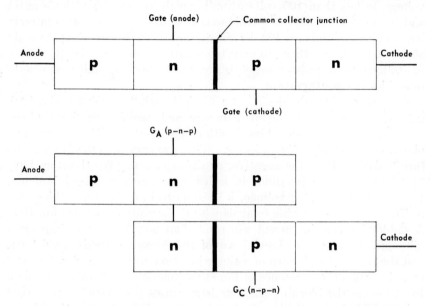

FIG. 15-2. The p-n-p-n transistor switch and two-transistor equivalent [5].

connections from the control regions) a p-n-p-n transistor or transistor switch and the three-lead device a controlled rectifier.

If we now consider the p-n-p-n transistor switch, we find that it may be represented as two ordinary transistor configurations connected together and sharing a common collector junction, as shown in Fig. 15-2. Note that the two equivalent transistors are unsymmetrical, that is, one is an n-p-n and the other a p-n-p. Note also that we are forced to use the terminology *anode*, *cathode*, and *gate* for regions which cannot be uniquely termed collector or emitter, since that depends on the direction of bias. We have, in addition, a wide latitude in the choice of the resistivity ranges for the several regions. In fact, theoretically, we should be able, by leaving one or more gate connections open, to match the performance of any four-region device starting from this configuration. To a degree, these things are possible, but the purpose of each type is different. Thus, the *p-n* hook transistor is designed for large current and power gain at relatively low absolute levels, the controlled rectifier for moderate to heavy power applications with high PIV ratings, and the *p-n-p-n* switching diode for fast, relatively high power switching. To perform some of these functions economically and to achieve still further flexibility for the needs of computers and control applications are the objectives of the *p-n-p-n* transistor switch.

Transistor switches are low-level devices that are limited in inverse voltage to less than 100 volts overall (much less on the cathode gate) and to currents of 100 ma (continuously) and that operate at relatively low frequencies (about 100 kc maximum). They can dissipate only about 300 mw, and they are very temperature sensitive. Despite these limitations, the transistor switch is a versatile, low-cost unit that closely resembles the gas thyratron.

We first consider the device as a whole (anode to cathode) with both gates disconnected. If the anode is now made positive, we find that the *p-n-p* transistor is forward-biased with respect to the emitter and reverse-biased with respect to the collector. This also applies to the *n-p-n* structure looked at from the negative, or cathode, end. We therefore have two forward-biased junctions in series with a reverse-biased junction. If we now increase the voltage, a breakdown in the forward (blocking) resistance will occur owing to avalanche at the mutual collector junction. A sudden increase in current will result, but now there is an important difference between the breakdown of the forward blocking resistance and the reverse breakdown of a single junction, as in a rectifier. In this case, because of the presence of the other junctions, the voltage required to overcome the blocking characteristic causes the overall α to exceed unity. This is due to the effect of increased emitter currents and efficiencies on the α's of both individual transistors. This results in a further decrease in resistance, reverting to a typical forward-biased characteristic for all current values above the critical value, known as the holding current. If the current is made smaller than the holding current, the device reverts to the forward blocking portion of the characteristic. This behavior is represented in Fig. 15-3.

This blocking behavior is typical of switching operations. Note that this is an unbiased performance curve, starting at the origin and proceeding through the nonlinear portion of the characteristic. In such regions, the value of α is known to be a function of the current. Obviously, for switching purposes, the device is on when the current exceeds I_H and effectively off if the current falls below I_H. All of this is typical of the four-region structure.

We have not, however, considered the role of the gates. These effects may be additive, although one is usually of low current capability for very small signals and the other of higher power for larger signals. Under suitable conditions of bias on the anode-cathode leads, these gates can be made to turn the switch on or off, either singly, differentially, or acting

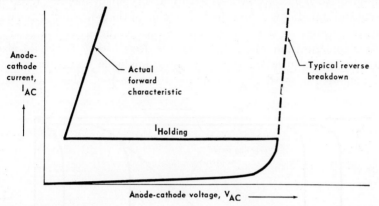

FIG. 15-3. Forward-biased p-n-p-n transistor switch showing blocking characteristic (both gates open-circuited).

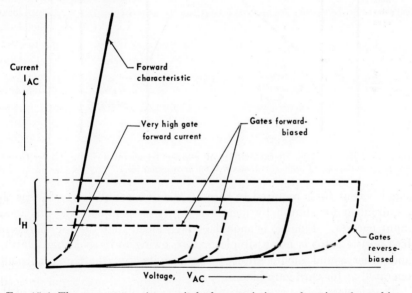

FIG. 15-4. The p-n-p-n transistor switch characteristic as a function of gate bias.

together. Some idea of the effects producing turn-on and turn-off conditions can be seen from the typical curves in Fig. 15-4. Note that for large enough forward gate currents the blocking action disappears and the unit resembles a normal diode in forward characteristics.

Considering the wide range of values possible for I_{AC}, I_H, and V_{AC}, it is obvious that positive switching with almost any degree of blocking can be achieved. For example, a value of V_{AC} may be selected and the unit so biased. Then, by change of bias on one or both gates, either turn-off or turn-on may be effected by a very small signal. (Note that the control signal now enters the base region of the p-n-p and n-p-n equivalent structures, so that the parameters are expressed for a grounded-emitter configuration when this device is described.) It should also be noted in this connection that the p-n-p analogue transistor has been designed with a small gain compared to the n-p-n. Therefore, only a

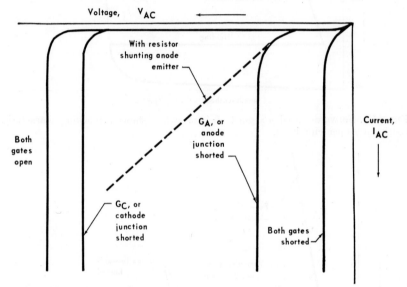

FIG. 15-5. Reverse-bias characteristics of p-n-p-n transistor switch. (*After General Electric Company, "Transistor Manual," by permission* [6].)

small bias at G_C is required to turn off the unit completely, whereas the requirement for shutoff by means of G_A involves almost the full anode input current. Finally, it should be mentioned that the use of reverse bias on the gates is not a simple matter, owing to internal regeneration and other signal modifications. These units are, therefore, normally operated at forward gate biases.

To complete this survey, a word may be said about the behavior of the unit under overall reverse bias, that is, with the anode negative, and typical values of the most important parameters under four conditions of limiting gate biasing. Figure 15-5 shows typical reverse-bias responses for several conditions, including the cases where G_A and G_C are shorted to the anode and cathode, respectively. This is equivalent to shorting out the emitter junctions in the corresponding analogue transistors. Also shown is the effect of shunting the anode gate with a resistor from the anode. Note that, as might be expected, most of the reverse voltage is sustained by the reverse-biased emitter junction at the anode, or low-gain transistor element.

Finally, listed in Table 15-1 are some typical values of selected parameters for a commercially available silicon transistor switch. (It should be mentioned that these parameters, unlike those tabulated in Chap. 13 for a two-junction transistor, are not standardized, and the nomenclature will vary from one manufacturer to another. Therefore, we may not regard them as fixed representations of the measured quantities.)

TABLE 15-1. GATE CHARACTERISTICS FOR A p-n-p-n TRANSISTOR SWITCH*

Parameter	Gates open	G_C shorted	G_A shorted	Both gates shorted
Forward blocking, volts...............	≈20	70	50	80
Reverse breakdown, volts.............	80	80	4	0.5
Holding current, ma..................	Small	0.5–1	2–3	2–3
Forward breakover current, ma........	Small	0.5	2	4
Current to turn on G_C, μa..........	≈1		10–60	
Volts to turn on G_C................	0.4–0.6		0.5–0.8	
Current to turn on G_A, ma..........	0–0.008	0.2–1.0		
Volts to turn on G_A................	0.4–0.6	0.6–0.9		

* After General Electric Company, "Transistor Manual," by permission [7].

15-4. THE p-n-p-n CONTROLLED RECTIFIER [3,8,9]

From the preceding discussion, the four-region, three-junction controlled rectifier (having only three leads, that is, with only one control gate, and therefore one region inaccessible from outside the device) may be readily understood. It is necessary to recall that the controlled rectifier is a power device, and it will therefore usually have large-area junctions, good thermal control, injection for large forward currents, and high-voltage capabilities. In most cases the frequencies will be low, owing to large capacitance at the collector junctions and thick, high resistance base regions.

As before, we may approximate the behavior of the p-n-p-n controlled rectifier by means of a two-transistor analogue, again with two regions shared. However, since we are now dealing with a rectifier, the important characteristics are those concerned with large current flow at low forward bias and small current flow at high reverse bias. In Figs. 15-6 and 15-7 we show the resulting breakdown for a p-n-p-n structure and the current multiplication factors α for each analogue transistor. (The control lead to one of the base regions is omitted for simplicity.) Note that here again there is a common collector junction, at the circuit point J, and that this junction is normally reverse-biased when the analogue transistors are forward-biased. There will therefore be a reverse current

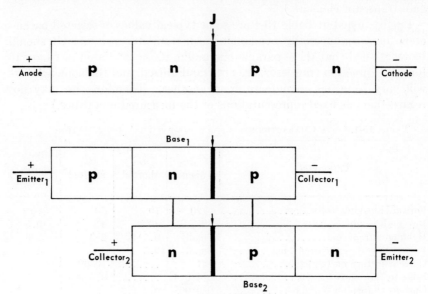

FIG. 15-6. Two-transistor analogue of p-n-p-n controlled rectifier.

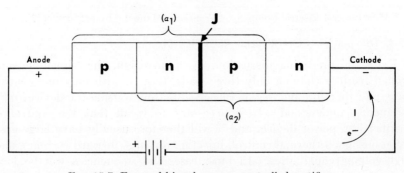

FIG. 15-7. Forward-biased p-n-p-n controlled rectifier.

component, resulting from the inequality between the reverse currents for the two individual transistors, at the collector junction.

From Fig. 15-7 we may write, by analogy with a normal transistor,

$$\alpha_1 = \frac{i_{c_1}}{i_{e_1}} \quad \text{holes} \tag{15-1}$$

and

$$\alpha_2 = \frac{i_{c_2}}{i_{e_2}} \quad \text{electrons} \tag{15-2}$$

Then, if we consider the current vectors at the point J and represent the total or net reverse current by a single value I_{SAT}, which is the algebraic sum of the two currents due to the emitters plus the reverse leakage at the reverse-biased junction, we may write

$$I_J = I\alpha_1 + I\alpha_2 + I_{SAT} \tag{15-3}$$

and since

$$I_J = I \tag{15-4}$$

$$I = \frac{I_{SAT}}{1 - \alpha_1 - \alpha_2} \tag{15-5}$$

Since I_{SAT} for a p-n junction is very small, the above expression for I can become large only if $\alpha_1 + \alpha_2$ is nearly unity. As seen before, the

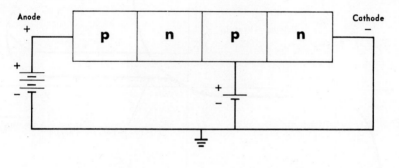

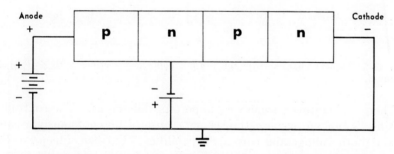

FIG. 15-8. Gate bias arrangements for p-n-p-n controlled rectifier.

values of α for this situation (unbiased junction) are current-dependent, increasing with i_e. Also, α is a function of V_{CE}, owing to eventual avalanche breakdown. Therefore, the p-n-p-n structure can become conducting if the V_{CE} values are increased or the overall V_{AC} values are increased. In the latter case, we have an analogy to the p-n-p-n transistor switch with both gates open. To increase the emitter current, we may forward-bias one base-collector junction as shown in Fig. 15-8, in which case the remaining structure resembles a forward-biased p-n rectifier of the injecting variety. (Note that for the complementary n-p-n-p structure, the bias polarities in Fig. 15-8 are simply reversed.)

In the absence of bias on the control base (or gate) we therefore obtain a characteristic similar to Fig. 15-9. This is very similar to the transistor switch in the preceding section, but it would show much larger power and voltage capabilities if the scales were indicated. As before, we may employ the gate current to modify the forward characteristic, as in Fig. 15-10. The qualitative similarity between this characteristic and that for the p-n-p-n transistor switch is obvious.

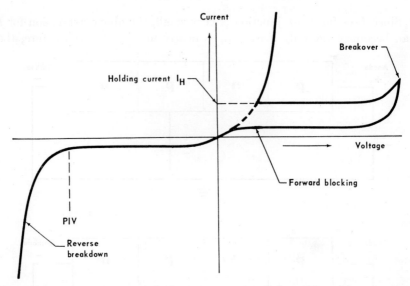

FIG. 15-9. Typical characteristic for p-n-p-n controlled rectifier with gate open-circuited.

There are, however, some very important differences. The controlled rectifier is not an oversize transistor because of its enormously greater power (both voltage and current) capabilities. Further, dimensionally, the base regions are heavy to achieve lower α's and high-voltage capability. As a consequence, however, the frequency response is poor, being limited in some cases to about 400 cps at rated load. If derated to compensate for heating, units can be made to respond to frequencies of about 20 kc. As rectifiers and control units, they are versatile and require very small signals to trigger them into forward conduction. This trigger signal may be of the order of 10^{-5} of the forward current compared with 10^{-2} for a power transistor.

Also, as a switch, the controlled rectifier exhibits the same thyratron-like characteristic as the transistor switch, except that here the only way to turn off forward conduction is to reduce I_{AC} below I_H. (Once conducting, the gate has no effect.) Finally, the use of the device as an amplifier over a limited range of applications is obvious.

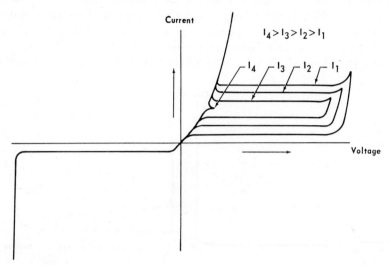

FIG. 15-10. Increasing gate bias versus forward current (p-n-p-n controlled rectifier). (*After General Electric Company, "Silicon Controlled Rectifier Manual," by permission* [10].)

15-5. THE p-n-p-n SWITCHING DIODE [11]

This is the last of the four-region, three-junction devices to be described. Here again we shall observe behavior very much like that of an ordinary thyratron. In this case there are only two leads and the inner regions are inaccessible. It is different from the preceding examples in geometry and doping level in order to produce a unit of high ON-OFF resistance ratio, a high reverse voltage capability, and fast switching times. The usual geometry calls for one very thin region (to achieve speed, or frequency response) and a fairly thick one adjacent thereto to support a high PIV. Finally, the unit is voltage- rather than current-controlled, resulting in a characteristic as shown in Fig. 15-11. The breakover voltage V_{BO} is carefully controlled, and it may reach values of 30 to 60 volts. The sustaining voltage V_S required to maintain the switch in the ON position can be as low as 0.7 volt, while the PIV has values of 400 volts or more. The ratio of ON to OFF resistance is of the order of 10^6, and switching times are less than 1 μsec, corresponding to a frequency in the megacycle range.

15-6. THYRISTORS [12]

These diffused-base transistors (usually of germanium, so far) derive their name from the thyratron-like behavior contributed by a special collector contact. In many respects they are similar to the p-n-p-n structures previously described, except that by use of diffusion and mesa geometry the frequency response is greatly enhanced and the signals required to turn on or off are made very small. As might be expected, they are strictly small-signal devices having a very limited power-handling

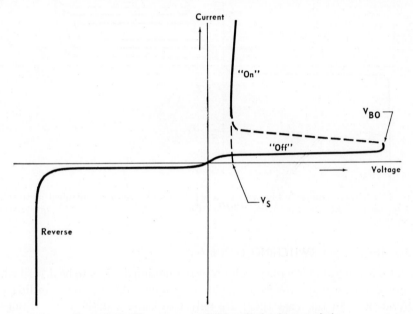

FIG. 15-11. The p-n-p-n switching diode characteristic.

ability (of the order of 100 mw). When used without the base lead for control, their behavior is very similar to that of the p-n-p-n switching diode described above, and it is voltage-controlled. Under operation as transistor switches, they resemble the four-region device, but they are not current-controlled or capable of significant power control. In addition, only one gate is provided.

The unique constructional feature of the thyristor is illustrated in Fig. 15-12. The upper diagram shows the grounded-emitter connection, most commonly used for this device, while the lower cross section emphasizes the similarity between this device and the ordinary diffused mesa transistor.

The unusual feature is the p-p^+ contact region at the collector lead.

Under small-signal conditions, this behaves like any other collector and simply gathers in emitted holes from the base region. However, since the doping level of the entire structure is rather high (the p-type collector is only of the order of 1 ohm-cm), an increase in current in the forward direction results in a space-charge buildup in the p-type collector region due to large hole emission currents across the emitter junction into the base. This charge buildup creates a nonequilibrium asymmetry, and electrons then flow from the collector contact through the p^+ region and

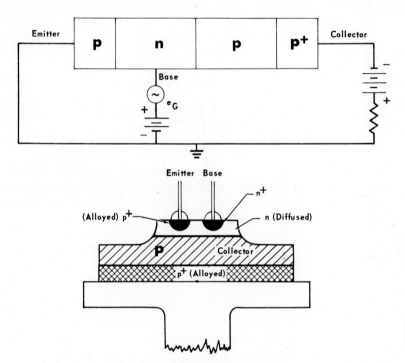

Fig. 15-12. The thyristor structure.

collector into the base, where they are destroyed by recombination. This large increase in current results in a negative-resistance characteristic of the type shown in Fig. 15-13 (compare with Fig. 15-11) for the case of open-circuited base lead and for two values of bias current on the base lead.

We thus have a very familiar characteristic, with OFF and ON zones separated by a negative-resistance region. We also have the unfamiliar effect of a p-p^+ region acting as a hole collector at ordinary currents and as an electron emitter at high currents owing to the space-charge effect. This duality of function is another example of the astonishing variety of behavior patterns possible for semiconductor devices with two or more junctions, depending on doping levels, geometry, biases, etc.

No simple, quantitative treatment of the electron-injecting p^+-p collector region (including the metal contact) is available. The difficulty involves the behavior of the metal-p^+ alloy region. The tunneling concept proposed by the originators of the device gives a fairly good correspondence with observed electron injection efficiency as a function of current. Another possible mechanism involves injection by the metal

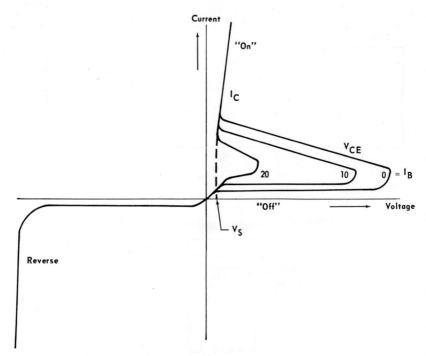

FIG. 15-13. Characteristic current-voltage diagram for a thyristor.

side of the metal-semiconductor contact. This will be further outlined in a later chapter. Qualitatively, the metal-p^+ interface is forward-biased (for electron flow) as a result of normal collector junction reverse bias. Therefore, if the reverse-bias barrier at the junction is modified by space charge due to emitter injection, the contact bias could become the controlling factor, and electron injection would take place. Unfortunately, the basic contact parameters are not too well understood, and there is therefore no simple means of determining the actual mechanism involved.

Typical performance data show frequency response in excess of 100 mc, turn-on and turn-off times of less than 0.1 μsec, a sustaining voltage V_S

of about 0.5 volt, and a reverse breakdown of the order of 65 volts. Even faster units are now in experimental production.

REFERENCES

1. J. J. Ebers: Four Terminal p-n-p-n Transistors, *Proc. IRE*, **40**: 1361 (1952).
2. J. L. Moll, M. Tanenbaum, J. M. Goldey, and N. Holonyak: p-n-p-n Transistor Switches, *Proc. IRE*, **44**: 1174 (1956).
3. J. F. Cleary (ed.): "Transistor Manual," 6th ed., chap. 19, General Electric Company, Syracuse, N.Y., 1962.
4. A. K. Jonscher: Physics of Semiconductor Switching Devices, in A. F. Gibson et al. (eds.), "Progress in Semiconductors," vol. 6, pp. 145–197, John Wiley & Sons, Inc., New York, 1962.
5. Ref. 3, p. 339.
6. Ref. 3, p. 341.
7. Ref. 3, p. 348.
8. G. W. A. Dummer and J. W. Granville: "Miniature and Microminiature Electronics," pp. 63, 64, John Wiley & Sons, Inc., New York, 1961.
9. F. W. Gutzwiller (ed.): "Silicon Controlled Rectifier Manual," 2d ed., pp. 1–10, General Electric Company, Auburn, N.Y., 1961.
10. Ref. 9, p. 4.
11. Ref. 8, pp. 42–44.
12. C. W. Mueller and J. Hilibrand: The Thyristor—A New High-speed Switching Transistor, *IRE Trans. Electron Devices*, **ED5**: 2 (1958).

16: Other Special Devices

16-1. INTRODUCTION

In the preceding chapter we described and examined four examples of a basic four-region, three-junction structure in different applications and the thyristor, which has properties so similar as to warrant its inclusion. Although this group probably includes the largest number of specialized multijunction devices used commercially, there are a number of other important species to be considered. In some cases these are unique for their particular application, whereas in others they are representative of new developments which promise to become important as the technology progresses.

16-2. AVALANCHE DEVICES

We have already considered a case of avalanche switching behavior in connection with the avalanche transistor described in Chap. 14. In this case a reverse-biased junction was made to avalanche by increasing the field, resulting in a forward flow of carriers and a decrease in overall resistance and switching behavior. The decrease in resistance was governed in this case by the base resistance, which limited the increase in current and resulted in a relatively high voltage drop in the ON condition. There is, however, another example of avalanche breakdown to be considered; it improves markedly on the behavior of the avalanche transistor, and it achieves very high switching speeds with only a moderate sacrifice of switching efficiency compared to the best four-region switches.

We are familiar with avalanche breakdown at a reverse-biased p-n junction. We now wish to consider the parallel phenomenon at an entirely different type of junction, namely, the n-n^+ or p-p^+ junction, under so-called "reverse bias," which means that the n^+ region is made positive and the n-type region (with a relatively higher concentration of holes) is made negative. At these junctions, the current flow is dominated by majority carriers under the influence of a field, and it is therefore a drift current. To a first approximation, the junctions may be said to be impenetrable for minority carriers in either direction, and so-called "injection" accompanying the avalanche effect is actually a majority carrier phenomenon. A further peculiarity of these structures

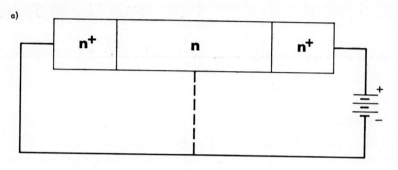

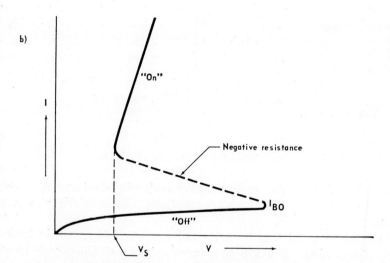

FIG. 16-1. The n^+-n-n^+ avalanche diode principle.

for low voltages and current densities lies in the fact that the so-called "reverse current" is higher for a given voltage than the forward current. (This is a second-order effect, but one of real significance in avalanche devices [1,2].)

16-2.1. Avalanche Diodes. The avalanche diode (really a switch) is an n^+-n-n^+ (or p^+-p-p^+) structure of the form shown schematically in Fig. 16-1. As shown in Fig. 16-1a the device is usually made up of three regions biased as shown by the solid line. For a consideration of the avalanche effect, however, we shall be concerned only with the effect on the right-hand junction, as though the bias voltage were connected only to the central n-type region as shown by the dotted line.

As the voltage is increased across the space-charge region arising from the inequality of doping levels at the n-n^+ junction, a value at which avalanche generation of hole-electron pairs takes place is reached. The minority holes flow into the n-type region under the influence of the field and set up a space charge there. This is countered by an influx of majority electrons flowing in from the external circuit. The conductivity of the n-type region increases, and the field becomes localized more at

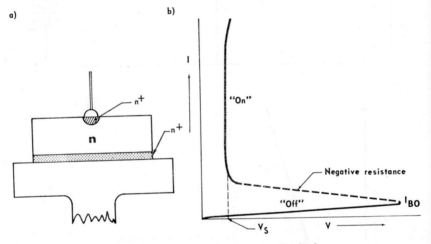

Fig. 16-2. Typical n^+-n-n^+ avalanche diode.

the n-n^+ junction, causing an even higher avalanche effect and eventually resulting in the familiar negative-resistance characteristic shown in Fig. 16-1b.

It should be noted that both V_S and I_{BO} are higher than in typical four-region devices. One way to reduce I_{BO} is to limit the current physically by change in geometry. This approach is shown in Fig. 16-2. In this structure the prebreakdown current (low-voltage majority carrier drift current) is limited by the area covered by the very small n^+ region of the alloyed wire contact. This also concentrates the field nearest to the small area, where it will be most effective in reducing the sustaining voltage, as shown in Fig. 16-2a. The resulting characteristic, Fig. 16-2b, shows the improvement in V_S and I_{BO}.

In their optimum form, these devices have fast switching times of the order of 0.01 to 0.1 μsec (100-mc range), which makes them comparable with the very best of the four-region devices of much greater cost. Unfortunately, there is still some disadvantage in terms of V_S and I_{BO}, with consequent power-handling problems.

16-3. INTRINSIC-REGION DEVICES

To overcome some of the early limitations of junction devices, a number of structures incorporating so-called "intrinsic" regions evolved. These were originally developed in an effort to improve the reverse voltage capability and frequency response without sacrifice in power gain. The earliest successful units were produced by combinations of diffusion and alloying techniques, starting from the highest-purity semiconductor

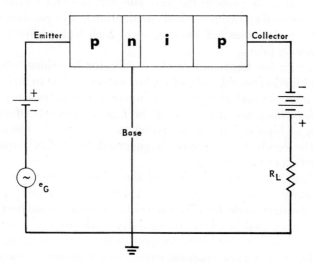

FIG. 16-3. A p-n-i-p transistor.

material then available. More recently, ion drift and epitaxial techniques have superseded the older method, making it possible to work with material of ordinary purity. As a result, many devices today incorporate a layer of relatively high purity material and function by virtue of the principles first demonstrated with intrinsic-region devices. In most cases there is no longer any reference to the original nomenclature.

We shall consider two early examples, both of which found limited commercial application. Their contemporary counterparts are of considerable practical importance and are so similar in operation as to require no further elucidation.

16-3.1. The p-n-i-p Transistor. This is a four-layer, three-junction structure in which there is an intrinsic (or near-intrinsic) region between the base and collector. This is shown schematically in Fig. 16-3 connected in the grounded-base configuration. The intrinsic layer acts like an additional dielectric at the collector junction, permitting the use of a narrow base region of low resistivity, reducing the collector-to-base capacitance (by effectively increasing the width of the depletion layer), and at the same time increasing the PIV for the reverse-biased collector junction. The net effect is to allow high-frequency response (up to 1,000 mc) without sacrifice in current gain or power-handling ability, although there are some limitations due to charge storage in the wide depletion region and the physical distance carriers must move to reach the collector after leaving the emitter. Fabrication was originally difficult (especially for the complementary n-p-i-n configuration), since it was necessary to protect the intrinsic region from contamination. With the success of epitaxial methods production is now routine, although the intrinsic region does not have to be of such high purity to achieve the same results [3].

The quantitative treatment of these collector junctions follows that of the step junction considered earlier, but so modified as to include either a layer of intrinsic material (with net impurity concentration producing fewer carriers than would be present because of thermal effects alone) or a region of high resistivity compared to the base or collector. The shape of the conductivity profile is governed by design requirements and fabrication economics [4].

16-3.2. The n-i-p Rectifier. This designation (or the reverse) represented an early (1952) attempt to provide a model of a power, or injection-type, rectifier made by alloying contacts onto germanium [5]. At that time the diffused-junction conductivity-modulated rectifier (usually of silicon) had not emerged as a competitor for germanium units, which were developed to a very practical commercial level even though forced cooling was necessary. Because of the thickness of the body and limited penetration by the alloy, as well as the relatively high purity of the bulk germanium, it was postulated that these structures could best be represented by an n-i-p scheme such as shown in Fig. 16-4. The representation at (a) assumes some diffusion beyond the contact region, while that at (b) limits the doping to the n^+ and p^+ regions. Both are shown without bias voltage. It was assumed that both junctions emitted majority carriers in the forward direction and that these recombined in the intrinsic region. Connections to the external leads were assumed to be nonrectifying and capable of supplying majority carriers from the external circuit. These models gave reasonable agreement with theory for the configuration they represented. Such rectifiers, however, were inefficient and suffered the temperature limitation characteristic of all germanium devices. There were also other limitations due to storage effects in the wide space-charge (intrinsic) region.

With the development of the diffused silicon device, it was possible to take advantage of its higher intrinsic resistivity (better reverse blocking or higher *PIV* for a thin layer) and decreased temperature sensitivity, ending the need for forced cooling. In addition, because of the narrower blocking region, the forward voltage could be lower and the efficiency higher. A further improvement was due to the better uniformity and reproducibility of diffused structures.

A comparison of the model of Fig. 16-4 with that described in Sec. 6-7 will show that, except for the assumption regarding two junctions, the

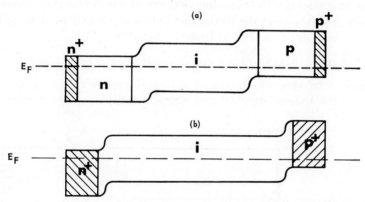

Fig. 16-4. The *n-i-p* representation of a power rectifier.

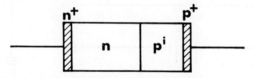

Fig. 16-5. Schematic representation of a diffused-junction power rectifier.

n-i-p unit is similar to the diffused-junction power rectifier with conductivity modulation. This latter structure is assumed to be a *p-n* junction with one region heavily doped compared to the other and with large-area, low-resistance lead attachments capable of injecting. If we let the heavy surface doping of a diffused structure (suitable for lead attachment) be represented as p^+ and n^+, respectively, and let the high-purity side of the junction approach intrinsic purity, but still be of definite type p^i, we could then represent the situation as shown in Fig. 16-5. This involves only one semiconductor junction, but it is essentially equivalent to the earlier *n-i-p* concept and consistent with conductivity-modulated diode theory.

Although the original manufacturing methods for the *n-i-p* diode are not currently used, some of the principles involved find application in

240 Semiconductor Junctions and Devices

epitaxial structures, usually of relatively low power capability. The growth of an intrinsic (or near-intrinsic by comparison with the balance of the structure) layer on a low-resistivity substrate is relatively straightforward by these techniques. The final layer, or p^+ in Fig. 16-5, is then made by alloying or diffusing to the epitaxial growth. The epitaxial method provides control of geometry and resistivity profile and makes possible the realization of diode properties expressed in terms of forward and reverse voltages as well as response times which are free of the limitations of ordinary diffusion or alloying combinations. In the case of diodes, as was seen with transistors in the preceding section, it is becoming customary to abandon the distinction between p-n and p-i-n structures, since the boundary line can at times be indistinct [6].

16-4. THE DOUBLE–BASE DIODE OR UNIJUNCTION TRANSISTOR [7]

This is a two-region, three-lead semiconductor device of wide utility owing to the unique stability and reproducibility of its properties, both

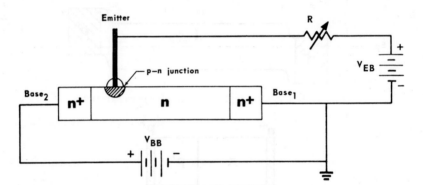

Fig. 16-6. The double-base diode, or unijunction transistor, schematic.

as functions of environmental conditions (temperature, aging, cycling, etc.) and, for any of the numerous standard models, from batch to batch during manufacture. In addition, the cost is modest. The double-base diode is another form of controlled switch, with a negative-resistance characteristic of great stability, virtual immunity to pulse damage, and controlled firing voltage at a very low current.

The geometry is shown in Fig. 16-6, which depicts the construction of an unsymmetrical diode with two "base" leads but without collector junction, collector region, or collector lead. Again the problem of nomenclature shows up, and we find the device called (properly) a double-base diode and (improperly but popularly) the "unijunction transistor."

If we now connect the bias supplies as shown, we shall have a fixed voltage V_{BB} appearing across the length of the n-type bar between the ohmic contacts at Base₁ and Base₂. The bar will appear to have a resistance of 5,000 to 10,000 ohms. If there is no voltage applied in the emitter circuit, the bar acts like a voltage divider and I_E, flowing through the reverse-biased emitter, will be typical of the I_{SAT} for a reverse-biased p-n junction. If V_{EB} is slowly increased, it will eventually forward-bias the junction, causing injection of holes and a nonequilibrium situation. This is then countered by electron flow through the Base₁-Base₂ external circuit, further decreasing the resistance between the emitter and Base₁. This effect pyramids until the voltage at the emitter falls as the current increases, resulting in a negative resistance. The characteristic is shown in Fig. 16-7. In this plot a normal forward characteristic is shown by the dotted line. If the Base₁-Base₂ circuit is open, increasing V_{EB} will trace

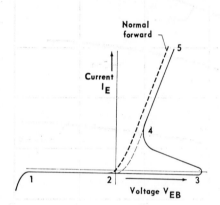

FIG. 16-7. Emitter characteristic for double-base diode. (After General Electric Company, "Transistor Manual," by permission [9].)

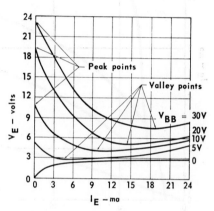

FIG. 16-8. Current-voltage characteristic for unijunction transistor. (After G. W. A. Dummer and J. W. Granville, by permission [8].)

the curve 1245, almost the same as a normal forward characteristic except for bias. If a fixed bias V_{BB} is applied across the two base leads, the curve 12345 is traced as the emitter voltage V_{EB} increases.

A further interesting property of this device is shown in the characteristics plotted in Fig. 16-8. This plot shows the current-voltage relationship between I_E and V_E as a function of V_{BB}. Note the points labeled as peak and valley points. To the left of the peak points (I_E very small) the emitter junction is reverse-biased and only the I_{SAT} flows. To the right of the valley points the n-type region is saturated with injected carriers and only the dynamic or positive resistance (about 10 ohms) enters the picture. Between these points on each curve there is a region of negative resistance. A more conventional characteristic is shown in Fig. 16-9.

242 Semiconductor Junctions and Devices

These devices are widely used for medium power switching, oscillation, firing circuits for controlled rectifiers, timing circuits, etc. Their biggest advantage in applications lies in the stable negative-resistance region of their characteristic. In most other respects their performance is not specialized, but covers a broad range of general interest and utility.

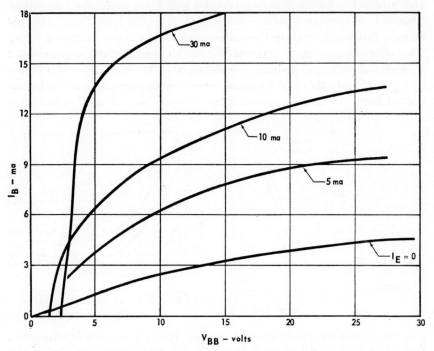

FIG. 16-9. Base characteristic for different emitter bias (double-base diode). (*After General Electric Company, "Transistor Manual," by permission* [9].)

16-5. THE JUNCTION TETRODE TRANSISTOR

The junction tetrode transistor was an early development of the grown-junction variety aimed at improving the frequency response at high power gain and high operating temperatures [10]. Units made by a combination of growing and diffusing techniques [11] are now on the market.

The chief enemy of high-frequency operation at high power is the product of the base resistance and collector capacitance $r_b C_{cb}$. To reduce the base resistance, or the resistance of the path through all parts of the base region to the base connection, a smaller geometry or cross section would be desirable. Similarly, the collector capacitance may be reduced by decreasing its area. To decrease the area in practical terms by reducing the physical dimensions is difficult beyond a certain point for reasons of fabrication (particularly since the collector contact should be large for low collector resistance) and also because reduction in the mass of material aggravates the heat-dissipation problem and limits the power which may be handled.

Therefore, to avoid a dumbbell geometry and thereby retain good thermal cross section, an electrical field is used to reduce the effective base area. Such an arrangement is shown in Fig. 16-10 for a very simple geometry. Other geometries are used, but all have in common the above structural and thermal assets while, at the same time, greatly decreasing the base area. In this configuration the base has at least two leads, one of which is biased to oppose the forward emitter bias. Therefore, since the reverse bias on the emitter constrains the junction to emit only over

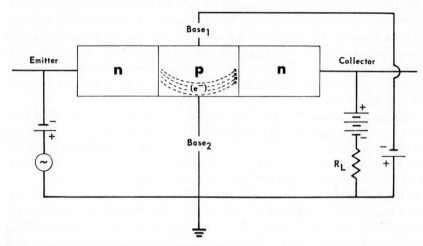

FIG. 16-10. A junction tetrode transistor (the base region is usually very narrow). (*After D. M. Warschauer, by permission* [12].)

a small area near lead Base$_2$, the effective base area is reduced and its resistance is lower, the collector capacitance is down, and high-frequency operation is possible. Further, there is an adequate thermal sink, so that high power levels can be handled.

As in all other design problems, some advantages are won at the expense of others. In this case, the current handling capability is smaller than might be expected for the average bar cross section, and the input resistance, or required emitter bias, is higher. Also, these units are quite expensive.

16-6. FIELD-EFFECT DEVICES

Up to this point, it has been repeatedly stated that the bulk of semiconductor device performances and characteristics are dependent on minority carrier effects. This type of reasoning has served nicely to

244 *Semiconductor Junctions and Devices*

explain junction phenomena, where the change in conductivity of the structure is due to the presence of excess minority carriers in the region of interest (even though they are majority carriers in the region from which they come). The principal departures from this strict pattern of reasoning involve reverse breakdown of junctions as in rectifiers, zener and avalanche diodes, and the small-signal forward response of esaki, or tunnel, diodes. Even in these cases, the physical interpretation of the observed behavior is easily reconciled with the flow of minority carriers at or near p-n junctions.

We now come to an entirely different type of behavior which involves only majority carriers. (There must always, of course, be some effect due to minority carriers so long as we are dealing with semiconductors. However, in the phenomena to be reviewed in the concluding sections of this chapter, the conductivity changes are due to majority carrier variations, and we know, from the p-n product theory, that such concentration changes will be so large compared with the corresponding concentration modulation of minority carriers as to permit us to completely ignore the latter.)

We shall first examine two devices which depend on the so-called field effect. This is analogous to the junction field effect resulting from space-charge accumulation at junctions. In this class of devices, however, the resulting fields are less intense and are subject to external modification to a degree not possible with junction geometries.

16-6.1. The Current-limiter Diode. This device (which is almost the only remaining example of the use of the term "varistor") has not yet been widely used for reasons which will appear later. It represents an early effort to accomplish current limitation in a manner as simple and elegant as that of the zener diode for voltage regulation. We shall briefly describe an early version of such a device and indicate its shortcomings [13].

It is known that if an asymmetrical junction is formed, there is a deeper penetration of space charge into the highly purified material. Theoretically, by proper choice of doping levels and external bias, this charge region could be enlarged to any reasonable dimension. Further, since the space charge is always such as to repel majority carriers, it should be possible to set up a variable barrier for majority carrier flow. Such a structure would be that shown in Fig. 16-11. Note that the depletion layer will extend principally into the n-type region by virtue of the doping levels and that the junction is reverse-biased by a variable voltage V.

If the reverse-bias voltage is zero, the space-charge region does not extend clear through the n-type zone. If a small reverse voltage is applied as shown, we shall obtain a flow of electrons (arrows) through that portion of the high-resistivity n-type region which is not in the space charge. With increasing voltage, we would expect a linear increase in current. However, since the voltage is also acting to reverse-bias the junction, the space-charge region will spread out toward the surface of the horizontal n-type region. This decreases the cross section for current flow and has the effect of increasing resistance. Eventually, with increasing voltage, the space-charge region would reach the surface, resulting in a greatly reduced current increment for an increment in voltage. In the ideal case, the resistivities and geometry would be so selected that, after a critical voltage was reached, the effect of increased bias on the flow of electrons would be exactly balanced by the spread in space charge until, finally, avalanche breakdown at the junction would supervene.

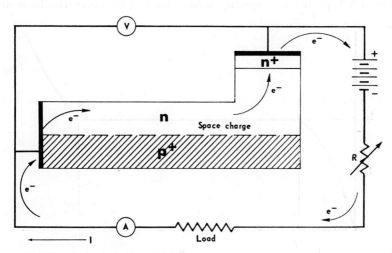

FIG. 16-11. Schematic of varistor, or current-limiting diode. (*After G. W. Dummer and J. W. Granville, by permission* [14].)

This ideal behavior is shown in Fig. 16-12. In this figure V_P is the so-called "pinch-off voltage" and I_P the "pinch-off current," while BV is the reverse breakdown voltage for the junction. In practice, V_P should be low in relation to I_P to avoid power loss and heating. Also, the I_P curve should be truly invariant with voltage. Typical units have V_P of about 5 volts for I_P values of 5 ma, which is not too good a ratio. Initial resistance is high (200 ohms), which means loss of power. In some units, however, the I_P characteristic is reasonably flat—a change of 1 volt producing less than 10^{-4} amp change in current [14].

These devices have inherent problems. The geometry is uneconomical; the presence of the ohmic contact on the left end shorts across the junction; and the exposed surface of the n-type channel invites many surface problems such as dirt, smoothness, and ambient atmosphere.

246 Semiconductor Junctions and Devices

16-6.2. The Unipolar Transistor. This device, also called the field-effect transistor, is another majority carrier device very similar in conception to the current limiter. However, it avoids construction problems and exposed surface problems and can be made reproducibly. It is not only a current limiter, but also a modulator or amplifier. As will be seen shortly, it has a sharp voltage breakdown but no negative-resistance characteristic; so that to remain within the power dissipation limit, the device must not be allowed to reach a voltage near breakdown.

A typical structure is shown in Fig. 16-13. The upper drawing is a schematic view corresponding rather closely to the geometry of one of the first units built at Bell Laboratories. The lower drawing is a section through a typical contemporary commercial unit, where the large-area

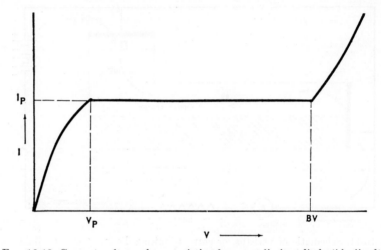

FIG. 16-12. Current-voltage characteristic of current-limiter diode (idealized).

contacts, the source completely surrounding the gate and drain, and the thin, layer-built structure give optimum transconductance, high breakdown voltage, and high input impedance.

Note that the two junctions forming the channel are reverse-biased and that the source is connected in common with the gate. The drain is forward-biased for majority carrier flow, since this is the conduction mechanism. The space charge is shown at or near pinch-off for the source-drain current.

An actual plot of current-voltage characteristics at various gate bias values V_{GS} is shown in Fig. 16-14 for a typical device of the diffused variety. Note that the current increases sharply with V_{DS} at low voltage. After reaching a value of pinch-off voltage V_P, a further increase in voltage causes no further increase in current up to V_{MAX}, the rated safe breakdown voltage. If V_{MAX} is exceeded, the unit breaks down and the current increases very steeply, with severe heating and damage. These units must therefore be protected against runaway.

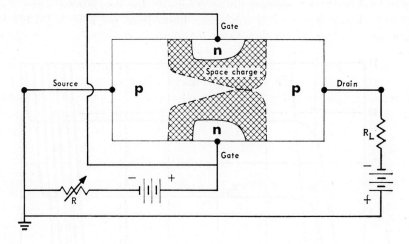

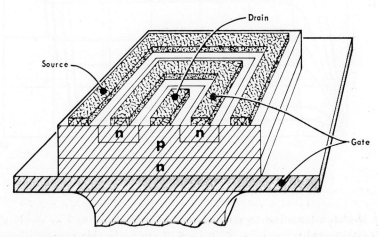

FIG. 16-13. The p-channel unipolar or field-effect transistor. (*After Amelco Semiconductor, "Field Effect Transistors," by permission* [15].)

The mechanism of pinch-off is best understood by considering the cross section of the conducting channel available for current flow. As the gate bias is increased, the cross section becomes smaller and the current I_D falls. If the V_{GS} is fixed and V_{DS} is increased, the drain current increases up to the pinch-off point. At this current value the cross section is fully used for current flow, and further increase in the voltage merely increases the length of the pinch-off path, causing an increased IR drop and resulting in current limitation. Since the gate junctions are reverse-biased under pinch-off, the breakdown when V_{DS} exceeds V_{MAX} is a typical avalanche breakdown for a reversed-biased diode.

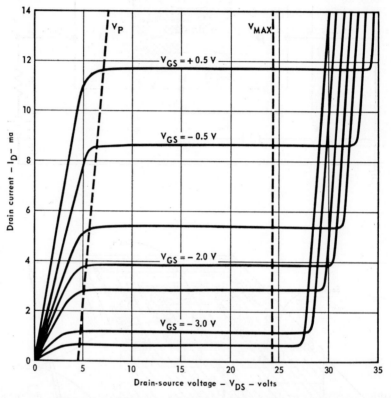

FIG. 16-14. Common-source drain characteristics of field-effect transistor. (*After Texas Instruments Incorporated, Bulletin DL-S 622727, by permission* [16].)

It is also interesting to note (Fig. 16-15) that $V_P \approx V_{DS} + |V_{GS}|$ for the small-signal range ($V_{GS} \leq 0$, $I_D \leq 3$ ma). In this case, V_P is taken as that value of V_{DS} (with $V_{GS} = 0$) at which the current reaches a steady value and is no longer increased if V_{DS} increases. For the device of Fig. 16-15, this gives $V_P = 5$ volts and $I_D = 1.75$ ma. This results from the fact that, to a first approximation, the reverse bias on the gates near the source is essentially $|V_{GS}|$, while at the drain end it is approximately $V_{DS} + |V_{GS}|$. Since pinch-off occurs first at the drain end, the relationship between V_P and $V_{DS} + |V_{GS}|$ is not difficult to understand [15].

The input circuit is effectively looking at the impedance of back-biased junctions, and typical values are of the order of 1 to 10 megohms. This is unique among devices we have so far considered, and it makes the field-effect transistor popular as a vacuum-tube replacement. The frequency response is well up into the 10- to 100-mc range, and the voltage controlled characteristic of these devices emphasizes the analogy to vacuum tubes. Also, since these are majority carrier devices, noise due to carrier recombination is very low; and as an unusual plus value, the devices are resistant to radiation damage. (This arises because loss of

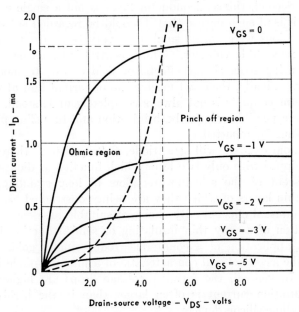

FIG. 16-15. Small-signal pinch-off characteristic of field-effect transistor. (*After Amelco Semiconductor, "Field Effect Transistors," by permission* [15].)

minority carrier lifetime due to radiative creation of recombination centers is not important.)

As mentioned earlier, the danger of thermal runaway is ever present. Aside from this, these units can withstand storage up to 300°C and will operate to 175°C, although with sharp loss of efficiency, as might be expected from junction theory. Finally, despite their advantages, commercial versions of this unit are very costly and have had chiefly military support to date.

16-7. TUNNEL OR MAJORITY CARRIER INJECTION DEVICES

This group, the last to be included in this discussion of multiple-junction configurations, represents an area of great current interest. We shall describe one classic example of these newest developments, the thin-film transistor. Before undertaking this specific description, a word should be added to alert the reader to this rather new and potentially important area, since the basic principles are not those of the junction device as we now know it. There are, however, enough areas of similarity to attract our interest and to make it possible to understand their behavior in terms with which we are already familiar.

Many differences in the basic concepts will appear even in this brief discussion. To begin with, these are majority carrier devices, using injection. Second, the semiconductor layer is not a single crystal, and frequently it is of such nature that it would be regarded as an insulator if the dimensions were not so small that very real currents are caused to flow by reasonable voltages. Third, control is by means of an insulated gate using space-charge effects, or field effects, to modulate the carrier flow. Finally, the exact mechanism of current flow at the metal-semiconductor contacts is not always completely understood, and there seems to be a possible role for both injection and tunneling phenomena in devices thus far reported.

Although our example will be a triode, or "transistor," there are other devices equipped with only two leads which function as diodes. (These are reminiscent of the selenium rectifiers of long standing, but the extremely thin layers of which they are being constructed has produced some very unusual effects.) By choice of suitable base electrode material, formation of a very thin insulating or semiconducting layer, and subsequent application of a counterelectrode, usually gold, properties similar to the asymmetrical characteristic of a junction diode have been observed. In general, the forward voltage drops are higher and the reverse saturation current is often larger than in the highly efficient single-crystalline silicon devices.

Despite these current problems, the interest in these new diodes is very keen. Metal-oxide systems such as $Al\text{-}Al_2O_3$, $Ta\text{-}Ta_2O_5$, $Ti\text{-}TiO_2$, and others of equally unlikely nature have been successfully employed for the electrode-"semiconductor" portion of the structures. To obtain desirable performance, the oxide (or sulfide) layers and counterelectrodes are usually prepared as thin films by evaporative or sputtering techniques, although some reports suggest that less costly techniques may also become important. Because of the metal-insulator or metal-semiconductor contacts involved, the polycrystalline nature of the "active" layer, and the wide diversity of experimental approaches, the underlying theory is not well worked out. As will be seen from the discussion of such heterogeneous contact systems in a later chapter, there are serious gaps in the present experimental picture, even for relatively simple and widely used examples. If an attempt is made to comprehend these newer systems, the problems simply multiply. It seems likely that both majority carrier injection at the contacts and tunneling are responsible, in part at least, for the observed behavior, but there is no consistent interpretation.

16-7.1. The Thin-film Transistor. The example to be discussed in this section also suffers from a lack of satisfactory physical theory. However, the controlled and reproducible techniques used to fabricate the device

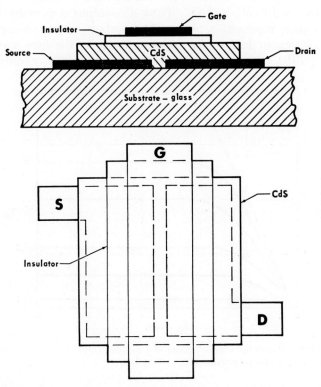

Fig. 16-16. Thin-film transistor structure. (*After P. K. Weimer, by permission* [17].)

and the good agreement of much of the performance data with the elementary theory thus far advanced strongly recommend the thin film transistor as one of the best examples of the type so far reported.

This device, first reported by Weimer in 1961 [17], is illustrated in Fig. 16-16 with a cross section of one of the simpler modifications and a plan view to indicate the approximate proportions involved. The structure is prepared by evaporating a suitable metal, such as gold, through an appropriate mask onto a clean glass substrate to form the source and drain, with a thickness of 15 to 50 μ. The metal must be able to make a nonrectifying, low-resistance contact to the chosen semiconductor. The spacing of the source from the drain is as small as possible, usually 5 to 25 μ.

This extremely narrow current path is one of the essentials of the design. The other essentials, both controlled by the evaporation process, are the thickness of the semiconductor layer (which is less than 1 μ) and the insulator (originally, evaporated SiO) with a thickness in the micron range but sufficient to avoid shorting or dielectric losses at gate voltages of 10 volts or less. These dimensions, achievable by thin-film techniques, represent a potential advantage, since spacings of the order of microns or less are in the same range as the base widths and junction widths for ordinary transistor structures after diffusion or

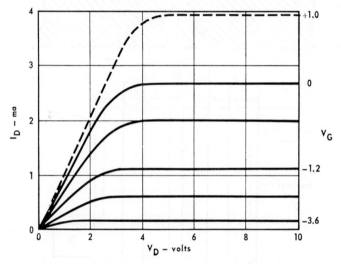

FIG. 16-17. Current-voltage characteristic of thin-film transistor with negative bias. (*After P. K. Weimer, by permission* [17].)

epitaxy, both of which operations are considerably more involved than the thin-film approach.

When we come to the performance of the TFT (thin-film transistor), we find another peculiarity of this structure. The active element, CdS, is a polycrystalline, n-type material which functions in this device as a majority-carrier-dominated region. Therefore, if we reason by analogy to the unijunction, or field-effect, transistor, we should expect maximum source-drain current for a zero gate bias and a decrease to pinch-off as the gate becomes more negative. Actually, this is seen for many units, particularly at low drain currents and low bias voltages. This characteristic is shown in Fig. 16-17, where pinch-off is near -3.6 volts on the gate. Note, however, the dotted curve corresponding to $+1.0$ volt and showing a higher I_D characteristic. This behavior is different from that of any device we have thus far considered, and it represents an increase in current for a positive gate bias.

Note that there is no electrical contact between the gate and the semiconductor. This effect is therefore a field effect in the truest sense. The field here is not the result of space-charge build-up at a junction, but it is a true induced field from an insulated electrode.* Operation in this fashion is known as *enrichment* when the gate is charged to attract the injected majority carriers and *depletion* when oppositely charged. The unipolar units described in the preceding section are true junction devices and operate almost entirely in the depletion mode.

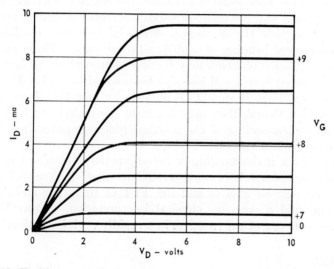

FIG. 16-18. Enrichment mode operation of thin-film transistor (source grounded). (*After P. K. Weimer, by permission* [17].)

Operation of the TFT in the enrichment mode provides a much larger range of amplification, as shown in Fig. 16-18. Here, the large change in I_D as a function of V_G is apparent. Note that this unit (not the same one as the previous example) would show operation in the depletion mode only at very low I_D, if at all.

The fact that the control gate of the TFT is insulated from the semiconductor and provides for operation in the enrichment mode has two very important consequences. First, the control currents are extremely low, making the device a voltage-controlled analogue of a vacuum tube, and second, direct coupling between amplifier stages is possible. Finally,

* Several other examples of this effect due to an insulated control gate have been reported, but they do not yet have commercial significance.

these units have shown frequency responses up to better than 10 mc, high input impedance, and excellent switching response at speeds of less than 0.1 μsec.

The physical process occurring in the TFT may be regarded as an extension of field-effect phenomena. The choice of a wide gap material such as CdS (gap = 2.4 ev) reduces the thermal generation of carriers. The device is therefore dependent on contact-injected majority carriers. Exactly how the current flows through the polycrystalline semiconductor and is modulated by the gate is not yet clear. The phenomena of operation in both the enrichment and depletion modes, with current saturation for a given V_G, are not fully understood. At least a portion of the problem is due to the very small dimensions, the contact behavior, possible tunneling effects, and unexplained variations from one unit to another.

We now conclude the discussion of special transistor configurations under the general heading of multijunction devices. There are many more varieties, some of which may become important in the future, just as some discussed herein will lose significance and be replaced. Except for the last example, it should be obvious by now that junction phenomena, varied though they may be, form the underlying basis which permits an understanding of the physical processes governing the performance of all these specialized units.

In time, if our understanding of heterojunctions between metals and semiconductors (especially polycrystalline materials) continues to improve, we may be able to add the TFT to the other multijunction devices and find a common physical interpretation. Meanwhile, the TFT represents one of the frontiers in a rapidly evolving technology.

REFERENCES

1. A. K. Jonscher: "Principles of Semiconductor Device Operation," pp. 130–132, G. Bell & Sons, Ltd., London, 1960.
2. A. K. Jonscher: Physics of Semiconductor Switching Devices, in A. F. Gibson et al. (eds.), "Progress in Semiconductors," vol. 6, pp. 166–173, John Wiley & Sons, Inc., New York, 1962.
3. H. C. Theuerer, J. J. Kleimack, H. H. Loar, and H. Christensen: Epitaxial Diffused Transistors, *Proc. IRE*, **48**: 1642 (1960).
4. J. M. Early: p-n-i-p Junction Transistor Triodes, *Bell System Tech. J.*, **33**: 517 (1954).
5. W. C. Dunlap: "Introduction to Semiconductors," p. 344, John Wiley & Sons, Inc., New York, 1957.
6. C. D. Root: A Look Inside the Diode, *Electron. Design*, **10**: 58 (May, 1962).
7. J. F. Cleary (ed.): "Transistor Manual," 6th ed., pp. 191–201, General Electric Company, Syracuse, N.Y., 1962.
8. G. W. A. Dummer and J. W. Granville: "Miniature and Microminiature Electronics," p. 65, John Wiley & Sons, Inc., New York, 1961.
9. Ref. 7, p. 192.

10. R. L. Wallace, L. G. Schimpf, and E. Dickten: A Junction Transistor Tetrode for High Frequency Use, *Proc. IRE*, **40**: 1395 (1952).
11. *n-p-n* Grown-diffused Silicon Tetrode Transistor, *Bull. DL-S* 961, *Type* 3N35, Texas Instruments Incorporated, Dallas, Texas, 1958.
12. D. M. Warschauer: "Semiconductors and Transistors," p. 135, McGraw-Hill Book Company, New York, 1959.
13. R. M. Warner, W. H. Jackson, E. I. Doucette, and H. A. Stone: A Semiconductor Current Limiter, *Proc. IRE*, **47**: 44 (1959).
14. Ref. 8, pp. 47, 48.
15. Field Effect Transistors, Theory and Application Notes No. 1, pp. 1–5, Amelco Semiconductor, Mountain View, Calif., June, 1962.
16. *p*-Channel Planar Diffused Silicon Field-effect Transistors, *Bull. DL-S* 622727, *Types* 2N2497, 2N2498, 2N2499, p. 2, Texas Instruments Incorporated, Dallas, Texas, 1962.
17. P. K. Weimer: The TFT—A New Thin-film Transistor, *Proc. IRE*, **50**: 1462–1469 (1962).

17: Metal-Semiconductor Contacts

17-1. INTRODUCTION

In developing the basic physical concepts of semiconductor-device function, attention has so far been focused on phenomena within the semiconductor lattice. When it was necessary to provide connections to external circuit elements, it has been assumed that leads could be attached in an appropriate fashion and that there were no electronic effects at the point of attachment: that is, such connections and leads were of negligible resistance and had no influence on circuit parameters. In a few cases it has been mentioned that an n^+ or p^+ layer was a desirable condition at a point of connection to the external circuit, and in connection with the thyristor and other majority carrier injection devices there was specific mention of a metal-semiconductor interface, or contact, which had significant properties in terms of device performance. In these cases, desirable characteristics were achieved by means of special configurations and fabricating techniques.

Unfortunately, the picture presented thus far represents a considerable simplification. It is true, from a practical point of view, that low-resistance ohmic contacts between semiconductor regions and conventional wire leads are necessary to the successful fabrication of all commercial devices. It is not, however, true that such connections are as simple to make or as inherently free of undesirable effects as ordinary metal-to-metal connections. To appreciate the problems involved in the manufacture of devices wherein contacts are required to be of low resistance necessitates a brief consideration of surface phenomena and their effects on the solid-state surface.

The subject of metal-semiconductor contacts has produced a vast literature [1]. From the days of the cat-whisker galena crystal radio receiver (a form of point-contact diode) the development of semiconductor technology and theory has been plagued by the problems of affixing leads in order to make measurements. At first, progress was largely empirical and good contacts were achieved by carefully following a procedure which had been demonstrated experimentally but not understood theoretically. The development of a satisfactory physical model was beset by experimental problems of a type which were new in the fields of electronics and metallurgy. In addition, there were so many conflicting observations and uncontrolled variables that much of the early literature was based on erroneous experimental conclusions.

It is not within the scope of this discussion to present the evolution of the present concepts and technology. There are still many unresolved questions, and fabrication methods are not always based on a well-understood theory. This chapter will simply outline the conditions which are believed to exist at metal and semiconductor surfaces in close contact. We shall start with a simple metal-metal system and consider the band energy diagram for each material. From this, we shall proceed to the case involving a semiconductor and metal, from which we will be able to deduce the rectifying or nonrectifying properties of an ideal contact. Finally, the effects of surface states and surface contamination, such as oxide, will be introduced to point out the role they can play in terms of contact resistance and other properties.

In a previous discussion of the electron band structure it was shown that electron energies ranged between a high (negative) value at the nucleus to a reference point at infinity E_∞, where the value of the energy reached zero, corresponding to the free electron in space. This latter condition represented a high absolute energy for an electron, since electrons must increase in energy to move farther away from the nucleus. Since we were then concerned only with changes in the electron energy between the bands of the solid, we represented the energy simply as E, increasing toward the value for a free electron. Later, in discussing this model of a solid, we introduced the Fermi level, or the average energy of electrons in partly filled bands. If we now combine these two representations (for the case of a metal) and omit all bands below the conduction-valence band, we arrive at the diagram shown as Fig. 17-1, in which E_c represents the energy of an electron at the lower edge of the combined valence-conduction band, E_F is the Fermi level energy, and E_∞ is the energy of the free electron in space (vacuum) no longer under any influence from the parent metal structure. All the electrons in the partly filled band above E_c are able to conduct.

Figure 17-1 is based on the energy interval between the free electron and the electron at the Fermi level E_F. This interval is classically defined as the work function W, a fundamental property of metals and semiconductors, and represents the work necessary to remove an electron from its average energy E_F within the band structure to a point in space

E_∞, where no further attraction for the lattice is experienced [2]. The electron work function is also defined as the negative of the electrochemical potential energy, which we recall as $(-qV_F) = E_F$ [3]. The work function is always given in terms of electron volts and is positive. It may be represented in terms of the energy scale of Fig. 17-1 by

$$W = E_\infty - E_F \qquad (17\text{-}1)$$

where $E_\infty = 0$ and E_F has a negative value.* In a system consisting of an ideal metal or semiconductor (without "surface states," cf. Sec. 17-5) in a vacuum environment (Fig. 17-1) it is assumed that the surface of the solid is free of all foreign materials and that an electron in crossing the surface will be influenced only by the asymmetry of the surface bonds,

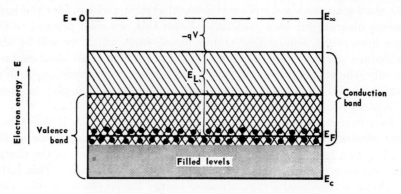

FIG. 17-1. The band energy system of a metal surface.

owing to the abrupt termination of the lattice. This effect is electrostatic, since the unsatisfied bonds of the surface atoms attract the emerging electron back to the solid. It is referred to as the surface or dipole layer, since the surface atoms may be regarded as positive ions and the emerging electron as the other end of a dipole. The value of the surface potential is given by the symbol χ, and the energy in electron volts becomes $-q\chi$. It is also significant that these effects normally occur in a very small distance at the surface, of the order of 10^{-7} cm, less than the mean free path of an electron [4,5].

The effect of a surface dipole layer such as that described above is predictable when the sample is in a vacuum. A question that arises, however, is what the result might be if two solids are in contact. A

* The energy of the electron at E_∞ is the true zero, or reference point, for the general case—a fact often overlooked in calculations involving only changes in energy within a solid system. When contacts or other interphase boundaries are involved, it is obviously necessary to work with E_F, E_∞, and W.

surface layer only 10^{-7} cm deep is of the order of interatomic dimensions and it is unlikely that two mechanically prepared surfaces would touch or overlap at more than a few points. However, since pressure, heat, or a low-melting solder is employed to improve the contact, it is usually assumed that dipole surface layers are present on both surfaces and affect the energy of an electron to the same degree as in the vacuum-solid system.

In addition to surface layers, there is the possibility that internal potential gradients (such as we have seen in the vicinity of junctions in semiconductors) may occur at or near the surface. These are due to space-charge effects and have much greater dimensions [5] (from 10^{-5} to 10^{-4} cm), so that surface dipole disturbances in a layer only 10^{-7} cm thick should have relatively little effect on them. Such potential gradients are known as inner potentials, represented by the term ϕ, and have energy equivalents of $-q\phi$. When these occur near the surface of a solid phase, they can also affect the energy of an emerging electron and must be considered part of the contact problem. The presence of surface contamination may add to or decrease the effect of either potential, χ or ϕ.

In order to relate these surface energy effects to the energy terms with which we are familiar, we recall that the average or Fermi energy of electrons is given by

$$E_F = \bar{\mu} = \mu - qV \qquad (5\text{-}6)$$

where the final term on the right is a potential term dependent on local conditions within the solid and μ is a kinetic term derived from consideration of an electron gas in a fixed volume. Near a surface the effects due to the potentials χ and ϕ must be part of the electrochemical potential energy $\bar{\mu}$ and are included in the term $-qV$. To be completely explicit, we should break the term $-qV$ into the potential energy corresponding to the lattice periodicity, local space-charge effects (as in the case of the p-n junction), a surface dipole term $-q\chi$, and a term $-q\phi$ reflecting inner potential effects near the surface. For real surfaces, we should perhaps include another term due to impurities such as oxide and adsorbed gases. Such detail is neither necessary nor practical, since independent measurements of these increments are not possible and since we are concerned in this discussion only with surface or contact phenomena. We shall therefore use a term E_L to represent the lattice or binding energy of an electron (sometimes called the chemical potential energy) taken to be the energy corresponding to the bulk material, including such potential-energy increments as are constant despite surface changes [6].

We may then write

$$E_L = \mu - qV \quad \text{bulk} \tag{17-2}$$

and
$$W = E_L + q\phi + q\chi = (E_\infty - E_F) \tag{17-3}$$

From these expressions, since E_L is a constant throughout the sample for a given temperature, W will depend on $q(\chi + \phi)$. For a clean metal surface at constant temperature, this term is also constant and W is a characteristic of a pure material as shown in Table 17-1 [7].

TABLE 17-1. WORK FUNCTION VALUES

Metal	Work Function, ev
Ag	4.73
Al	4.08
As	5.11
Au	4.82
Cs	1.81
Cu	4.38
Na	2.28
Ni	5.03
Pt	5.32

Approximate values for Ge and Si are 4.3 and 3.6 ev, respectively. Owing to surface effects, however, such measurements are rather uncertain.

17-2. METAL–METAL CONTACTS

If we now consider two unlike metals at the same temperature, separated by a gap or a vacuum, we obtain Fig. 17-2, where we now show only the important energy levels E_c, E_F, and E_∞. We have used the energy of a free electron as a common reference point with respect to

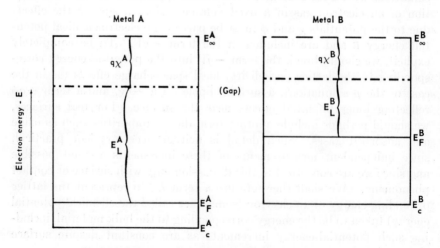

FIG. 17-2. Two unlike metal surfaces with vacuum gap.

both metal A and metal B. In this case, the work functions are [8]

$$W^A = (E_\infty^A - E_F^A) = (E_L^A + q\chi^A) \tag{17-4}$$
$$W^B = (E_\infty^B - E_F^B) = (E_L^B + q\chi^B) \tag{17-5}$$

where the ϕ potentials are not included, since in a metal we do not observe space-charge effects and have, for ideal surfaces, only the dipole potentials χ.

If these metals are now brought into equilibrium in a thermodynamic sense, we know that the Fermi level in the system as a whole, now consisting of both metal A and metal B, must be constant in the absence of a disturbing field. If the Fermi level is thus made constant throughout the system, we find from Fig. 17-2 that $E_\infty^A > E_\infty^B$.* In other words, if an electron in metal A overcomes the lattice energy E_L^A and the surface electrostatic energy $q\chi^A$, it must have acquired more energy than a similar electron in metal B. Therefore, if

$$E_F^A = E_F^B \tag{17-6}$$

we have from Eqs. (17-4) and (17-5)

$$(W^A - W^B) = (E_\infty^A - E_\infty^B) = (E_L^A - E_L^B) + q(\chi^A - \chi^B)\dagger \tag{17-7}$$

Equation (17-7) defines an energy difference whose potential equivalent is known as the Volta potential, or contact potential, V^{AB} for the system metal A in contact (thermal equilibrium) with metal B [8]:

$$-qV^{AB} = (W^A - W^B) = (E_\infty^A - E_\infty^B) = W^{AB} \tag{17-8}$$

Since W can be measured by direct experimental methods for an individual material, values of the Volta potential can be predicted (Table 17-1). In a two-component system involving a metal for which W is not known, for example, or in which surface effects may influence the value of χ (and W), the contact potential V^{AB} can be measured directly and a value for χ or W for the unknown material can be obtained. From a practical point of view these measurements are difficult, even in systems of pure metals, and the literature records a considerable spread of work

---
* Although $E_\infty = 0$ by definition for each individual metal, the use of a common energy scale to plot both sets of band energies has the effect of displacing the zero for one sample.

† The physical significance of the term $E_L^A - E_L^B$ will be appreciated if it is recognized that the chemical potential energy, also referred to as lattice energy or binding energy, is actually the half-cell potential energy of a standard electrode under equilibrium. The difference, $(1/q)(E_L^A - E_L^B)$, is the Galvani potential θ, or the net emf from a standard cell containing the two electrodes. Note that the Galvani potential can be measured directly and that it is a property of the bulk phase of a material. It is also the source of the thermocouple and Seebeck effects [9].

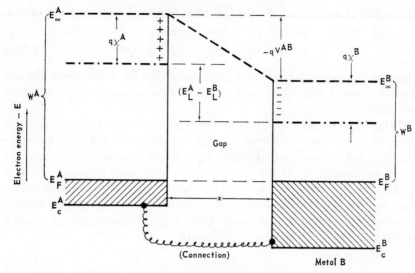

FIG. 17-3. Contact potentials in the two-metal system at equilibrium.

function values. At least part of the difficulty is due to surface effects peculiar to a given experiment which distort the measurements.

Figure 17-3 shows two metals in thermal equilibrium, with the resulting field across the gap balanced by space charges. By inspection we may write

$$-qV^{AB} = (E_\infty^A - E_\infty^B) = (W^A - W^B) \qquad (17\text{-}9)$$

and
$$(E_L^A - E_L^B) = -qV^{AB} + q(\chi^B - \chi^A) \qquad (17\text{-}10)$$

or from Eq. (17-8)

$$(E_L^A - E_L^B) = (W^A - W^B) + q(\chi^B - \chi^A) \qquad (17\text{-}11)$$

which is the same as Eq. (17-7).

Finally, in Fig. 17-4 we show the same system with the metal samples actually in contact and with the Volta potential balanced by the space charge at the boundaries of the solid phases. Note that in this case of clean metal surfaces the Volta potential difference is the algebraic sum of the difference in chemical potential and the difference in surface dipole potentials [Eqs. (17-7) and (17-8); Fig. 17-3]. The contact in Fig. 17-4, including the charge layers, is similar to the case of the p-n junction structure we have previously considered, except that the depth of charge penetration is almost zero, owing to the extremely large surface electron concentrations available in metallic conductors. As in previous cases, there is no current flow through the contact if the circuit is completed and all elements are at a single temperature.

The metal-metal contact also differs from a p-n junction in a semiconductor in another important respect. Despite the existence of a Volta potential, or charge barrier, such systems do not exhibit rectifying properties. There is, in other words, no significant difference in the resistance to current flow as a function of direction. This is another

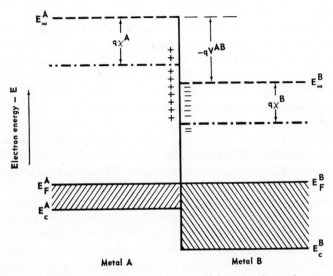

Fig. 17-4. Surface charge or contact potential at metal-metal contact.

result of the very high electron density in metals. A small bias in either direction produces an overwhelming flow of electrons across the barrier.

17-3. IDEAL METAL–SEMICONDUCTOR CONTACTS

17-3.1. Rectifying Contacts.
Having discussed the simple behavior of metal-metal contacts and defined the Volta or contact potential, we are now in a position to repeat the process for the case of the ideal metal-semiconductor contact. We shall assume that a simple model of a semiconductor surface (analogous to that of the metal surface) is valid and that the work function, the potential-energy components, and the bulk lattice energy are unaffected by contamination, etc. The more general and most frequently encountered case of semiconductor surfaces under various external disturbing influences will be considered later.

The principal difference which we shall encounter in the metal-semiconductor contact will be concerned with the low carrier density (by a factor of 10^3 to 10^4) in the semiconductor, even at relatively high doping levels. This will be found to have two related effects. First, whereas a charge layer sufficient to balance the Volta or contact potential in a metal can exist in a surface "skin" of the order of 10 A (10^{-7} cm) thick, an equivalent space charge in a semiconductor will penetrate 10^{-3} to 10^{-5} cm, depending on the doping level. This will produce changes in the band configuration very similar to those which have been observed in p-n junctions. Second, electron energy states that are able to trap, or immobilize, an electron or hole are believed to exist at the surface of a semiconductor. This action is similar to the behavior of donor and acceptor impurities, and since the trapped carrier is at an energy level in the forbidden gap, significant space-charge effects, or double layers, are created. These also cause changes in the band configuration. The theory of these "surface states" is rather complicated, even for clean surfaces. All semiconductor systems, including the ideal case, will therefore be sharply different from the metal system.

For the present we shall omit surface state effects and speak of clean semiconductor surfaces under the influence of surface charge layers (χ potentials) only. We shall also, as was usual in early work with impure materials, ignore variations of the Fermi level as a function of impurity content. This will simply divide semiconductors into two conductivity types with a Fermi level that is constant for all practical p-type materials and also constant, but at a different value, for n-type semiconductors.

If we now place a metal and n-type semiconductor (without surface states and therefore with no inner potential gradient) near to each other, with a vacuum gap between, we shall have a situation such as is shown in Fig. 17-5a (cf. Fig. 17-2). In this representation we introduce the new notation W_B^S, which refers to the work function of the semiconductor as measured in the bulk, or away from the surface under consideration. For this example we choose a metal and n-type semiconductor such that W^M is larger than W_B^S, since it will be shown that rectification depends on the type and work function of the semiconductor. In Fig. 17-5b (cf. Fig. 17-3) the two specimens are in thermodynamic equilibrium but not yet touching physically. Note that there is a field due to the Volta potential equal to V^{MS}/x between the surfaces. This field results from the migration of electrons through the external connection to the surface of the metal, leaving a positive charge on the semiconductor. In this case, however, the positive charge is not confined to the skin or surface layer. Owing to the low carrier concentration, there is a space charge producing a gradient in the inner potential ϕ just beneath the surface, which causes a shift in the band edges analogous to the situation during junction formation (Sec. 5-2, etc.). (The fact that both surface and bulk potential changes produce similar displacements of the band structure is an important aspect of the modern treatment of heterocontacts.)

From Fig. 17-5b, we obtain by inspection, as previously obtained for

Eq. (17-11) and Fig. 17-3 [10],

$$(E_L^M - E_L^S) = (W^M - W_B^S) + q(\chi^S - \chi^M) \qquad (17\text{-}12)$$

It is also apparent from the figure that in this system

$$-qV^{MS} \neq (W^M - W_B^S) \qquad (17\text{-}13)$$

since W_B^S is a bulk property, whereas the Volta potential V^{MS} depends

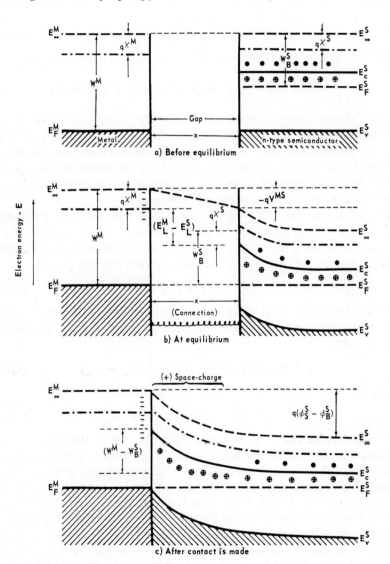

FIG. 17-5. Rectifying metal–n-type-semiconductor contact (no surface states).

on the effective surface work functions. If we let ϕ_B^S represent the value of the inner potential in the bulk of the semiconductor and ϕ_S^S the same value at the surface, we have

$$W_S^S = W_B^S + q(\phi_S^S - \phi_B^S) \tag{17-14}$$

in which W_S^S is the effective work function at a surface where a space charge has altered the inner potential term. It therefore follows that

$$-qV^{MS} = W^M - W_S^S \tag{17-15}$$

and $$-qV^{MS} = (W^M - W_B^S) - q(\phi_S^S - \phi_B^S) \tag{17-16}$$

Substituting in Eq. (17-12), we obtain, considering only potentials,

$$\frac{1}{q}(E_L^M - E_L^S) = -V^{MS} + (\chi^S - \chi^M) + (\phi_S^S - \phi_B^S) \tag{17-17}$$

It also follows, since the width of the gap will cause the space charge in the semiconductor phase to change, that both $\phi_S^S - \phi_B^S$ and V^{MS} are functions of gap spacing. In other words, when a metal-semiconductor contact (as defined in this section) is involved, the observed Volta potential is not dependent solely on the properties of the bulk materials, as it was in the case of the metal-metal system.

Yet we know that $E_L^M - E_L^S$ is dependent only on bulk properties. It is therefore a constant for a given system and independent of gap width. Also, for a given pair of electrodes, it is reasonable to assume that χ^M and χ^S are constant. We may then rewrite Eq. (17-17)

$$-V^{MS} + (\chi^S - \chi^M) + (\phi_S^S - \phi_B^S) = \text{const} \tag{17-18}$$

Since the term $(\phi_S^S - \phi_B^S)$ will become larger as the field increases because of the narrowing gap, V^{MS} must become smaller. Finally, at contact, $V^{MS} = 0$ and

$$(\chi^S - \chi^M) + (\phi_S^S - \phi_B^S) = \frac{1}{q}(E_L^M - E_L^S) \tag{17-19}$$

Then, from Eq. (17-7),

$$(W^M - W_B^S) = q(\phi_S^S - \phi_B^S) \tag{17-20}$$

as shown in Fig. 17-5c. The space charge is now at its greatest penetration, and the entire potential change takes place within the semiconductor. This potential term $\phi_S^S - \phi_B^S$ is the net sum of the chemical potential difference and the difference in surface dipole potentials (cf. Sec. 17-2 and Fig. 17-3). Note that there is no longer any significance to the terms V^{MS} or $W^M - W_S^S$ when the two phases are actually in contact. Strictly speaking, the barrier height should now be W^M, but it can be shown theoretically [10] that the upper part of the barrier is very easily penetrated by the tunnel effect (a postulate which is supported by experimental evidence), so that the effective barrier height becomes equivalent to the difference in bulk work functions, given by Eq. (17-20), and should be dependent on the properties of both the materials used for the contact. The forward electron current direction is from semiconductor to metal over a barrier equal to $(\phi_S^S - \phi_B^S) - V$, where V is the forward bias. In the reverse direction the barrier is $(\phi_S^S - \phi_B^S) + V_R$ as seen by the metal, giving rise to only a small saturation current. These conditions correspond to rectification.*

An analogous situation results for a p-type semiconductor-metal contact if the work function W_B^S is chosen larger than W^M. The final configuration is shown in Fig. 17-6. Under forward bias, electrons flow

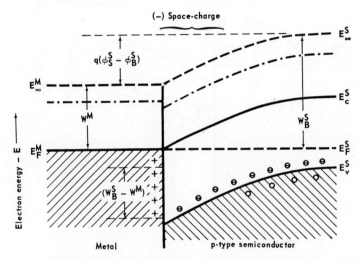

FIG. 17-6. Rectifying metal–p-type-semiconductor contact (no surface states).

from the metal into the semiconductor and holes into the metal, and rectification results because the holes cannot move down into the metal on reverse bias. This is the opposite of the n-type case considered above.

17-3.2. Nonrectifying Contacts. In the preceding section the examples were based on the facts that $W^M > W_B^S$ for the n-type contact and $W^M < W_B^S$ for the p-type contact. Under these conditions, rectification

---

* If we apply a current flow analysis to this contact and assign the symbols I_f and I_g to the forward and thermal generation currents, we shall find the reasoning analogous to earlier examples, except for the I_g component arising on the metallic side of the contact. Since there is no spontaneous generation of hole-electron pairs in a metal, the hole current I_g from the metal to the semiconductor is really not a hole current, but another electron current, and it is limited by the thermal hole generation process in the semiconductor. This leads to an equation for the current-voltage characteristic identical to that of the earlier rectifier equation.

is obtained. If these conditions do not hold, the contacts are said to be ohmic or nonrectifying.

The energy diagram of Fig. 17-7 illustrates this situation for the n-type case. Because $W_B^S > W^M$, there is no barrier and the system might be regarded as having a built-in forward bias. Schematically, this may be represented by stating that the lower edge of the conduction band dips below the Fermi level, which results in a rush of carriers to fill the depression. Since the metal is a rich source of electrons, the depression becomes completely filled and the contact resistance vanishes, no matter what external bias is applied. A similar argument prevails for the p-type

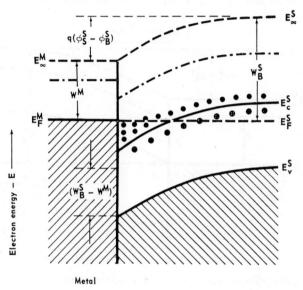

FIG. 17-7. Nonrectifying metal–n-type-semiconductor contact ($W_B^S > W^M$).

case (Fig. 17-8), except that in this case $W_B^S < W^M$ for a nonrectifying interface.

In recognition of this behavior it was postulated that the region adjacent to the interface, where there is a high electron density, might show a greater conductivity than either the metal or the adjacent semiconductor. The term "accumulation layer" was therefore proposed to characterize such a situation (by contrast to the familiar term "depletion layer"). It was recognized, of course, that such an accumulation layer could exist only if there was a higher local density of states available to the charge carriers than in either of the discrete solid phases.

This concept of heterocontacts is of considerable historic interest, since it was the mechanism proposed to explain the opposite rectification polarities observed with a single semiconductor but with different impurities. This led to the first recognition that there are two semiconductor types, and the n- and p-type designations referred to the sign of the voltage bias on the semiconductor side of the contact for forward current flow.

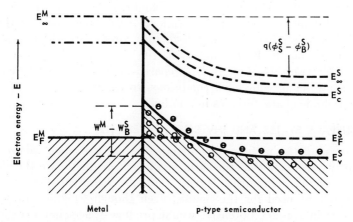

FIG. 17-8. Nonrectifying metal–p-type-semiconductor contact ($W^M > W_B^S$).

17-4. THE RECTIFICATION SERIES [11]

In the preceding sections it was found that rectification could be theoretically related to the work functions of the metals and bulk semiconductors comprising a metal-semiconductor heterojunction in the absence of surface states. The reverse postulate concerning accumulation layers was abandoned early, since the layers were never shown to exist by any unequivocal measurement. It was therefore reasoned that they were so thin by comparison with the bulk material on either side as to be undetectable.

Nevertheless, from the ideal theory above it follows that if metal-semiconductor contacts could be made up by using a series of metals of known work function, it would be possible to obtain the work function for the semiconductor as a function of type and perhaps impurity concentration by observation of the point at which rectification ceased. Experimental verification of these postulates was undertaken, and it was expected that a plot of rectification efficiency against values of the metal work functions would put the metals in the proper order and that the slope of the line would match that derived from simple theory. In some cases, such as those involving contacts to carefully prepared selenium surfaces, it was possible to arrange the metals in the proper order with respect to work function (for the cases when rectification was observed), but the variation in rectification efficiency with work function was too small by almost an order of magnitude. Similar partial success was achieved with copper oxide surfaces, but the method was generally unsuccessful.

These experimental results became even more difficult to interpret when samples of a single semiconductor varying in resistivity were used in an effort to understand the effect of band energy shifts. Even simple experimental measurements of thermionic work functions for the elemental semiconductors were unsatisfactory, producing widely scattered values, often on duplicate runs with the same sample. It was also common for two sets of contacts, prepared under what was believed to be the same set of controlled conditions, to exhibit rectification in one case and ohmic characteristics in another.

17-5. SURFACE STATES AND INVERSION LAYERS [12]

In studying the metal-semiconductor system described in preceding sections, we assumed that the surface of a semiconductor (in the absence of oxide or other impurity layers) could be represented by a model similar to that used for the metals, even though we were aware that differences in contact potential would produce space-charge regions of considerable depth in the semiconductor. In other words, we assumed that the inner potential values were unaltered for the semiconductor at a distance and that only upon approaching a metal (in thermodynamic equilibrium by means of an external connection) was a field created, causing a space charge and warping the band edges. However, the physical picture of a solid semiconductor surface, and particularly the deficiency of electrons at the point of discontinuity owing to incomplete atomic bonds, suggests that these unsatisfied bonds might trap electrons, causing an accumulation of charge at the surface. In the case of metals, the experimental evidence indicates that this is of little importance, since electrons are plentiful, the layer is very thin, and there is a rapid exchange of electrons with the bulk material. In this case the surface was treated in simple terms and the contact potential developed between two metals was successfully related to bulk properties (that is, work function). In the case of a semiconductor, however, we have an entirely different situation. It has been shown that charge accumulation at the surface and depletion regions just beneath the surface can occur in the presence of an external electric field and that the depletion layer penetrates to an appreciable depth within the bulk material. We have not yet considered the energy of the electrons at or near a semiconductor surface, although the position of their levels with respect to the Fermi energy level might produce layers with distinctly p- or n-type properties.

The study of these matters began in 1932 with the work of Tamm [13], who investigated some of the conflicting experimental work on semiconductors and specifically whether surface states (then called Tamm states) might have energies lying in the forbidden gap. Tamm and his coworkers assumed the surface to be electrically neutral unless perturbed by an electrostatic field. Such a field would, for example, involve any gradient used to make measurements. It was also postulated that in germanium, for example, one could expect at least one dangling, or unsatisfied, bond per surface atom, or about 10^{15} surface states per square centimeter. Then, since the unsatisfied bonds could form acceptor states when a field was present, electrons would be trapped and unable to move, so that p-type surface conductivity was predicted. This model assumed a clean surface with no disturbing influences due to the environment or surface contaminants.

Recently, measurements of work function, conduction, and field effect on "clean" surfaces prepared by the Farnsworth technique [14] have been reviewed by Handler [15]. The concentration of Tamm states was estimated to be in the range of 10^{13} to 4×10^{14} per cm^2. In addition, the strongly p-type nature of surface conductivity on an n-type sample and apparent lack of dependence of the work function upon bulk impurity level (that is, position of the Fermi level in the forbidden gap) all lent some support to these conclusions. In other areas, notably the interaction of the clean surface with an oxygen environment, the proposed interpretation was questioned. Furthermore, the realization of an atomically clean semiconductor surface without strains, defects, or other distorting influences is a most difficult task; and even if a theory were in good agreement for the perfect surface, it would be of little practical value until extended to real surfaces and phase boundaries.

For real surfaces the effect and importance of surface states was realized by Shockley and Pearson [16] early in the history of transistors (1948). When these workers tried to modulate the conductivity of a germanium rod by means of an induced field, the observed effect was much too small to be consistent with theory, and it was postulated that at least some of the induced charges were trapped in surface states and thereby rendered immobile and unable to carry a current. Other observations, such as the failure of the rectification series experiment and the apparent constancy of the work function of the semiconductor despite changes in doping level or type, gave impetus to a more general theoretical approach [17].

The explanation offered by Bardeen [18,19] and since amplified and broadened by many other workers is based on the concept of a spontaneous double layer at the surface. This arises in part from the existence of unsatisfied bonds but is usually modified by the presence of surface contamination, such as the oxide layer which is found on all samples of silicon or germanium under normal circumstances. In the Bardeen concept, electrons (or holes) from regions near the surface are attracted to acceptor (or donor) levels, forming immobile negative (or positive) ions. The region just beneath the surface skin then acquires a space charge resulting from repulsion of free carriers away from the charged skin or ion layer. These double layers are also called inversion layers if there is a change of type in the space-charge region compared to the bulk semiconductor, and they are regarded as normal properties of all surfaces including the clean surface. In the latter case, since a double layer is assumed to be present, the situation is different from that proposed by Tamm, who regarded the surface as neutral until it came under the influence of an external field or another solid phase at a contact.

To understand the implications of the double-layer or surface barrier hypothesis of Bardeen and coworkers, it must be pointed out that the surface properties of a semiconductor may be influenced by a variety of factors such as surface oxide, adsorbed moisture and other materials from the atmosphere which alter the properties of the oxide, lattice imperfections, diffused impurities in either semiconductor or oxide layer, and dipole effects. Any or all of these may, in general, contribute surface states resulting in an immobile charge layer. The surface barrier height is then the net algebraic sum of these positive and negative potentials, but it cannot be separated into its individual components. It is impossible to predict the manner in which such an array of surface states will be modified by an external field or contact with another solid.

It also follows, therefore, that the surface barrier may be zero (owing to opposing potential effects) without revealing any useful information about the surface charge layer composition. Finally, as a result of double-layer formation, the surface of any semiconductor will be more or less conducting than the bulk, depending on the number and energy levels of the carriers forming the space-charge region.

A qualitative picture of the effect of such double layers may be seen from Fig. 17-9, in which four extreme situations are compared. For an n-type sample (a) where the electrons are removed from the surface (by ion formation due to surface states acting like traps, or acceptors, or by an external field) there is a p-type region, or inversion layer, beneath the surface, as may be seen from the distortion of the bands and the Fermi level position. For a sample which is p-type (c), under the same influence as (a), there is no inversion, but there is a strong increase in p-type conductivity at the surface owing to the migration of holes to neutralize the negative charge. In examples (b) and (d) we find the opposite effects, resulting in inversion at (b) and increased n-type conductivity at (d).

Finally, the concentration or density of surface states as proposed by Bardeen may be as high as 10^{12} per cm^2 [21] (corresponding to the degenerate doping level of about 10^{18} atoms per cm³), with the result that they can screen work function variations due to shifts of the Fermi level as a function of doping. In much the same manner, the existence of a double layer or spontaneous surface barrier can mask the effects of different bulk material properties, such as chemical potential and work function. This is reflected in the insensitivity of the Volta potential. Calculated

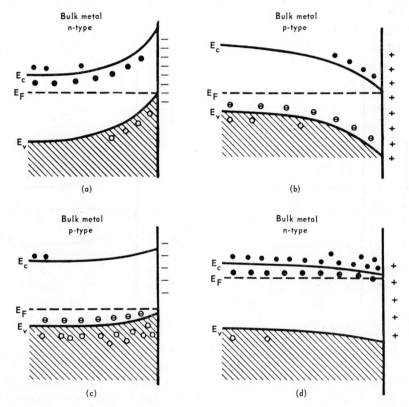

FIG. 17-9. Double layers at semiconductor surfaces. (*After L. B. Valdes, by permission* [20].)

values for the surface barrier height in low-resistivity (0.65-ohm-cm) n-type germanium reach values of 0.36 ev, or about one-half the forbidden gap width [22].

17-6. METAL–SEMICONDUCTOR CONTACTS

17-6.1. Equilibrium Conditions. The effects of surface states and double layers on real heterocontact systems are imperfectly understood. For example, no simple interpretation has yet appeared for structures which include oxide layers or which involve local variations in impurity concentration on the semiconductor side in the vicinity of the interface. Many commercial devices are therefore dependent on empirical techniques, particularly rectifiers of the copper oxide type. Even in the area of simple nonrectifying contacts, which are of prime importance in commercial production of devices, the behavior of a particular combination of metal and impurity semiconductor can be predicted only if the fabrication process is able to duplicate the assumptions involved in the theoretical treatment. In most cases, the techniques finally selected are based on a mixture of theory and experience and reflect a lack of

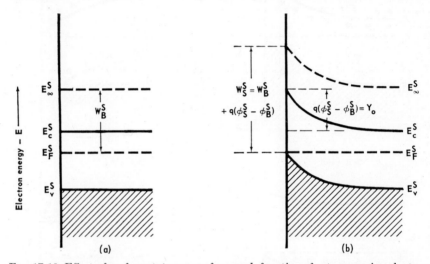

FIG. 17-10. Effect of surface states on surface work function of n-type semiconductor.

understanding in the overall area. In the following paragraphs, a first-order approximation of the metal-semiconductor contact in the presence of surface states will be compared with the model of the preceding sections.

Since the controlling factor will be barrier height and since no method is now available to evaluate the individual contributions of such things as oxide, moisture, and polarizable adsorbed layers, we shall limit the surface effects to a simple double layer of charge (due to carriers trapped in surface states) and a space-charge region in equilibrium. We further require that there be no foreign materials on the surface and that the impurity concentration be uniform throughout the semiconductor near the contact.

For an n-type specimen, Fig. 17-10 compares the band structure at the surface with and without surface states. In Fig. 17-10a we have the same band structure as that of the sample in Fig. 17-5a, while in (b) the effect of surface states on the surface work function W_S^S is shown.

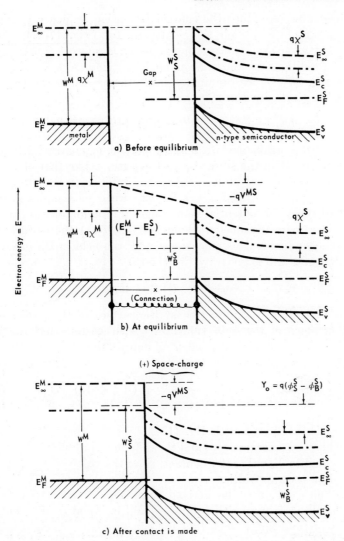

FIG. 17-11. Metal–n-type-semiconductor contact with surface states.

The spontaneous double-layer field has increased the inner potential by an amount $(\phi_S^S - \phi_B^S)$ and produced a similar change in the work function as compared to the bulk value W_B^S. Note that the situation in Fig. 17-10b is similar to that of Fig. 17-9a, where an inversion layer is present. At the moment, except to recall that inversion layers correspond to high surface fields, the question of inversion is secondary, since our primary concern is the height of the surface barrier given by $q(\phi_S^S - \phi_B^S)$ and designated as Y_o.

We may now reproduce the contact formation sequence of Fig. 17-5 but with surface states on the semiconductor (Fig. 17-11). For the sake of clarity, we shall choose a metal with a large work function; otherwise, the conditions are the same as in the previous series.

From Fig. 17-11b we obtain by inspection a value for $(E_L^M - E_L^S)$ which is identical with Eq. (17-12):

$$(E_L^M - E_L^S) = (W^M - W_B^S) + q(\chi^S - \chi^M)$$

where all values again refer to bulk properties, or characteristics of the material rather than the surface. In this case, unlike that of Fig. 17-5, the presence of surface states screens the interior from external fields (such as the potential drop across the gap with the samples in equilibrium), and the Volta potential is no longer a function of the gap width. For the ideal case (assumed to hold in Fig. 17-11) the surface barrier on the semiconductor does not change with the field in the gap, so that V^{MS} is constant and is given by Eq. (17-15)

$$-qV^{MS} = (W^M - W_S^S)$$

where the effective work function at the surface of the semiconductor is used in place of the bulk value. We also assume that the barrier height Y_o (Fig. 17-10) is independent of gap width. We may then write

$$W_S^S = W_B^S + Y_o \quad (17\text{-}21)$$

and $\quad -qV^{MS} = W^M - (W_B^S + Y_o) \quad (17\text{-}22)$

or $\quad (W^M - W_B^S) = -qV^{MS} + Y_o \quad (17\text{-}23)$

This may be compared with Eq. (17-20) for the case without surface states, where V^{MS} became zero and the difference in bulk work functions became equal to Y_o only after contact, since the value of Y_o in that case depended on the gap width.

A further result may be derived from Fig. 17-11. As before, the barrier height seen from either phase should be equal to W^M, but again the effect of tunneling is such that the effective barrier height is simply Y_o, which is a characteristic only of the surface barrier layer of the semiconductor and is independent of both bulk work functions. This explains, in a simple manner, the failure of the rectification series experiment, where surface states were not allowed for in the underlying theory.

17-6.2. The Effect of Bias. It is now possible to make a few general statements concerning the rectifying or nonrectifying properties of metal-semiconductor contacts with surface states (Fig. 17-11c). An applied voltage will displace the Fermi levels on opposite sides of the interface, and this will appear as an increment in the surface barrier height Y_o. A simple explanation is that when a current flows as the result of applied bias, the largest voltage drop will appear across the region of highest resistivity, which in this case is the space-charge region of the semiconductor. A simplified sketch of the equilibrium and biased cases is given in Fig. 17-12. Note that the barrier, as seen from the metal, is independent of bias but that the barrier as seen from the semiconductor is proportional to the bias voltage V_R. (This behavior was also true for the

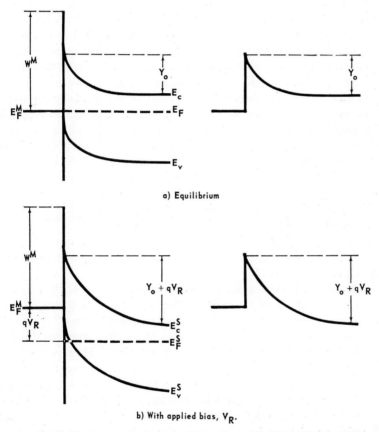

FIG. 17-12. Metal-semiconductor contact at equilibrium and under bias.

contact without surface states, Fig. 17-5, but the barrier height was dependent on the work functions in that case.)

The current under forward bias (from the metal to the semiconductor in Fig. 17-11) will be limited by Y_o at low bias and by body resistance thereafter. In the reverse direction, the current will be limited by $Y_o + qV_R$, diminishing as V_R increases and saturating at a relatively low value of V_R. This is exactly the behavior to be expected from a junction rectifier, and, in some cases, the rectifier equation may be used to describe the behavior of a rectifying metal-semiconductor contact with surface states.*

A further analogy may be recognized if we consider the possibilities of forward injection, as in conductivity-modulated rectifiers described in Sec. 6-8. The metal is a ready source of electrons, and the space-charge region may be considered as the high-resistivity or blocking side of the rectifier. According to this model, an injecting metal contact is readily understood and would be rectifying in the reverse direction.

Perhaps the more important consideration, however, is the need for nonrectifying, low-resistance contacts for device purposes. Based on the present model, a double layer will always exist, giving rise to a surface barrier, which would tend to impart a rectifying characteristic. If the barrier were eliminated, there should be no tendency for the contact to exhibit resistance in either direction. The general approach to this problem is to use a solder or welding material which contains an impurity which will impart the same type of conductivity (that is, n- or p-type) as is already present in the semiconductor. The effect of this, after the impurity diffuses into the semiconductor surface, is to destroy the rectifying property of the contact by creating an impurity gradient which opposes the field set up by the double layer. This is frequently accompanied by pretreatment of the semiconductor at the desired point of contact by adding a high local concentration of impurities, which narrows the space-charge region until it is small compared to the mean free path of an electron.

17-6.3. Impurity Variations in the Semiconductor.

There are several ways in which the contact-forming properties of a semiconductor should, at first glance, be influenced by its impurity concentration (type, position of the Fermi level, etc.). In the junction these effects are readily understood and yield to quantitative interpretation. However, although the work function of a clean semiconductor was expected to be directly dependent on the Fermi energy level, so that a different value was predicted for n- versus p-type material, this has not been observed. According to the surface-state hypothesis of Bardeen, such changes in impurity level (with or without junction formation) also produce changes in the $\phi_S^S - \phi_B^S$ term which compensate for the change in Fermi energy level. Furthermore, when there is surface contamination in addition (as there is except under the most rigorous techniques of ultra high vacuum), the surface barrier value may vary, depending on the contaminants, over a

---

* This was also found to be true in the hypothetical case of contacts without surface states (see footnote, Sec. 17-3.1). Neither theoretical model is able to predict the observed behavior of many practical systems. However, the surface-state representation, with barrier height dependent on Y_o rather than bulk work functions, has proved more powerful in interpreting many surface phenomena.

range at least as large as the Fermi level variation from the intrinsic or midgap position to the onset of degeneracy.

It is generally assumed, therefore, that contact and surface phenomena are nearly independent of the impurity concentration over the range from intrinsic to near-degeneracy (about 10^{17} impurities per cm^3). From a practical point of view, however, control of contact properties for high-purity semiconductors is much more difficult than for those which have been highly doped—perhaps because the semiconductor approaches the behavior of a metal more closely as the carrier concentration increases. It is likely, although not certain, that the most important change with high doping levels is one which affects the surface and the subsurface space-charge region, rather than any change in the Fermi energy level.

REFERENCES

1. A recent review article by A. R. Plummer, The Semiconductor-Gas and Semiconductor-Metal Systems, in P. J. Holmes (ed.), "The Electrochemistry of Semiconductors," pp. 61–140, Academic Press Inc., New York, 1962, contains an excellent bibliography. See also Ref. 11.
2. C. Kittel: "Introduction to Solid State Physics," 2d ed., p. 266, John Wiley & Sons, Inc., New York, 1958.
3. A. R. Plummer: The Semiconductor-Gas and Semiconductor-Metal Systems, in P. J. Holmes, (ed.), "The Electrochemistry of Semiconductors," p. 82, Academic Press Inc., New York, 1962.
4. Ref. 2, p. 239.
5. L. B. Valdes: "The Physical Theory of Transistors," p. 239, McGraw-Hill Book Company, New York, 1961.
6. E. Spenke: "Electronic Semiconductors," p. 337, McGraw-Hill Book Company, New York, 1958.
7. "Handbook of Chemistry and Physics," 36th ed., pp. 2342–2344, Chemical Rubber Publishing Company, Cleveland, 1954.
8. Ref. 3, p. 100.
9. D. K. C. MacDonald: "Thermoelectricity: An Introduction to the Principles," p. 44, John Wiley & Sons, Inc., New York, 1962.
10. Ref. 3, pp. 101–104; 122–129.
11. Ref. 6, pp. 336–372.
12. Ref. 3, pp. 72–84.
13. I. Tamm: Über eine mögliche Art der Elektronenbindung an Kristalloberflächen, *Physik. Z. Sowjetunion*, **1**: 733 (1932).
14. R. E. Schlier and H. E. Farnsworth: Low-energy Electron Diffraction Studies of Cleaned and Gas-covered Germanium (100) Surfaces, in R. H. Kingston (ed.), "Semiconductor Surface Physics," pp. 3–22, University of Pennsylvania Press, Philadelphia, 1957.
15. P. Handler: Electrical Properties of a Clean Germanium Surface, in R. H. Kingston (ed.), "Semiconductor Surface Physics," pp. 23–51, University of Pennsylvania Press, Philadelphia, 1957.
16. W. Shockley and G. L. Pearson: Modulation of Conduction of Thin Films of Semiconductors by Surface Charges, *Phys. Rev.*, **74**: 232 (1948).
17. W. E. Meyerhof: Contact Potential Difference in Silicon Crystal Rectifiers, *Phys. Rev.*, **71**: 727 (1947).

18. J. Bardeen: Surface States and Rectification at a Metal-Semiconductor Contact, *Phys. Rev.*, **71**: 717 (1947).
19. W. H. Brattain and J. Bardeen: Surface Properties of Germanium, *Bell System Tech. J.*, **32**: 1 (1953).
20. Ref. 5, p. 241.
21. J. Bardeen, R. E. Coovert, S. R. Morrison, J. R. Schrieffer, and R. Sun: Surface Conductance and the Field Effect on Germanium, *Phys. Rev.*, **104**: 47 (1956).
22. E. O. Johnson: Simplified Treatment of Electric Charge Relations at Semiconductor Surfaces, *RCA Rev.*, **27**: 525 (1957).

18: Point-contact Devices

18-1. INTRODUCTION

We have deferred consideration of point-contact devices, despite their importance as the first successful devices, for two reasons. First, the theory underlying their function is imperfectly understood even after almost a century (Sec. 1-2), and second, they involve active metal-semiconductor contacts of a highly specialized nature. It is therefore logical to discuss them at this point after junction devices have been explored and after some of the problems with heterojunctions have been outlined.

Despite some very considerable disadvantages, point-contact diode devices are still actively marketed. The manufacturing process is deceptively simple, but since much of it involves the empirical know-how of the fabricator, the true variables are almost impossible to isolate or study. The cost of these units is quite low, and their frequency response is excellent; but except for high α in the transistor, the other properties are generally inferior to those of junction configurations. We shall look more closely at the exact properties when specific examples are discussed. It may be mentioned here, however, that although the very nature of these units limits them to small power capabilities, the concept of small-signal behavior, in the sense of the term when applied to junction devices, is meaningless, since there is no region of operation wherein equilibrium or theoretical performance is observed. Point-contact devices may therefore be described as sharply nonlinear under all operating conditions.

18-2. THE RECTIFIER–DETECTOR

The use of the cat whisker with crystals of galena (PbS) or iron pyrite (FeS_2) was one of the earliest applications of semiconductors in a metal-semiconductor rectifier combination. In early radio terminology they were known as "detectors," and they served to detect the audio-frequency modulation of the high-frequency carrier by performing a half-wave rectification [1,2]. Since the audio part of the circuit could not react to the higher frequency, the modulating signal was thereby made audible, after amplification. In practice, it was often necessary to readjust the contact on the crystal until a new area of maximum reception (rectification efficiency) was found. Since the crystal was a polycrystalline material and since its surface condition was not under any kind of control, it is not surprising that local effects could be important. Factors such as surface impurities (water, oxygen, etc.), local heating, vibration, and changes at the other contact all were uncontrolled and not appreciated. It is small wonder, therefore, that despite the improvements of the

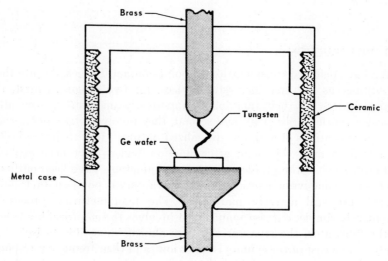

FIG. 18-1. Schematic view of point-contact diode.

period, the much more reliable and reproducible vacuum tubes became a universal replacement as soon as they were available.

There was, however, one area in which the point-contact device excelled. The frequency response was very high, in the kilomegacycle range, a region in which conventional tubes were almost useless. Therefore, with the advent of radar, at microwave frequencies, the point-contact configuration was revived and the modern semiconductor diode was created [3,4]. Many problems were solved. The semiconductor became single-crystalline germanium or silicon, the surfaces were rigorously cleaned so that the back contact was stable and of low resistance, while the point contact could be made to almost any part of the exposed face without loss of rectification efficiency. Finally, once the contact was made and tested, the point (now of tungsten for better hardness and stability) was permanently fixed in place for the life of the unit. Cooling was greatly improved, and the effects of stray heating were minimized. The typical structure represented in Fig. 18-1 has a germanium element. Case materials could be all glass, meta insulated as shown, or all metal, depending on the frequency and application. In addition, the free space in the case was often filled with a waxy substance known as potting compound to seal out moisture, air, etc.

The peculiarity of the point contact lies in its very good rectification ratio and, because of the small contact area, very low capacitance, resulting in good high-frequency response. The conductivity mechanism is controlled by the spreading resistance which is observed when a small point contacts a wafer of a thickness at least 10 times the contact diameter. It is then found that, owing to the hemispherical equipotential lines radiating outward from the point and constantly increasing the area of conduction in proportion to the square of the distance, the resistance is independent of the size and shape of the wafer (if not too thin) and is controlled exactly by the volume of the point-contact region. It may be shown, for material of resistivity ρ, that the resistance r to forward flow of current at the contact is given by

$$r = \frac{\rho}{2d} \tag{18-1}$$

where d is the diameter of the point. As d becomes small, r increases inversely. The reverse flow of current, on the other hand, is proportional to the area of contact, or $1/d^2$. Therefore, by making the contact smaller, the rectification ratio is improved, even though the forward drop rises. Also, a smaller contact results in less capacitance and therefore a higher frequency capability. The contact diameter is chosen as a compromise, being as small as possible (frequency, rectification efficiency) and yet large enough to provide reasonable forward voltage drops. As might be expected from these considerations, the current-carrying capacity is sharply decreased.

Typical commercial diodes of either germanium or silicon are used for operation at frequencies beyond 50 kmc and show excellent switching properties, but for very small signals. In the audio range, germanium units have shown PIV values as high as 150 volts.

18-3. THE BONDED RECTIFIER

It was mentioned earlier that these units are most likely to be regarded as junction devices as a consequence of the method of manufacture but that they have the properties of point-contact devices. Bonded rectifiers are constructed in much the same way as point-contact diodes are, except that the point is a metal or alloy of conductivity type opposite to that of the basic wafer. Such diodes are made of both silicon and germanium with points or wires of aluminum or gold-gallium alloy (for n-type substrates). After the unit is assembled, heat or a current pulse is used to alloy or bond the wire to the substrate, producing a small area of p-type silicon beneath the contact. Figure 18-2 shows a schematic of such a unit in which the size of the p-type alloy region is greatly exaggerated for the sake of clarity. This type of connection has been discussed before with reference to mesa transistors and other structures requiring small emitter or base regions. Here, however, the technique is applied to a diode, the spreading resistance formula holds, the contact-forming techniques are similar, and the properties are similar to those of a point-contact device. Wire diameters are as small as possible; the wafer also is

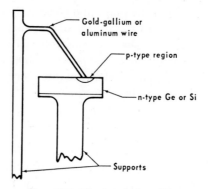

FIG. 18-2. The bonded rectifier.

small; and the emphasis is directed toward the small-signal, low-capacitance range.

The chief advantage over the point-contact device lies in the existence of a true p-n junction of very small dimensions. Units made of silicon with aluminum wire show rectification ratios of 10^7 to 10^8 with PIV values of about 100 volts. Reverse currents at 1 volt are as low as 0.5 na with corresponding forward currents of 50 ma. Their switching capability is therefore most attractive, and high-frequency performance is good [5].

In another version a gold-gallium wire is bonded to a germanium wafer. (The presence of gold is frequently an asset to high-speed operation, if the geometry is optimized, since it tends to promote recombination of carriers and avoids storage and delay effects.) In these units the emphasis is on speed, in the high kilomegacycle range. PIV ratings of 100 volts can be achieved, but at some sacrifice in speed. Once again, these devices are capable of very small power handling (of the order of 0.05 to 20 mw) [6].

18-4. THE TYPE A TRANSISTOR [7]

This device deserves a special mention, since the original disclosure of this transistor by Bardeen and Brattain in 1948 may truly be regarded as the beginning of the present era of semiconductor devices. Though rather crude and poorly understood, it was the basic model for future devices, including those of commercial significance for several years. By demonstrating the basic principle of transistor action, it led to the announcement two years later of Shockley's junction transistor. With the advent of the latter configuration, the point-contact transistor has become almost extinct, although certain types may still be obtained for replacement service.

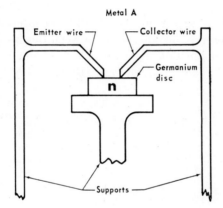

FIG. 18-3. The point-contact transistor configuration.

The general arrangement of the elements of a point-contact transistor is shown in Fig. 18-3. Although there were literally dozens of variations in structure, the essential character did not vary. The germanium wafer, usually n-type, was fastened to a base by means of a nonrectifying, low-resistance solder joint. The two pointed wire elements were then placed on the wafer as close to each other as practical, usually with a contact-to-contact spacing of less than 0.1 mm if possible. The wires themselves were very fine, ranging up to 0.08 mm in diameter. The amount of pressure on the wires and the surface condition of the germanium were empirically fixed. After assembly, the contacts underwent a "forming" process during which high-current pulses were passed through the contact points from base to collector (reverse) for periods of some seconds or fractions of a second to improve the α of the structure. Similar treatments were given the emitter point. Correlations of results, in terms of current density, polarity, duration, wire material, semiconductor type, surface preparation, method of forming and cleaning the points, pressure on the points before and during forming, and the ambient atmosphere (that is, "forming gas"—an arbitrary mixture of nitrogen or argon with hydrogen) were never satisfactory. By control of as many variables as possible and by careful techniques designed to exactly replicate from run to run (as a means of controlling unknown variables) it was possible to produce commercial units [8–10].

Theoretically, these effects have never been systematically explained. A further problem arises if we consider the composition of the wires, or points, before forming. In the type A transistor, both wires were of phosphor bronze, presumably n-type, as was the germanium wafer. In later versions the emitter became a p-type alloy. Other improvements in technique fo lowed. A brief review of the theory will be presented in the next section.

Before leaving the type A device, it will be of interest to tabulate the parameters and observe the differences between this historic ancestor and the typical modern junction device of which so much has been said in earlier chapters [cf. Table 9-2 and Eqs. (10-5) to (10-8)]. In Table 18-1

TABLE 18-1. POINT-CONTACT TRANSISTOR PARAMETERS*

r Parameter	Internal Resistance
r_{11} = 530 ohms	r_e = 240 ohms
r_{12} = 290 ohms	r_b = 290 ohms
r_{21} = 34,000 ohms	r_c = 19,000 ohms
r_{22} = 19,000 ohms	r_m = 34,000 ohms

$$\alpha = 1.8$$

* After W. Shockley, by permission [11].

are the r parameters for a typical 1948 unit, as well as the internal resistances. The tabulated values are for a grounded-base configuration with R_G = 500 ohms, R_L = 20,000 ohms, and a total power output of 5 mw. It is clear that the output impedance is low, with about the same input impedance, and that the power capability is very poor in comparison to a junction device. On the other hand, α is greater than unity, the operating power gain is 47 (17 db), and the frequency response is in the 3- to 5-mc range, despite high noise levels which were such as to make a determination of the cutoff frequency somewhat doubtful in the early units.

In retrospect, after years of intensive technological advance and the substitution of an entirely different physical structure for which we have a reasonably well understood theory (which might or might not apply to the type A transistor if we knew enough about it), it is remarkable to compare the values in Table 18-1 with those of more current interest. Despite its mechanical shortcomings, the transistor, as exemplified by the device of Bardeen and Brattain, was launched into semiconductor science as a remarkably well developed prototype, quite capable of setting off the electronic revolution which followed. Through a happy coincidence of chance and inspiration, the original unit possessed properties of great promise. Indeed, some of these, such as the speed of response and high α, have been rather painfully acquired in the junction types, despite their flexibility.

18-5. COMMERCIAL TRANSISTORS

No truly defined line can be drawn between the type A and so-called commercial point-contact transistors. The process of the next ten years was one of empirical optimization which produced a brief period of wide acceptance, only to lose out as the economics and reproducibility of the junction devices took over. The attempts to improve the techniques to match the performance of other types were very largely empirical, and many evolutionary improvements were of benefit to both. Today, the point-contact transistor has outlived its usefulness, and it would be regarded as completely obsolete if it were not for replacement needs in elaborate circuits which are still functioning and which cannot be adapted to the newer units on an economical basis.

Before looking at the theory of these remarkable triodes, it will be instructive to enumerate briefly the ultimate advantages and disadvantages which, taken in a total evaluation, caused junction units to supplant the point-contact transistor. Considering first the electrical parameters in their simplest form, we would find that, although the input resistance could be lowered, it was never possible to achieve the high r_{22} value of the junction unit [12]. As a practical consequence, the reverse saturation current I_{CBO} was at best about 150 μa compared with 5 μa for a similar power rating and reverse voltage. This was a serious drawback, since the PIV values for the point-contact collector were not too far from those of the junction units (50 to 100 volts) [10,12]. In frequency response, the junction devices, such as mesa transistors and others described herein, were eventually upgraded to meet an α cutoff in the low megacycle range. More important, however, was the noise level of the point-contact structure, which was never satisfactorily reduced [13]. Inability to handle higher power levels was another limitation of the point-contact device, but it was not a decisive drawback, since there is a large area of very low power applications where speed and low capacitance are needed and where the two physical types met on a common basis. Indeed, the large α of the point-contact type (until the advent of the three-junction types and a decrease in their cost due to mass production) gave it a very real advantage. Values of α were 3 or greater in the desired frequency range, and in this regard were clearly better than for the junction device [8].

More important were two other characteristics of the point-contact device which could not be eliminated and which were absent from the junction version. The nonlinearity of the type A transistor was recognized at an early date, and much was attempted to improve it [14]. However, as an example, Fig. 18-4 represents a "good" characteristic for a point-contact device. Note that the large value of α may be inferred from the increasing values of I_C for a fixed I_E as V_C is raised. The low output impedance (or collector resistance) is evident from the appreciable

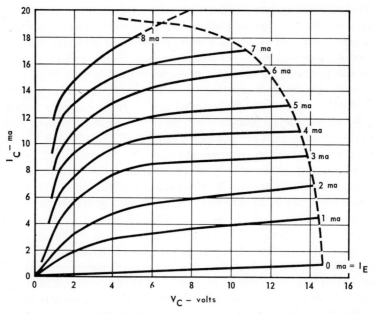

FIG. 18-4. Current-voltage collector characteristic of point-contact transistor. (*After L. B. Valdes, by permission* [14].)

slope of the curves. Finally, the curves do not extend to zero for $V_C = 0$. A definite voltage is required to start the conduction, owing to drift field effects (see below). This means, for switching, that the saturation region will give discontinuous currents as a function of V_C or I_E. In addition to the nonlinearity problem was that of reproducibility from unit to unit [12,13]. It is very likely that, in the absence of a compelling economic motive, circuit designers would have learned to live with characteristics like those of Fig. 18-4. Unfortunately, it was not possible to design to normal tolerances because of the nonreproducible behavior of the units themselves. Even where the manufacturer did a very careful testing, sorting, and rejecting job, the shelf life effects and in-service aging, as well as temperature sensitivity [15], were unpredictable and not at all understood by either the producer or the consumer. Taken together, these factors spurred the development of junction techniques, although if any single cause could be said to have been most compelling, opinion would select nonuniformity from unit to unit.

The theory of the point-contact transistor is made very complicated by lack of fundamental knowledge concerning the area of the contact and the few microns of original semiconductor surrounding it after the forming process. In cases where the emitter wire was of a different conductivity type than the wafer, a p-n junction is plausible. At the collector, however, and indeed at the emitter if an n-type wire is used for both, we are forced to contemplate some form of n-n^+ junction. At the emitter it injects, as we have seen for the metal-semiconductor case, while at the collector it seems to both inject and collect if we are to account for the increase in α beyond unity. (This may be somewhat similar to the behavior of the metal-semiconductor p-p^+ collector junction we have seen in the case of the thyristor, but in the present instance the transition is not as abrupt and it becomes difficult to base a plausible explanation on such an effect.) The simplest, and under the circumstances possibly the best, theory involves the small area of the contact and the marked effect on transistor gain (or α) if the spacing is increased by a small factor. For example, if the intercontact spacing is increased by a factor of 5 from 0.05 mm (truly a minimum) to 0.25 mm, the gain falls by a factor of 100, almost a third-power dependence on the distance. All of this suggests a field effect, particularly since the field at a very small contact is expected to be very high in the local region [7].

It is therefore postulated that the emitter serves its normal purpose and emits holes into the n-type substrate. This is accomplished, despite the n-n^+ configuration of the "junction" at the emitter, because it is forward-biased (like a p-n-p emitter) or made positive with respect to the base. Owing to the intense field at the point, the minority carriers (holes) are forced into the n-type region against a concentration gradient. They thus behave like a drift current rather than the diffusion current we are used to. Once they are in the substrate, the intense field of the reverse-biased collector pulls them to it by drift effect, and they are there collected. At the same time, however, the n-n^+ junction at the collector, now biased in favor of electron injection, also injects electrons, some of which are involved in hole annihilation but many of which flow to the base, thereby increasing I_C over I_E and creating an α greater than unity [14]. (Just why the emitted electrons at the collector are not in turn collected at the nearby positively biased emitter is not clear, although there is a tremendous difference in field strengths if the relative bias potentials are examined—the difference is perhaps 100 to 1.) Finally, it is obvious, from the experimental data that the emitter current I_E

governs not only its own contribution to I_C but also the extent to which I_C is increased by electron flow to the base. The situation may be represented as in Fig. 18-5 if the above gross mechanism is correct.

Obviously, this is without quantitative meaning. If we knew more of the nature of the formed contacts (and indeed, more about metal-semiconductor contacts in general) it would be a start toward a more accurate interpretation. Unfortunately, the very small dimensions involved, the very high field intensities, and the surface problem have thwarted experimental investigation. We must therefore be content to suggest a mechanism in fairly simple terms and to reason as best we can by analogy

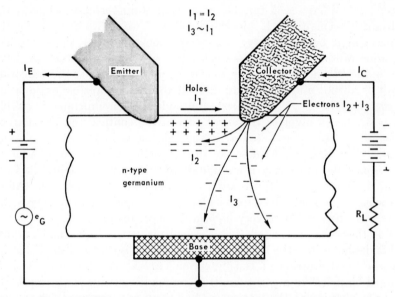

FIG. 18-5. Electron-hole flow scheme for point-contact transistor.

to the junction behavior, which is now familiar. There is at least some justification for such an approach, since the outward behavior of the point-contact transistor is almost startlingly similar to that of the junction variety, and the small-signal equivalent-circuit analysis of Chaps. 10 and 11 may be used for most point-contact devices in regions of stable operation within the d-c and low-frequency range. Unlike junction devices, instability for values of α greater than unity is a serious problem in designing circuits for use with point-contact transistors.

REFERENCES

1. A. Coblenz and H. L. Owens: "Transistors: Theory and Applications," p. 3, McGraw-Hill Book Company, New York, 1955.

2. H. C. Torrey and C. A. Whitmer: "Crystal Rectifiers," pp. 5, 6, McGraw-Hill Book Company, New York, 1948.
3. G. W. A. Dummer and J. W. Granville: "Miniature and Microminiature Electronics," p. 39, John Wiley & Sons, Inc., New York, 1961.
4. Ref. 2, pp. 7–11.
5. G. L. Pearson and B. Sawyer: Silicon p-n Junction Alloy Diodes, *Proc. IRE*, **40**: 1348 (1952).
6. Ref. 3, p. 42.
7. W. Shockley: "Electrons and Holes in Semiconductors," pp. 34–53, 101–108, D. Van Nostrand Company, Inc., Princeton, N.J., 1950.
8. Ref. 7, pp. 108–116.
9. Ref. 1, pp. 212–215.
10. Ref. 3, p. 59.
11. Ref. 7, p. 41.
12. J. A. Morton: Present Status of Transistor Development, *Proc. IRE*, **40**: 1314 (1952).
13. R. D. Middlebrook: "Introduction to Junction Transistor Theory," p. 8, John Wiley & Sons, Inc., New York, 1957.
14. L. B. Valdes: "The Physical Theory of Transistors," pp. 350–352, McGraw-Hill Book Company, New York, 1961.
15. J. R. Tillman and F. F. Roberts: "An Introduction to the Theory and Practice of Transistors," p. 197, John Wiley & Sons, Inc., New York, 1961.

19: Device and System Reliability*

19-1. INTRODUCTION [2]

Despite the impressive record of reliability, virtually unlimited life (in excess of 10^5 hr), and a very high operating stability which have materially added to the impact of semiconductor devices in the electronics industry, a surprisingly intensive effort is still devoted to improving device and system reliability. There are three basic reasons for this. First, the reliability demonstrated in specific areas by selected devices has led the industry to expect the same or better reliability for all other types. Second, environmental and service conditions have become increasingly severe, and finally, the consequences of a single failure, even though it might fall well within acceptable limits in ordinary service, have become intolerable.

In applications such as satellites, undersea cable amplifiers, and weapons-guidance systems there is an increasing need for new, high-performance device systems designed to have a zero failure rate, even under these extreme service conditions. It is therefore not surprising that the scope of reliability test programs has continued to enlarge and that there is a persistent gap between the desired performance and the technological ability to meet the demand.

The factors which comprise a comparative reliability determination in areas involving semiconductor devices are now broad and complex. The individual component must not only meet the demands of performance, service life, and resistance to environmental influences but also be compatible with such other factors as circuit elements to which it is connected, assembly procedures, testing methods both as a discrete device and as a part of the overall circuit, storage time, and mechanical handling.

* The parts of this chapter dealing with device reliability, including data and illustrations, are adapted from the General Electric Company's "Transistor Manual," by permission [1].

For the purpose of this discussion, we shall arbitrarily distinguish between two areas of reliability evaluation. In the first we shall deal with the isolated device subject only to the testing and evaluating normally performed by the manufacturer. The end result of these operations is a device which meets the standards of the supplier and is delivered to the user with data supporting the specifications. The second phase will concern such a device from the moment it is first handled outside its original package by the user and will include testing or screening, assembly, mechanical handling, and an entire spectrum of electronic circumstances outside the limits of predelivery evaluation. This latter problem is more generally a system reliability evaluation, and it depends on the various factors which influence the device-system interaction.

To draw a clear line between the device and system reliability concepts, we begin by clarifying the term "semiconductor device" as it is used in this discussion. By analogy to the vacuum tube, and because the condition of the unit at the point of sale should represent a reproducible situation, we elect to define a semiconductor device as a unit in its as-delivered condition, fresh from the manufacturer. This will mean, in general, that the unit will comprise an enclosure or can of glass, plastic, ceramic, or metal, varying in size from slightly larger in diameter than the wire leads to large structures containing rectifying wafers more than an inch in diameter. (These are stacked in series, in some cases, for high-voltage performance, but the external holder is merely mechanical.) The enclosures may contain one or more junctions (in some cases several interconnected junction structures in one or several chips of silicon) and will be provided with two to eight or more leads. The metallic enclosure is frequently a lead, particularly with rectifiers, where heat removal makes this type of construction desirable.

Whatever its nature, the device cannot be subdivided (as by disconnecting external joints) without destroying the integrity of the device closure as made by the manufacturer. It constitutes, in short, a complete circuit element (just as a vacuum tube does), carries its unique terminology, and is sold as a standard item of electronic hardware. It has been tested and found to comply with the rating data supplied by the manufacturer, and the test methods used were carefully selected to produce no permanent or lingering deleterious effects. At this point, the semiconductor device is potentially at the peak of its condition, backed by the manufacturer's experience and evaluation.

19-2. RELIABILITY AND PRODUCTION

Planning for device reliability must begin at the design stage. Proper design is a combination of theory, experience, and empiricism. A theoretically perfect design may not be possible in actual practice under the current state of the art; conversely, experience may dictate that best results are obtained by the deliberate selection of a design which is not fully understood, as in the case of the metal-semiconductor injection contact on which the commercial success of the thyristor is based. The ultimate design may be arrived at by trial-and-error methods, particularly in a new device. For example, an ideal transistor structure might require base region thicknesses of the order of angstrom units. Yet, from experience, we know that the practical problems are not under control, and a compromise must be effected. Similarly, if a unit has a new configuration or involves a different lead attachment technique, the final selection may depend on the observed behavior of the structure as a whole or with reference to the methods of lead fastening.

Most design parameters are fairly obvious, even if the solution to the problem may not be capable of complete justification on theoretical grounds. No comment is needed with respect to mechanical ruggedness and good ohmic contacts unaffected by normal age and heat. Thermal stability demands matched expansion coefficients, good thermal conductivity from all internal parts to the case, stud, or other heat sink connection, and temperature-resistant hermetic seals for closure and leads.

A much more difficult area is surface protection [3], particularly at surfaces near or overlying a junction. If surface changes occur, the junction may be shorted or caused to leak current at the surface. Two supplementary approaches are used in connection with surface problems. First, the surface is rendered as clean as possible and then deactivated, either by an added layer such as evaporated oxide or by simple chemical passivation. Second, to further protect the sensitive surface, the entire unit is hermetically sealed in an inert-gas atmosphere (rarely a vacuum) plus an adsorbent or chemical getter to remove and retain traces of moisture at all normal operating temperatures. As an ultimate precaution, the entire finished unit is frequently baked after sealing to be sure the getter has absorbed all the gas it can and that equilibrium is now established so that lower-temperature behavior will be stable and independent of environmental changes. Baking times are of the order of hours to weeks.

Design considerations, however, are of no value unless they can be carried out in the manufacturing operation. Obviously, material quality, workmanship, and the manufacturing environment (that is, clean room conditions) are vital, and tooling is important for reproducibility. However, by far the worst hazard is the human being. Many of the most elaborate control systems exist only to render it more difficult for human error to influence the results. A great part of in-process control and inspection is aimed at this one hazard. For example, a skilled worker, without most of the expensive equipment, can produce high-quality devices under adverse conditions because of his basic comprehension of the variables which must be watched. If the same operation is attempted on a production basis with less skilled workers, the production of a single high-quality device becomes a notable rarity.

Despite these numerous variables and unresolved theoretical questions, today's design and manufacturing techniques have progressed to a remarkable state of reproducibility. When one considers the result, in terms of the reliability which is routinely achieved, it emphasizes the magnitude of the technological revolution of the past decade.

19-3. DEVICE TESTING AND FAILURE MECHANISMS [2,4]

To be able to design and produce units of high reliability, we must first be able to evaluate devices in terms of performance. Let us suppose that we have a target reliability of no more than 0.001 per cent failures per 1,000 hr, with failure defined for every parameter by which we characterize normal operation. In other words, we shall test devices under controlled conditions, starting with a representative sample, and we shall determine the number (and per cent) of failures as a function of elapsed time. A failure will be recognized when one or more parameters being monitored reach a degraded value chosen for the test or when some other malfunction of the unit renders it impossible to maintain the original test conditions. We might be concerned with parameters such as power gain, voltage gain, current gain, temperature effects, leakage, vibration, humidity, frequency response, breakdown voltages, and noise levels, to name only a few. All of these are important from a reliability viewpoint, and we need accurate data for each failure mode at the reliability level required.

In obtaining such data, we must recognize that a test period of 1,000 operating hours is 6 weeks at least, and for intermittent service it might be years. If every unit were to be tested, the inventory would become formidable, and many units might be partially degraded by the testing before they could be sold. Second, for a target of 0.001 per cent the sample size becomes [2,5] enormous (nearly 10^6 units), and there is an irreducible error in measurement, testing, handling, etc., which is estimated to be about 0.001 per cent. When all these factors are considered for each parameter to be tested, the cost becomes excessive, the results become unreliable because of accumulative errors, and much useful life of the device is consumed. The obvious recourse is a sampling technique and tests which can be extrapolated.

The most common test program is to determine the suitability of a device for a given application. To do this, test periods of 6 to 12 months are used with fairly small (a few per cent) samples, and sometimes under mildly accelerated conditions. This program is continued through many production cycles, and the results are checked against actual operation in the field. If the tests are suitably chosen, it is found possible to relate the data to the application and to predict a reliability level from the failure rates observed. (Note that there may be other applications, not yet tested for, wherein the same device might have a poor rating.)

The results must be corrected for measurement errors, test equipment malfunction, and accelerated test conditions (that is, stresses higher than those to be expected). From such failure rate data, however, the broad picture of performance builds up, and, provided the testing continues throughout the period of manufacture of the device, serves as an absolute control on the process and also as a means of evaluating reliability. Meaningful figures in the 0.1 to 0.01 per cent per 1,000 hr range are common [6]. This, then, is the true reliability test, and it should be noted that it is tied to the application in the sense that the test parameters are defined and limited in terms of the proposed end use.

Such a program, however, is not too practical for normal production testing, since the delay is frequently prohibitive. For this purpose it is assumed that the production line is stable, and only enough testing is done to show up statistical trends. If, on the basis of a small sample of a lot and an overall test of not more than 1,000 hr, there is no trend away from the norm, the lot may be shipped. If a trend shows up, a more extensive program will be required to evaluate it. If it is only one bad lot, it may relate to an out-of-control process step, and it may be easily corrected. In general, if such a situation arises, shipments are made from stock and no more lots are approved until normal conditions are restored.

All of the above contemplate a truly representative sample. In a later section we shall discuss the occasional freak and the screening of every device which is used to guard against such a rare but highly undesirable unit. Before discussing typical results obtained from comprehensive reliability evaluation studies, it will be instructive to look briefly at some of the causes of failure and the types of test stress which reveal them.

1. Structural flaws are such things as cracked or weak parts, poor connections, and loose particles in the device; they are usually revealed by mechanical tests such as strength, vibration, shock, or pressure. They can obscure results of temperature behavior and, by short circuits or heating, change the electrical characteristics. These are usually errors of design or poor production control, with the latter effect frequently being a random occurrence.

2. Thermal fatigue, or inability of the device to function in a stable manner when subjected to temperature variation, is revealed by temperature cycling and power dissipation effects. These tests will be more specifically described in a later section.

3. Encapsulation flaws result in a nonhermetic seal, permitting impurities (moisture or oxygen) to affect the active element or, in some designs, to disturb the heat-removal mechanism. Various methods of leak detection are applied to the finished device. Some such direct method is almost a necessity to distinguish between a leaking closure and enclosed contaminants in a leak-free enclosure, since these effects are often hard to distinguish by thermal or electrical measurements.

4. Entrapped gaseous contaminants can be simply trapped through failure to displace them or through an overloaded or nonfunctioning getter. The effects are typical of surface contamination, except that there may be an induction period before the gas reaches the surface, owing to grease, oxide, adsorption, etc. Usually high-temperature storage will show this up.

5. Ionized impurities present at the time of encapsulation on the surface act as a short-circuiting mechanism. If the leakage current is measured as a function of temperature, the increase will be proportional to the temperature until recombination of ions, owing to thermal mobility, overtakes the static leakage and the current decreases. For some silicon devices this can be as high as 150 to 200°C.

6. A partially passivated surface may be modified by entrapped moisture and cause surface-state effects, which show up as trapping in the base region. This behavior is, of course, temperature sensitive, and therefore a measurement of h_{fe} as a function of temperature will detect it.

7. Finally, we consider the case of semiconductor material flaws, usually electrical in nature. From the detailed discussion of earlier chapters it is not too difficult to envision the type of failure to be found at junctions made by different techniques or to be ascribed to nonhomogeneous starting material. In general, these failures are detected by high-voltage or high-current operation. The breakdowns are often irreversible. Even if the circuit is limited to prevent catastrophic breakdown, parameters such as h_{fe} and BV_{CEO} will show anomalous behavior. Complete short circuits are not uncommon.

19-4. FAILURE DISTRIBUTION DATA [2]

For each of the above failure mechanisms there will exist a simple relationship between failure rate and time for any commercial transistor. In Fig. 19-1 we have plotted such a probability of failure as a function of time to failure in a generalized form. The probability (or per cent failures per 1,000 hr of test) is plotted against elapsed time, in hours. The graph is divided into regions A to F wherein the failure rate is varying as indicated. The first peak in the failure probability represents "workmanship" failures, or those associated with one or more of the design or manufacturing failure variables previously discussed. Wear-out failures after a long interval are believed to be typical for all semiconductor devices, although few have been tested long enough to reach this region.

Region A, labeled "delay," is due to the definition of failure, which almost always involves a safety factor beyond the initial parameter value. In other words, it may take some time for the parameter under discussion to degrade to the failure point after the start of testing. In region D, the drop-off in workmanship faults is balanced by the increase in wear-out failures. The overlap is indicated by the dotted lines.

Note that this typical curve is for a single parameter and is a generalized case presented to show the normal behavior of transistors and other

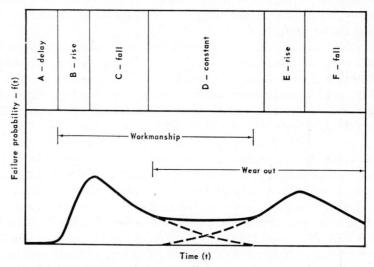

FIG. 19-1. Generalized failure-time relationship. (*After General Electric Company, "Transistor Manual," by permission* [7].)

devices if a large variety of types and test conditions are averaged. Some varieties show deviations from this behavior pattern, but it is a good statistical picture. As concrete examples, we include Figs. 19-2 to 19-6, taken from published data of a leading manufacturer covering commercial device production in the period 1959 to 1961. These examples will illustrate the meaning of failure rate studies, but they should not be taken as typical of the most reliable devices available today.

Figure 19-2 shows the failure rate as a function of time for a p-n-p germanium alloy transistor. The delay lasted beyond the first 125 hr for all 1,200 units, followed by a sharp rise and gradual decrease. After 5,000 hr the failure rate was essentially constant. (Total failures from 3,000 to 25,000 hr were only three, but the sample size was reduced to 400 or fewer units.) In Fig. 19-3 the same unit is shown but with the

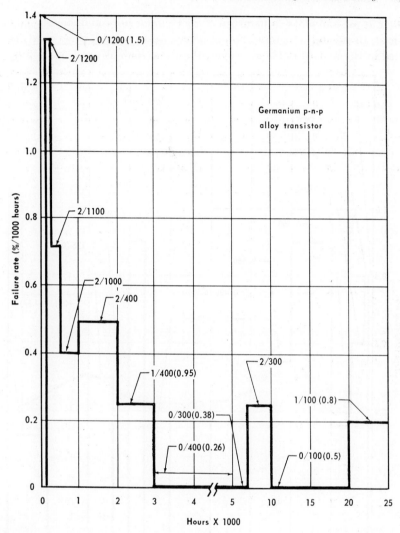

FIG. 19-2. Failure rate at normal conditions. (*After General Electric Company, "Transistor Manual," by permission* [8].)

test run at 100°C, the maximum rated temperature. The delay was seen again, followed by a less precipitous decrease and by a total of 13 failures in the 3,000- to 25,000-hr interval. In this case, the effect of higher temperature was still seen even beyond the 5,000-hr region. It does not seem desirable to operate this germanium device at higher temperatures.

A diffused silicon device with the *n-p-n* configuration is shown in Fig. 19-4; failures resulted from cycled operations up to 30 per cent of rated

300 *Semiconductor Junctions and Devices*

power at an ambient temperature of 25°C under normal operating conditions. In this case there was no observed delay period, since eight units failed in the first 125 hr. However, the behavior was good, with failures

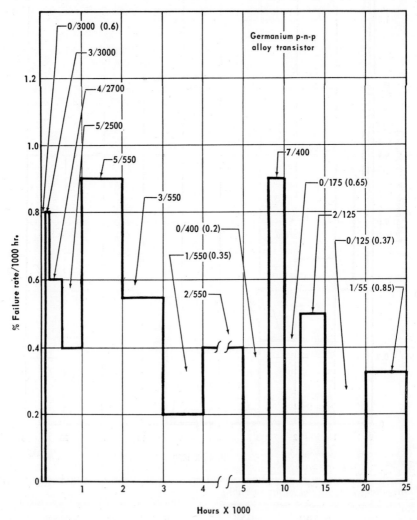

FIG. 19-3. Failure rate at maximum rated temperature. (*After General Electric Company, "Transistor Manual," by permission* [9].)

declining rapidly and approaching zero after 5,000 hr. Only four failures occurred after 3,000 hr, but again the sample size had been reduced.

The behavior of a silicon *n-p-n* diffused planar passivated transistor under maximum power operation is shown in Fig. 19-5. As might be expected, the initial failure rate was high, but it fell off rapidly as the accelerated stress weeded out the failure-prone units. After 1,000 hr the rate was down to zero, and it remained there. In other tests of this unit (not shown) a very low failure rate was observed out to 7,000 hr. No delay time was observed when the first readings were made, after 100 hr.

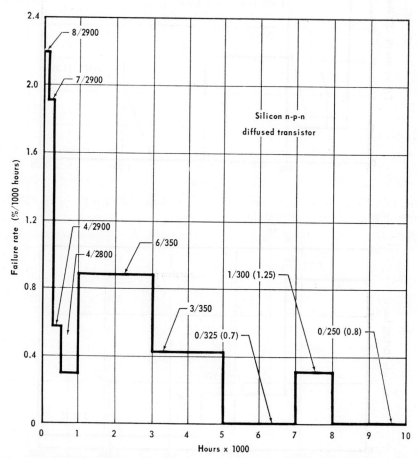

FIG. 19-4. Failure rate at 30 per cent rated power, 25°C ambient temperature. (*After General Electric Company, "Transistor Manual," by permission* [10].)

Finally, in Fig. 19-6 data are shown for a germanium *p-n-p* mesa transistor under reverse-bias testing at temperatures between 55 and 75°C. Earlier tests on this device showed a persistent failure rate at elevated temperatures (100°C) out to 10,000 hr. With improvements in the manufacture, and by limiting the temperature to 75°C maximum, the failure rate was seen to converge rapidly in the first 1,000 hr, with no change in the sample size. Although it was not followed beyond the

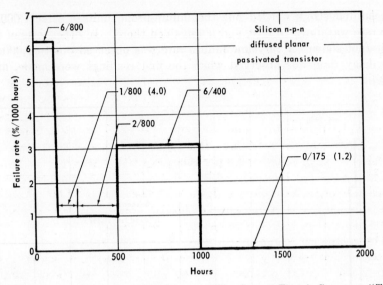

FIG. 19-5. Failure rate at maximum power. (*After General Electric Company, "Transistor Manual," by permission* [11].)

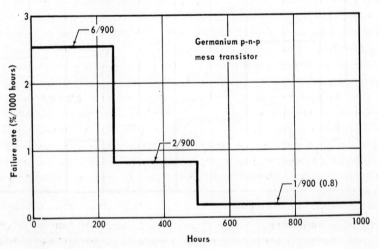

FIG. 19-6. Failure rate by reverse bias. (*After General Electric Company, "Transistor Manual," by permission* [12].)

first 1,000 hr, the failure rate was already appreciably below that seen after 10,000 hr in the preceding test.

The data show calculated failure rates corresponding to zero at certain times for Figs. 19-2 to 19-5, but we cannot say that these are equivalent to 0.01 per cent per 1,000 hr or better, unless we account for reduced sample sizes, total failures up to that time, etc. The fractional designations on the graphs show the number of failures in the numerator and the approximate remaining sample size in the denominator for each time interval. If we consider Fig. 19-4, for example, we see that eight transistors out of 2,900 failed in the first 125 hr, or a failure rate of $(8/2,900)(1,000/125) = 2.2$ per cent per 1,000 hr. During the next 125 hr, the failures dropped to seven, for a rate of 1.9 per cent per 1,000 hr, followed by a decrease to four, but this time for twice the time interval, or 0.57 per cent per 1,000 hr. During the first 500 hr there had been 19 failures out of 2,900, or 1.3 per cent per 1,000 hr average.

The sample was then reduced by 100 units and during the next 500 hr there were only four failures in 2,800, or 0.3 per cent per 1,000 hr for this time interval. The overall rate for the first 1,000 hr was 0.8 per cent per 1,000 hr, representing 23 failures out of 2,900 units.

The sample size was again reduced by 2,450 units, and only 350 remained on test. The remaining units behaved as shown, with six failures in the next 2,000 hr (0.86 per cent per 1,000 hr); three failures in the next 2,000 hr (0.43 per cent per 1,000 hr); no failures in the next 2,000 hr; and one in the interval from 7,000 to 8,000 hr (0.33 per cent per 1,000 hr). From 8,000 to 10,000 hr there were no additional failures in the remaining units, and the test was then ended after more than 13 months.

From a reliability standpoint we need a basis for evaluating the probable accuracy of this and other zero rate values. If we consider the method of calculation, it is seen that the failure rate will increase directly with the number of observed failures and inversely with the length of the test interval and the size of the sample. Therefore, if the test interval is short, and the sample small, the failure rate will be large for a given number of failures. Statistically, however, the larger the sample and the longer the time interval the more meaningful will be the result and the greater the confidence that it represents a true statement of the probability of failure. Therefore, we make use of a statistical approach to determine the interval or range of failure rate values, containing the observed value, within which there is a specified probability that the true failure rate value lies. These intervals are called confidence intervals [13], and the limits are known as confidence limits for the desired probability. We elect to compute 80 per cent confidence intervals, such that any interval will contain the true value of the failure rate in eight out of ten cases; in two out of ten the true value will be either higher or lower.

Since we are primarily concerned with the lowest possible failure rates in seeking high reliability, we have shown the upper value of the 80 per cent interval in parentheses for all cases of a single or zero failures in Figs. 19-2 to 19-6. (The lower values are, of course, zero or very near to it.) From these data, the zero failure confidence ratings range from 0.2 per cent per 1,000 hr (Fig. 19-3, 5,000 to 7,000 hr with a sample size of 400 units) to 1.5 per cent per 1,000 hr (Fig. 19-2, 0 to 125 hr with a sample size of 1,200 units). For a single failure the values range from a low of 0.35 per cent per 1,000 hr (Fig. 19-3, 3,000 to 4,000 hr with a sample of 350 units) to 4.0 per cent per 1,000 hr (Fig. 19-5, 125 to 250 hr with a sample of 800 units).

From these and other values shown on the graphs, it is apparent that the goal of 0.01 per cent per 1,000 hr failure rate was not in sight at 80 per cent confidence with these 1959–1961 production units. An order-of-magnitude improvement was needed, with the ultimate goal a further factor of 10 beyond that. In the next sections several additional testing procedures will be discussed, because they have resulted in significant improvements in device reliability as well as pointing the way to major improvements in production and quality controls.

Before proceeding to these recent methods for reliability improvement, two consistent patterns demonstrated by the failure rate–elapsed time curves are worth noting. In all cases presented in this section (and in the vast majority of all other such data) there has been no clear indication of wear-out or the rising failure rate after long service predicted in Fig. 19-1. Some of these tests were followed for 25,000 hr, or about three years, without any evidence that the units were failing owing to old age or wear-out. In the cases of Figs. 19-2 and 19-3 there appears to be an increase in failure rate during the last 5,000 hr. In both cases, however, this is confined to one unit, and we do not know when in the last 200 days of operation this particular failure occurred. The other tests were stopped at 10,000 hr or before, without even a minor increase.

It would therefore seem that the value of 10^5 hr quoted at the beginning of the chapter might be realistic as a useful life estimate and that wear-out will not be observed unless tests continue beyond 10 years or are conducted under circumstances which will cause accelerated wear due to abnormally high stresses. Under such testing conditions the phenomenon of wear-out may eventually be observed and its causes investigated. To date, there is little reason to be concerned about such effects as a reliability hazard, since very few evaluations or applications are concerned with service lifetimes of 10 years or longer.

Another common characteristic of the curves discussed above is the relatively high initial failure rate and the tendency for this to decrease appreciably in most cases. This is directly in line with the theory as exemplified by Fig. 19-1. It is also the basis for a powerful means of attacking unreliability. In four of the five tests shown in this section, the initial failure rate dropped to a relatively low value after 1,000 to 5,000 hr. In Fig. 19-2 the initial failure rate up to 3,000 hr was about 0.5 per cent per 1,000 hr, while for the period 3,000 to 25,000 hr it fell off to 0.1 per cent per 1,000 hr. For the transistor in Fig. 19-4 the corresponding values were 0.68 per cent per 1,000 hr prior to the 5,000-hr mark and 0.07 per cent per 1,000 hr thereafter up to 10,000 hr; for the unit in Fig. 19-5 the values were 1.6 per cent per 1,000 hr prior to the elapse of 1,000 hr and "zero" from there out to 2,000 hr; and for the transistor of Fig. 19-6, the failure rate was 1.6 per cent per 1,000 hr up to 500 hr, after which it fell to 0.2 per cent per 1,000 hr up to the end of the test at 1,000 hr.

In the high-temperature test of the germanium transistor in Fig. 19-3 the results were not so encouraging. Up to 5,000 hr the failure rate was 0.5 per cent per 1,000 hr compared with a relatively high value of 0.3 per cent per 1,000 hr for the period 5,000 to 25,000 hr. Here again we see the deleterious effect of high temperature on this unit.

With the exception of the last case, all failure rates decreased at least fivefold after an initial high value and remained low for the balance of the test. This behavior suggests that workmanship flaws be eliminated by preservice operation (also known as burn-in) over a long enough interval that the bulk of the failures are removed before reliability data are taken and before the device is sold or put into active service. Further, if the weaker units (which presumably will fail at some point in their service life) could be eliminated by accelerated or overstress testing, the situation might be improved still more. These techniques are described in the next sections.

19-5. SCREENING TESTS AND BURN-IN

All of the device reliability tests so far considered are based on the assumption that the sample chosen for the test is representative and that a small (of the order of a few per cent) group of units will therefore accurately reflect the entire behavior of the production run. In general, this assumption is valid, provided that freak units are somehow detected and removed before sampling. The term "freak" actually means any device which would pass a very casual inspection (such as packing) but which is widely different from the normal production run. The occurrence of freaks is entirely random, and it is not safe to assume that their frequency and behavior will be covered in the normal testing for long-range reliability. Nor is it correct to include them in the data from lots to be sampled, since long-term testing, which is both expensive and vital to the manufacturer, could be invalidated by only one or two freaks in a sample of several thousand.

The procedures for elimination of freaks are called *screening*, and they are applied to every unit. It is therefore important that they be as simple as possible and yet be effective. One very important aspect of such an operation is an understanding of the product involved and therefore some foreknowledge of what to look for. It is, in addition, desirable that the tests be fast and relatively inexpensive.

The following have been proposed [1] as preliminary screening criteria to be utilized as needed for a particular device. The first group consists of inspections and short-term measurements.

1. Visual inspection for gross faults, that is, missing or loose leads, crooked closures, failure to cover with paint or plating.

2. h-parameter check on automatic curve-tracing equipment, or more elaborate go–no-go automatic testing and sorting machinery.

3. Visual oscilloscope check on stability of BV_{CBO} and I_{SAT}, looking for irregular or jittery behavior.

4. Diode leakage stability I_{CBO} over an interval of about one minute.

5. Low-temperature I_{CBO} for entrapped moisture.

6. X-ray fluoroscope for internal arrangement.

7. Leakage of case by standard leak-detector methods.

More time consuming, but necessary in many cases, are some or all of the following test procedures.

8. Vibration and shock—can be mechanized.

9. High reverse bias at high end of rated temperature range; usually followed for a period of hours or days. Can be taken to avalanche breakdown in some cases without damage, and it can thereby replace other high-current or high-power thermal effects.

10. High collector current at low V_{CB}; also can be run as power check by varying V_{CB}.

11. Thermal stress, usually run by rapid cycling of power levels over periods of days up to a week, depending on the device.

12. Ambient temperature effects, more or less important, dependent on the device involved and the rating.

Finally, there is the phenomenon of burn-in, or actual operation, to eliminate the high initial failure rate (usually assumed to be poor workmanship) and, where possible, to detect potential failures by overstress operation and thereby improve both the initial failure pattern and the steady-state level. Burn-in always follows the above screening process, and it not only improves the reliability of devices actually sold but also furnishes information about the design and manufacturing steps, which leads to overall improvements. This is a relatively recent development, and it has been greatly dependent on economics and the degree of reliability required. In any case, there is always the problem of reducing useful life of good units by overly long burn-in.

The conditions of burn-in can be quite varied. In some cases the parameter most likely to cause failure is simply continued on test, sometimes at an augmented level of stress, until a specified interval has passed. Units which have failed are then removed, and the balance are regarded as the output for that production run. In one specific case involving the silicon device shown in Fig. 19-4 burn-in consisted of operation at a power level of 75 per cent of rated value for one week (168 hr). The acceptable units were then tested at the same power level for an additional 2,700 hr. The failure rate was reduced by a factor of 8 compared to units without burn-in. We recall that this device had dropped to an average failure rate of 0.07 per cent per 1,000 hr for the 5,000- to 10,000-hr period during a cycled power test up to only 30 per cent of full power, without burn-in. The implied result is therefore encouraging, and the range of failure rates below 0.01 per cent per 1,000 hr begins to seem more realistic. As the industry comes to apply burn-in phenomena more widely, target reliability values should be more easily realized.

In summary, the device reliability evaluation program described above consists of screening (with or without burn-in), sampling, and extensive applications testing. Results are typical of Fig. 19-1 for the average case with reference to a single parameter. Additional spot testing on smaller samples is used to keep control of production variables on a short-term basis. The results obtained permit continuing evaluation of the reliability values for a given device in a given application. The burn-in technique has found important application as part of the screening process, but it is not yet universally used, mainly for economic reasons. Ideally, an effective, short burn-in interval on a device of very long life would be the best situation, both economically and in terms of reliability.

19-6. DEVICE–SYSTEM INTERACTION [2]

In the preceding discussion we considered the reliability problem only up to the point where the finished, tested unit was packaged and sold. The results of the evaluation programs, screening techniques, and burn-in pretreatment lead us to expect reliabilities (at 80 per cent confidence interval) of the order of 0.01 per cent per 1,000 hr or better for critical components. We now wish to consider the larger problem which exists when a multiplicity of devices such as those considered here are assembled to form a system. (By "system" we mean a complete electronic circuit or functioning equipment, which includes all the individual components of varying individual reliability ratings as well as the connections, structure, etc., deriving from the assembly operation.)

Let us first propose that all the components in such a system are so chosen as to have reliability ratings of the required level and that the system is carefully constructed. Then where is the reliability problem? It would seem logical that such an approach would ensure a successful outcome. Yet it is a fact that actual circuits, made as described above, are frequently far less reliable as a whole than would be expected from the combined effects of the most failure-prone units which make them up. Furthermore, if the individual components are derated and are used to only a small fraction of their rated capacity in an effort to avoid failures, the overall circuit is not, in general, any more reliable (owing to the necessity for additional units or stages), and it will almost surely be bulkier, heavier, slower, and more costly.

Before examining this situation more closely, it is instructive to inquire why these matters have become so much more serious with the advent of semiconductor devices. During the era of vacuum tubes, the failure rate of tubes alone was high enough to constitute a limitation on overall reliability. As semiconductors entered the picture, other components, such as capacitors, became poor by comparison. Improved manufacturing techniques reduced failure rates, and soon all components were showing low individual failure rates. Then, by simple arithmetic, it was possible to propose circuits of a new order of reliability, and electronic failures were to become virtually a thing of the past. Unfortunately, this was not borne out in actual experience, although there were always a number of tantalizing examples of predicted performance being achieved or exceeded. In most cases, however, failures were earlier than they should have been, unpredictable, and without a meaningful pattern. Indeed, much to the chagrin of the device makers, an uncomfortably large portion of the failures were apparently traceable directly to the most "reliable" devices themselves.

Obviously, the original concept of unit reliability as a measuring stick was not good enough. Other factors associated with the creation of a "system" were introducing the controlling variables. Effects due to mechanical assembly, choice of associated components, unanticipated variations in ambient conditions (including undesirable stray signals from failure of other units), and connections between circuit elements were larger and more random and were therefore able to vitiate the predictions based on unit reliability data. For example, connections in miniature circuitry are particularly difficult to evaluate, since a poor connection may open up, produce heat, or introduce stray resistance or intermittently do all three.

Ideally, we might require that each device be rated in advance with respect to each of these hazards as part of the overall evaluation procedure. In so far as these influences can be anticipated, it is at least theoretically possible to predetermine how a given component will perform and, if testing is sufficiently detailed, to arrive at a meaningful reliability figure for device-system interactions. Yet it is obviously impractical to test every device against every known mechanical risk. It is equally unrealistic to assume that any workable evaluation program can anticipate all combinations of electronic hazard to which a unit may be subjected. Consider, for example, the thousands of circuits in which a device may be quite properly used and the infinite variety of potential stresses which may be imposed simply by small differences in design.

We are therefore face to face with an apparent paradox. A vastly complicated problem involving an almost infinite number of possible variables and test situations has, in some cases at least, been resolved rather successfully despite the lack of a systematic means of measurement or evaluation. In the discussion to follow, we shall attempt to describe some of the techniques which have contributed to improved system reliability, recognizing that the problem is still very far from solution and that we do not yet know how to take full advantage of individual component reliabilities.

19-7. SYSTEM RELIABILITY IMPROVEMENT

Since there is no general method of attack on the overall problem of system reliability, the most workable approach is to examine selected aspects of the interaction between the device and the system in terms of a limited number of parameters. All others are eliminated by making certain assumptions, such as a zero failure rate for other parts of the system and 100 per cent effectiveness of precautions during assembly. It is then possible to examine such a simplified system from a reliability standpoint and to arrive at an evaluation in terms of the hypothetical model. If the proposed system should fail to match the criteria set for it, we conclude only that the device (and therefore the system) failed for reasons outside the conditions of evaluation.

It is as if we were asked, for a simple analogy, to evaluate an incandescent light bulb. We might determine the effects of vibration, ambient temperature, leakage of the glass-metal seal, loose connections, intermittent versus continuous service, age, and perhaps many other factors. We would not, however, consider it necessary, for today's domestic applications, to test for 100 per cent overvoltage, the peril of falling tree limbs, or thermal shock. Therefore, if a sampling of typical units were to be destroyed by such an accident to the system or surroundings, we would not consider it a reflection on the light bulbs or an error in the designated reliability evaluation. The difference between this example and the semiconductor device problem is only one of degree. Instead of a few unexpected stresses, such as uncontrolled line voltage, we may have ten or even a hundred, some of which are not only hard to diagnose but even harder to detect, occurring sometimes singly and sometimes in random patterns.

We shall approach the problem by regarding the device as a known quantity and accumulate data concerning the interaction between it and selected external system variables. This approach leads to knowledge of small fractions of a system, each of which must then be considered in relation to larger arrays. This procedure is also compatible with a growing trend toward a physical interpretation of failure data which attempts to relate the statistical results to specific mechanisms and to correct weaknesses by changes in design, manufacturing, or testing [14]. To illustrate these concepts, we shall look briefly at two aspects of the device-system reliability problem. First to be considered will be the problem of maintaining the device in its as-delivered form by protective measures, both electronic and mechanical. After this, we shall consider device selection and the manner in which this can influence reliability.

19-7.1. Device Protection. From an electrical point of view, a device becomes subject to possible damage as soon as it is connected into a circuit of any kind. This includes testing, troubleshooting, or other steps taken during manufacture as well as abnormalities of the external circuit during service. The sources of uncontrolled aberrations reaching the device in the form of signals outside the normal range are widely varied. Typical circuit malfunctions include short circuits, stray signals, overloads, harmonic frequencies, high-resistance connections producing local heating, and open bias circuits permitting uncontrolled response by the device. Many of these are traceable to a single flaw or failure elsewhere in the system, but some may be a complex product of a number of interacting abnormalities.

Our previous discussion of device testing, rating, and failure mechanisms provides the background for understanding the impact of these stresses on the device and the many ways in which a failure may result. (Note that this may be only a single link in a chain of failures started at some remote point in the circuit and continuing on to other units.)

We may know, for example, that the device has a PIV rating of 60 volts and that the circuit, as designed, will never reach more than 50 volts. However, should a failure elsewhere produce a surge of 100 volts, we expect breakdown and, unless the current is limited, almost certain permanent damage. We accordingly hope that the circuit is such as to limit the reverse breakdown current to a safe value. Note, however, that even if the unit in question fails, it has not failed within its normal test parameters. A burnout due to a 100-volt uncontrolled surge is no reflection on the device in question; it is a system failure which produced an abnormal set of circumstances for which the device (and circuit) were not designed. In other cases, we may find devices subject to conditions for which there are no test data. Consider the effect of a high-voltage, high-frequency transient at the collector, a phenomenon which could arise in the control circuit of a power supply. If this is at all likely to occur, it would be necessary either to accumulate test data for the application or make provision to protect the device against such a surge. Finally, connections may fail in a variety of ways and so produce electrical overloads or stray signals.

These examples have been concerned with more or less devastating electronic mishaps, and not with failure due to long-range service under normal or near-normal conditions. In Sec. 19-7.2 we shall consider the matter of device selection as another element in system reliability.

In addition to electronic mishaps, devices can suffer mechanical damage from a number of causes. The simple act of connecting the device to the circuit may introduce flaws of several different types. Leads are bent and often cut in the process of assembly; bending can destroy the hermetic seal at the point where the lead emerges from the enclosure; and excessive flexing can strain internal connections. Even the cutting operation has been found to be deleterious if not so performed that no axial acceleration is transmitted through the lead to the seals and internal connections. A smooth shearing action has been found to minimize such hazards. Excessive heat for uncontrolled time intervals during soldering operations is another variable which is present in device utilization, but not in the manufacturer's test procedure. There are also handling accidents in assembly, resulting in harmful impacts due to dropping onto a hard surface, etc., and cleaning procedures, such as ultrasonic methods, which may set up resonant vibrations within the device. These can result in internal mechanical damage to the device structure.

To achieve a level of device protection compatible with the reliability of the device places a considerable burden on system design and manufacture, as well as on the other elements comprising the circuit. Each mechanical accident to any component during assembly constitutes a risk for every other device, since a failure at some remote part of the system can produce damage and cause another component to fail. For example, to maintain a system reliability equivalent to that of the device (that is, a failure rate of 0.01 per cent per 1,000 hr), only one failure per 1,000 hr is permitted for each 10,000 possible points of failure. If we count each component as a single possibility and if each connection has the same potential as a source of trouble, we are proposing one failure per 1,000 hr in a hypothetical system of 2,000 components and 8,000 connections. This degree of perfection is understandably difficult to achieve.

Problems of a mechanical nature are generally approached in a straightforward manner. Handling and assembly operations can be automated and structural damage minimized, although there always remains an element of risk in that a few devices may suffer during a mechanical operation which is gentle enough for many thousands of supposedly identical units. In general, workable control tests are awkward, and the trend is to require maximum strength in devices as manufactured and to reduce the hazard in assembly by minimizing stress.

Connections are perhaps the most difficult problem. Reducing the number by use of multiunit solid circuit components and the use of fully automatic methods seem to be the best approaches. In this area, the solution must be one of optimization, recognizing that detection of the one potential failure in 1,000 or 10,000 operations will be difficult and expensive.

19-7.2. Device Selection. We come next to the question of selection of the correct device for the circuit. This is important for two basic reasons. First, the multiplicity of components with overlapping characteristics offers us a deceptively wide choice for a given application; yet a mismatched circuit element of the greatest reliability may prove very poor if the circuit makes demands outside its optimum design range. For example, if a chosen device fails in a circuit and it is found that it was forced to operate in a marginal manner, the selection was wrong and should be changed. The second reason is the image of the first. Since other devices are involved, a choice may be wrong because a signal can be generated without harm to the unit in question but can damage another unit whose reliability rating under normal circumstances is as good or better.

Two useful guides aid in the selection process. As a result of experience and a knowledge of manufacturing, it is possible to identify the device parameter having the greatest failure potential and to avoid dependence on this property in the chosen application. If this cannot be done entirely, the data will permit a selection of the device presenting the most favorable prognosis. Second, matching the application and environmental conditions to the device as a function of geometry and method of manufacture will greatly assist in making a good choice.

For example, consider a transistor application where switching characteristics of maximum stability over a long period of time are critical. Assuming that a junction triode is chosen, we know that it will have highly stable input and output current-voltage characteristics but that the effect of leakage currents will be dangerous, since they will affect the OFF resistance. We know that the common-base or common-emitter configurations are stable but that the BV_{CEO} is also a variable, owing to surface problems, and that the forward gain h_{fe} for the common emitter is therefore not as stable as the common-base parameter h_{fb}. Our search should therefore concentrate on devices (probably with some form of effective surface treatment or passivation) whose leakage currents are stable with time, and we should plan to use them in a circuit dependent on α rather than β. (Note that a 1 per cent change in α, from 0.98 to 0.99, corresponds to a change in β from 49 to 99.)

Again, suppose the application is to involve power operation at high temperature, with the possibility of thermal shock, or momentary excursions beyond the nominal range. We should, first of all, obviously select silicon. Next, we would require a rugged, massive construction, probably a diffused device with a good heat sink. We would also be concerned about the elevated temperature performance of lead attachments and case integrity. The most sensitive parameters to study have been found to be thermal resistance or temperature rise per unit power dissipation under steady-state conditions, the common-emitter forward current gain h_{fe}, and the saturation voltage $V_{CE(SAT)}$. In general, since temperature is the worst enemy of device performance and reliability, some form of cooling may be desirable when the service conditions approach such high failure regions. However, the overall system reliability is not improved unless the cooling means is at least as trustworthy as the units being cooled. It might well prove to be better to derate or increase the power dissipation rating of the circuit elements and avoid artificial cooling.

Considerations such as those in this and preceding sections are more a part of system design and proper interpretation of manufacturers' data than the properties of the semiconductor device itself. They are, however, essential first steps in achieving overall reliability with any system, since, if misapplied, the most reliable unit may become a liability and, if due care in making connections and mechanical handling is not taken, the "reliable" device may be ruined before current is first turned on.

REFERENCES

1. J. F. Cleary (ed.): "Transistor Manual," 6th ed., chap. 18, General Electric Company, Syracuse, N.Y., 1962.
2. H. E. Corey: Reliability: A Design Objective, and E. Bleicher, H. E. Corey, and R. M. Ryder: Principles of Reliability Characterization, in H. E. Bridgers, J. H. Scaff, and J. N. Shive (eds.), "Transistor Technology," vol. 1, chaps. 26, 27, D. Van Nostrand Company, Inc., Princeton, N.J., 1958.
3. M. C. Waltz: Laboratory Measurement of Transistor Reliability, and Transistor Reliability in Field Use, in F. J. Biondi (ed.), "Transistor Technology," vol. 3, pp. 191–198, 204–217, D. Van Nostrand Company, Inc., Princeton, N.J., 1958.
4. N. P. Burcham et al.: Germanium Alloy Junction Transistors, in F. J. Biondi (ed.), "Transistor Technology," vol. 3, chaps. 11, 12, D. Van Nostrand Company, Inc., Princeton, N.J., 1958.
5. I. Bazovsky: "Reliability: Theory and Practice," chap. 21, Prentice-Hall, Inc., Englewood Cliffs, N.J., 1961.
6. Partial Summary of Matrix I Data, Minuteman High-reliability Component Parts, *Rept. EM*-3066, *Electron. Components Group, Autonetics Div.,* North American Aviation, Inc., Downey, Calif., February, 1962.

7. Ref. 1, p. 313.
8. Ref. 1, p. 322.
9. Ref. 1, p. 323.
10. Ref. 1, p. 326.
11. Ref. 1, p. 332.
12. Ref. 1, p. 330.
13. Ref. 5, chap. 22.
14. M. E. Goldberg and J. Vaccaro (eds.): "Physics of Failure in Electronics," Spartan Books, Inc., Baltimore, Md., 1963.

Appendix: Conduction in Filled Bands

The conduction mechanism derived from the band model of solids was based on electrons (and holes) in the valence and conduction bands only. In the case of metals these bands overlapped, forming a single continuum of levels with no forbidden gap, while in the case of semiconductors conductivity was governed by the electron and hole concentrations in the conduction and valence bands, respectively. In both cases, no role was asigned to bands of lower energy (which are filled with electrons) or those of higher energy which are completely empty. Based on this model, the behavior of the free electrons in the transition region was predicted on the basis of the Fermi probability function, and the derivation of this expression involved only electrons in partly filled bands, which were assumed to be the only carriers free to move under the influence of an external field.

It is intuitively easy to see why an entirely empty band cannot assist in the conduction process, but the other limiting case, that of the filled band, is not so obvious. Why is it possible, for example, for the vacancies in the valence band of a semiconductor to move relatively freely and constitute a hole current, while the large excess of electrons in the same band is unable to contribute? Similarly, if the number of electrons in the valence band increases (and the holes therefore decrease) why does the conductivity due to this band decrease, finally reaching the point where the band is filled with electrons, but no current can be made to flow even under very high fields?

The precise theoretical explanation of this mechanism is beyond the scope of this discussion. Qualitatively, however, the following argument is useful. Consider a filled band, or one in which every available energy state, as determined by solutions of the wave equation and the Pauli exclusion principle, is assumed to be occupied by an electron. It is also assumed that forbidden gaps, in which there are no allowed states, exist on both sides of the filled band and that they are of an appreciable width (of the order of 1 ev or more). If a field is applied, normal conduction would require that the electrons attain a velocity (on the average) proportional to the direction and intensity of the field. The electron must therefore accelerate and in so doing must gain energy. The accelerated electron must now move to a state of higher energy; but if the band is full and the forbidden gap above it is large compared to the energy gained by the electron, this move is possible only if another electron falls to a lower energy state, leaving an opening. Thus for every electron which gains energy (acceleration) owing to the applied field, another must lose the same amount of energy. Therefore, there is no net energy gain within a filled band and no net current flow. The effect of the field can be thought of in terms of a circulating current within the band, owing to a continuous exchange of electrons seeking to accelerate and to reach the top of the band with others which give up the same amount of energy and fall into the lower energy levels. Such a current cannot be detected and does not influence the conductivity of the filled band.

In the case of the valence band of a semiconductor with acceptor impurities present, the average number of electrons which can accelerate is exactly equal to the number of vacancies. Since these vacancies behave like positive electrons, we call them holes, and the contribution to conductivity from the valence band is proportional to the hole concentration. The remaining electrons, which fill all the states available in excess of the hole concentration, are not able to take part in charge transport under the influence of a field, since they are governed by the same reasoning as was applied to the filled band. In the conduction band in a semiconductor containing donor impurities there are many available empty states, and every electron which reaches the band contributes to the conductivity. In both types of conduction the numerical value of the conductivity is proportional to the concentration of carriers (holes or electrons), the charge, and the mobility (Chap. 2).

This qualitative mechanism applies equally well to any filled or nearly filled band, starting with the valence band and including bands lying at lower energy levels in the structure. In these lower bands, however, the effect of an applied field is negligible, owing to the greatly decreased energy of the electrons corresponding to stronger bonds to the nuclei. The possibility of conduction by creation of a vacancy is therefore zero for all practical purposes. The only bands which can share in carrying a current are the highest nearly filled band and the first empty band above it. This redefines the transition region (in which the electrons obey Fermi-Dirac statistics) from another point of view and shows that a model based only on the valence and conduction bands is the correct one to describe the mechanism of conduction in metals, insulators, and semiconductors.

Author Index

All references are given in full at the end of each chapter. The following list gives page and reference number (in brackets) for each mention of an author's work. Items contained in the reading list of Chapter 1 are omitted.

Anderson, A. E., 204[6]

Bardeen, J., 3[5,6], 271[18,19], 272[21]
Basovsky, I., 295[5], 303[13]
Biondi, F. J., 96[14], 98[19], 294[3], 295[4]
Bleicher, E., 292[2], 295[2], 297[2], 307[2]
Bode, H. W., 150[4]
Brattain, W. H., 3[5,6], 20[2], 271[19]
Braun, F., 2[4]
Bridgers, H. E., 292[2], 295[2], 297[2], 307[2]
Burcham, N. P., 295[4]
Bylander, E. G., 96[17]

Chapin, D. M., 107[34]
Christensen, H., 238[3]
Cleary, J. F., 184[1], 185[1], 186[1], 192[3], 193[3], 195[4], 201[1], 220[3], 221[5], 224[6], 225[3,7], 240[7], 241[9], 242[9], 292[1], 298[7], 299[8], 300[9], 301[10], 302[11,12], 306[1]
Coblenz, A., 138[2], 139[1], 154[5], 161[1], 172[1], 174[1], 178[2], 282[1], 286[9]
Coovert, R. E., 272[21]
Corey, H. E., 292[2], 295[2], 297[2], 307[2]
Czochralski, J., 92[1]

de Ruyter, C., 204[4]
Dickten, E., 242[10]
Doucette, E. I., 244[13]
Dummer, G. W. A., 103[28], 106[32], 225[8], 229[11], 241[8], 245[14], 282[3], 284[6], 286[10], 287[10]

Dunlap, W. C., 78[1], 94[6], 94[8], 98[18,20], 238[5]

Early, J. M., 238[4]
Ebers, J. J., 94[9], 205[7], 206[7], 218[1]
Esaki, L., 103[29]

Faraday, M., 2[2]
Farnsworth, H. E., 271[14]
Frosch, C. J., 96[14]
Fuller, C. S., 96[14], 107[34]

Gibson, A. F., 122[3], 123[3], 211[10], 220[4], 235[2]
Goldberg, M. E., 310[14]
Goldey, J. M., 218[2], 220[2]
Granville, J. W., 103[28], 106[32], 225[8], 229[11], 241[8], 245[14], 282[3], 284[6], 286[10], 287[10]
Gray, D. E., 3[10]
Guillemin, E. A., 139[2]
Gutzwiller, F. W., 225[9], 229[10]

Hall, R. N., 93[4,5], 95[10]
Handler, P., 271[15]
Hannay, N. B., 8[1], 55[4], 80[2], 82[2], 87[2]
Hart, K., 204[4]
Hilibrand, J., 230[12]
Holmes, P. J., 256[1], 258[3], 261[8], 265[10], 266[10], 270[12]

Holonyak, N., 218[2], 220[2]
Hunter, L. P., 94[6]

Jackson, W. H., 244[13]
Johnson, E. O., 273[22]
Jonscher, A. K., 122[3], 123[3], 211[10], 220[4], 235[1,2]

Kingston, R. H., 271[14,15]
Kircher, R., 150[3], 166[2]
Kittel, C., 3[11], 8[1], 258[2,4]
Kleimack, J. J., 238[3]

le Can, C., 204[4]
Loar, H. H., 238[3]

McAfee, K. B., 72[12], 73[13], 74[13]
MacDonald, D. K. C., 261[9]
McKay, K. G., 54[3], 73[3,13], 74[13]
Maita, J. P., 26[3]
Mark, A., 96[16]
Meyerhof, W. E., 271[17]
Middlebrook, R. D., 9[2], 20[1], 52[2], 287[13], 288[13]
Miller, J. R., 196[5]
Miller, S. L., 123[4], 205[7], 206[7,8]
Moll, J. L., 218[2], 220[2]
Morin, F. J., 26[3]
Morrison, S. R., 272[21]
Morton, J. A., 287[12], 288[12]
Mueller, C. W., 230[12]

Nanavati, R. P., 76[16], 100[23]
Nussbaum, A., 35[1], 57[6]

Owens, H. L., 138[2], 139[1], 154[5], 161[1], 172[1], 174[1], 178[2], 282[1], 286[9]

Pankove, J. I., 95[11]
Pearson, G. L., 3[7], 20[2], 98[19], 107[34], 271[16], 284[5]
Pell, E. M., 99[21]
Pfann, W. G., 93[2]
Phillips, A. B., 63[9], 65[10], 120[2], 123[5], 206[9]
Pincherle, L., 100[26]
Plummer, A. R., 256[1], 258[3], 261[8], 265[10], 266[10], 270[12]
Prince, M. B., 96[14], 99[22], 107[35]

Roberts, F. F., 76[15], 100[24], 288[15]
Root, C. D., 240[6]
Ryder, E. J., 72[12]
Ryder, R. M., 150[3], 166[2], 292[2], 295[2], 297[2], 307[2]

Saby, J. S., 94[8]
Sawyer, B., 98[19], 284[5]
Scaff, J. H., 292[2], 295[2], 297[2], 307[2]
Schimpf, L. G., 242[10]
Schlier, R. E., 271[14]
Schrieffer, J. R., 272[21]
Seitz, F., 8[1]
Shive, J. N., 2[1], 4[13], 5[13], 8[1], 11[3], 13[4], 27[4], 28[4], 41[2], 43[3,4], 44[5], 45[6], 46[7], 56[5], 62[8], 84[3], 92[7], 94[12], 107[33,36], 108[36], 120[1], 126[6], 127[6], 292[2], 295[2], 297[2], 307[2]
Shockley, W., 3[7,8,9], 8[1], 50[1], 72[12], 271[16], 285[7], 286[8,11], 287[8], 289[7]
Smith, K. D., 96[14]
Smith, W., 2[3]
Sparks, M., 72[12]
Spenke, E., 4[12], 8[1], 59[7], 259[6], 269[11]
Stone, H. A., 244[13]
Sun, R., 272[21]

Tamm, I., 270[13]
Tanenbaum, M., 93[3], 218[2], 220[2]
Theuerer, H. C., 96[15], 238[3]
Thurmond, C. D., 96[14]
Tiley, J. W., 95[13], 96[13]
Tillman, J. R., 76[15], 100[24], 288[15]
Torrey, H. C., 282[2,4]

Vaccaro, J., 310[14]
Valdes, L. B., 74[14], 100[25], 102[27], 103[30], 104[31], 105[31], 203[2], 204[3], 258[5], 259[5], 273[20], 288[14], 289[14]
Veloric, H. S., 96[14], 99[22]

Wallace, R. L., 242[10]
Walston, J. A., 196[5]
Waltz, M. C., 294[3]
Warner, R. M., 244[13]
Warschauer, D. M., 136[1], 204[5], 214[11], 243[12]
Weimer, P. K., 251[17], 252[17], 253[17]
Whitmer, C. A., 282[2,4]
Williams, R. A., 95[13], 96[13], 102[13]

Zener, C., 72[11]

Subject Index

Acceptor, definition, 23
  density and Fermi level (*see* Fermi level, effect of impurities)
  energy level, 22–24, 45
  impurity, 21–32, 92–96
  ion, 22, 45, 49
Activation of carriers, 12–20, 81
  (*See also* Generation)
Active circuit (*see* Equivalent circuit)
Alpha (α), components of, 120, 206–212
  cutoff, 185–187, 287
  definition, 121, 151
  greater than unity, 207–212, 217–233
  (*See also* Current gain)
Atom model, 9–13
Audio frequency, 187n.
Avalanche (*see* Breakdown in p-n junction; Current gain)
Avalanche devices, 206–216, 234–237

Band, conduction, 15–17
  filled, 34, 39
  forbidden, 10–17, 24, 42
  valence, 13–17
Band energy, 10–17, 100–104
  and gap width, 102, 108
  in insulators, 14–17
  in metals, 13–17, 257–263
  in p-n junction, 51
  in semiconductors, 15–17, 263–279
  and surface states, 263
Band formation, 10–14
Band overlap, 15–17, 257
Band structure, 20–25, 51–53
Band theory, 9–25, 40–46
Barrier, formation of, 48–52
  height, 63–68, 273–276
  in p-n junction, 54, 58, 63
  space-charge, 53, 54

Barrier, surface, 272–279
  width (thickness), 62–65
  (*See also* Potential, barrier)
Base, 115
  conductivity of, 116
  contact, 115
  lifetime in, 116, 120, 123, 209
  resistance of, 138, 206, 242
  resistivity of, 115, 131, 206
  transport efficiency, 120, 123, 206
Base current, 124, 201, 206
  components of, 117–124
Base width, 116, 120, 123, 209
Beta (β) (*see* Current gain)
Bias, definition, 48, 77
  source, 133, 147–149
Black box, concept of, 139–147
  equations, 143, 162, 167, 181
  (*See also* Equivalent circuit)
Bonded diode, 283
Breakdown in p-n junction, 71–76, 82–89, 100
  avalanche, 73–76, 100–102, 120
  collector voltage in, 121–123
  field emission in, 72, 100–103
  incipient, 74, 120
  microplasma in, 74
  prebreakdown noise in, 74
  runaway, 205, 209
  for step junction, 100
  tunneling in, 75, 100–106
  Zener, 72–75, 100–103
Breakdown voltage, 71–76, 82–85, 120–123
Burn-in, 305–307

Capacitance, 60, 65
  barrier, 65–70, 185
  collector, 185, 238, 242

Capacitance, commercial symbols for, 185
  effect of bias on, 69
  emitter, 185
  formulas for, 66–69
  in p-n junction, 64–70, 106
    linear, 69
    step, 68
Carrier concentration (density), 25–32
  compensated, 24–27
  and conductivity, 19–25
  effect of temperature on, 25–29
  gradient, 52, 62–65
  and impurities, 21–25
  intrinsic, 18–21, 27–29
Carrier velocity, 19, 73
Carriers, activation of, 12–20, 81
  generation of, 18, 25, 70
Charge, 8
  density, 57–62
  motion in field, 9, 18–20, 53
Chemical potential, 259–267
Collector, 115
  bias polarity, 112–117, 204
  characteristic, 124–126, 203–207
  contact, 115, 230
  efficiency, 86, 87, 120
  multiplication, 73–75, 120–123, 206–212
  p-n junction as, 86, 112, 115
  resistance, 137
Collector capacitance, 185, 238, 242
Collector current, 116
  components of, 117–124
Collector voltage, 120–124
  maximum, 83, 185, 237
Commercial rating data, 188, 192
Commercial specifications, 184–198
  (*See also* Specification sheet)
Common-base, -collector, -emitter (*see* Grounded-base, -collector, -emitter connection)
Compensation, of impurities, 24, 25
  in junction space charge, 50
Complementary configuration, 220, 228
Conductance, surface, 271
Conductance parameters, 180, 181
Conduction, electronic, 5, 18–20, 34
  in filled bands, 34, 315
  by holes, 5, 18–20
  in insulators, 14, 15
  mechanism in solids, 3–17
  in metals, 13, 14
  in semiconductors, 15–32
Conduction band, 14–17
Conductivity, 5–19, 34
  bulk, 64, 83

Conductivity, compensated, 24
  of copper, 19
  equations for, 19, 28
  impurity, 21–32
  intrinsic, 18–21, 31
  of metals, 19–21
  n-type, 23–32
  p-type, 22–32
  of silicon, 19–21
  and temperature, 3, 25–29, 70
  type, bulk, 271, 278
    surface, 271
  (*See also* Resistivity)
Conductivity modulation, 90
  in rectifiers, 90, 238, 239
  in switch, 210
  in transistor, 122
Confidence interval, 303
Confidence limit, 303
Conjugate structure, 117, 121
Connections, failure of, 307–312
Contact, alloy, 236–240, 284
  injecting, 230–240, 249–254
  metal-metal, 256–263
  metal-semiconductor, 231, 249–254, 256–279
  nonrectifying, 267–270
  and rectification series, 269
  rectifying, 263–267
  (*See also* Work function)
  ohmic, 116, 268
Contact potential, 261–267, 276
Crystal, growth, 92–94, 97
  imperfections, 98
  single, 21, 92, 96
Current, base (*see* Base current)
  breakover, 225, 235–237
  collector (*see* Collector current)
  commercial symbols for, 186
  diffusion, 52–57, 79–82, 105
  drift, 53–57, 79–82, 86
  emitter (*see* Emitter current)
  equilibrium (*see* p-n junction)
  external, 121, 207
  forward, 54–57, 79–89, 98
  holding, 222–229
  internal, 121, 122
  leakage (*see* Rectifier; Switch; Transistor)
  loop, 148–153, 162–169
  maximum, 202
    absolute, 187, 192
  multiplication, 73–75, 120–123, 206–212
  pinch-off, 245–254
  reverse saturation, 82–90, 118–122, 226
  tunnel, 103–106

Subject Index 321

Current density, 8, 19
Current gain, 74, 114, 133
  alpha (α), 121–124, 137, 151
  avalanche in, 121–124, 206–212
  base transport efficiency in, 120, 123, 206
  beta (β), 138, 163
  collector efficiency in, 86–88, 120
  collector multiplication in, 73, 120, 206–212
  and comparison of circuits, 133–138
  dependence on emitter current, 210–212
  effect of base resistance, 206
  emitter efficiency in, 120–122, 206–210
  of grounded-base, 151
  of grounded-collector, 168
  of grounded-emitter, 162–164
  and load resistance, 168, 172–175
  maximum, 133–135
  in p-n-p-n hook transistor, 218–220
  short circuit, 151, 163, 168, 173–175
Current-limiter diode (see Rectifier)

Degeneracy, in p-n junction, 75, 102–106
  in semiconductor, 58, 102–106
Density of states, 59
Depletion layer (see Space charge)
Depletion mode, 253
Derating, 90, 308
Diffusion of carriers, 49–57, 82
Diffusion constant, 122, 209
Diffusion current, 52–57, 79–82, 105
Diffusion length, 122
Diode (see Rectifier)
Dipole layer, 258–263
Dipole potential, 261
Donor, definition, 24
  density and Fermi level (see Fermi level, effect of impurities)
  energy level, 23–25, 44
  impurity, 21–27, 31, 92
  ion, 23, 31, 44, 49
Double-base diode (see Transistor, unijunction)
Double layer, 271–278
Drift current, 53–57, 79–82, 86

Edge leakage, 74, 99, 197
Electrochemical potential, 39, 40, 59, 258
  (See also Fermi level)
Electron, charge on, 19
  conducting, 14–21
  excitation of, 12, 15, 18

Electron, free, 13, 34, 39, 49
  generation of (see Generation)
  mean free path of, 258, 278
  recombination of, 29–32, 49, 88
  valence, 12–17, 22–25
Electron concentration (density), 19, 31, 39
Electron energy, 10, 14, 57–59
  allowed values of, 10, 14, 38, 41
  average, 38, 45, 257
  binding, 11n., 259–266
  of impurity levels, 22–25, 44–46
  internal, 36, 58
  kinetic, 57–62
  in lattice, 259–266
  levels, in atom, 9–14, 33
    in solids, 39–41, 44–46
  potential, 12, 57–62
  states, 10, 14, 35–38
    density of, 59
    surface, 270–279
  surface, 259–264
Electron energy distribution, 14, 33, 59
  (See also Fermi function)
Electron gas, 39, 259
Electron-hole density product, 29–32
Electron-hole pairs (see Hole-electron pairs)
Electron spin, 10–12
Electron statistics, 33–40, 57
Electron traps, 264, 270–273
Electron tunneling, 100–106, 249–254
Emitter, 115
  bias polarity, 114–116, 204
  characteristic, 210–216
  conductivity of, 116
  contact, 115
  efficiency, 120–122, 206–210
  p-n junction as, 85–90, 104, 114
  resistance, 138
  resistivity of, 115
Emitter capacitance, 185
Emitter current, 114, 210
  components of, 117–124
Emitter voltage, 124
Energy, band (see Band energy)
  electron (see Electron energy)
Enrichment mode, 253
Epitaxial process, 96, 240
Equivalent circuit, 139
  active, 145
  for grounded-base, 143, 149–160, 174
  for grounded-collector, 167–170, 174
  for grounded-emitter, 161–167, 174
  passive, 145
  T network, 144
Esaki diode, 103–106

Failure of devices, confidence interval, 303
confidence limit, 303
distribution of, 297–305
general failure-time curve for, 295–305
material flaws in, 294, 297
mechanisms of, 295–297, 310–312
physical causes of, 296, 306–312
probability function for, 297
rate of, 296–298, 304–306
stress level and, 296
wear out and, 297, 304
(*See also* Reliability of devices)
Fall time, 186, 192
Fermi-Dirac statistics, 33–40, 57
Fermi energy, 35, 38
Fermi function, 33–46, 59
Fermi level, 33–46, 257–279
in compensated semiconductors, 46
in conductors, 40, 42
constancy of, 42, 51, 60
effect, of bias on, 79–81
of impurities on, 44–46, 85, 103
of temperature on, 40–44
in impurity semiconductors, 44–46
in intrinsic semiconductors, 40–44
in metals, 40, 42
in n-type semiconductors, 44
in p-type semiconductors, 45
in p-n junction, 51, 79, 103
Fermi potential, 40, 67
Field, critical, 72
electric, 13, 57–62
emission, internal, 72, 100–103
external, 244, 253, 270
internal, 71–73
and space charge, 53, 57–62
strength, 8, 60–62, 72, 102
Field effect devices, 243–254
Forward current transfer ratio (*see* Hybrid parameters)
Forward limiting resistance, 82–85
Forward transfer resistance (*see* Resistance parameters)
Four-layer devices, 217–233
Four-pole (-terminal) network, 140–143
Free electron gas, 39, 259
Free energy, and Fermi statistics, 37–40
and Gibbs function, 37–39
and Helmholtz function, 38
and Lewis and Randall function, 37n.
Frequency, commercial symbols, 185, 187
cutoff, 98, 185–188
Frequency response, 177–179

g parameters (*see* Conductance parameters)
Galvani potential, 261n.
Gap, forbidden, 10–17, 24, 42
Gate, 221, 246–254
anode, 221–225
cathode, 221–225
in thin-film transistor, 251–254
Generation, of carriers, 18, 25, 70
of electrons and holes, 20, 29–32, 107
of heat, 83–85, 90
rate of, 29–31
(*See also* Activation of carriers)
Gibbs function, 37–39
Gold-bonded diode, 283
Grounded-base, -collector, -emitter connection, 129–138
comparison with other connections, 133–138
equivalent circuit for (*see* Equivalent circuit)
phase reversal in, 130, 137
as switch circuit, 203–207

h parameters (*see* Hybrid parameters)
Heat, dissipation of, 202, 294
generation of, 83–85, 90
Helmholtz function, 38
Hole, 5, 18
charge on, 19
conducting, 18, 42
free, 49
Hole concentration (density), 19, 29–32
Hole-electron density product, 29–32
Hole-electron pairs, 19
combined, concentration of, 30
generation of, 20, 29–32, 107
recombination of, 29–32, 49, 88
Hole energy (*see* Electron energy)
Hole traps, 264, 270–273
Hybrid parameters (h parameters), 180–183
commercial symbols for, 186
conversion to resistance parameters, 181–183
forward current transfer ratio, 181, 183
and frequency, 183
grounded-base, conversion to grounded-emitter, 190
input impedance, 181, 183
output conductance, 181, 183
representation of, 181–183, 186
reverse voltage transfer ratio, 181, 183
for 2N332 transistor, 188–191
for 2N2193 transistor, 194–196

Subject Index   323

Impedance matching, 137
Impurity, 21–25, 92–96
  acceptor, 21–32, 92
  adsorbed gas, 259, 272, 274
  concentration of, 24, 31, 45
  concentration gradient of, 62–65, 68
  donor, 21–27, 31, 92
  and doping methods, 92–97
  effect, on Fermi level, 44–46, 85, 103
    on resistivity, 4, 21
  gold, 284
  immobile ion, 22–25, 48, 297
    and space charge, 50–52
  n-type, 21–27, 31, 92
  oxide layer, 259, 270–274
  oxygen, 282
  p-type, 21–32, 92–96
  surface, 259–279
Impurity compensation, 24, 25, 50
Impurity conductivity, 21–32
Impurity energy levels, 22–25, 44–46
Injection, of carriers, 52, 85–90
  of electrons, 112
  external electric, 74
  external photoelectric, 74
  of holes, 56, 88
  of majority carriers, 235, 249–254
  in metal-semiconductor contact (see Contact, metal-semiconductor)
  of minority carriers, 56, 112–118
  in p-n junction, 56, 85–90
  in rectifier, 98
  in switching, 208–210, 234–236
  in transistor, 111–115, 250–254
  (See also Conductivity modulation)
Inner potential (bulk, surface), 259, 270, 275
Input impedance, 249, 286
  of grounded-base, 135–137, 152, 176
  of grounded-collector, 135–137, 169, 176
  of grounded-emitter, 135–137, 164, 176
  and load resistance, 175–177
  (See also Hybrid parameters)
Input resistance (see Resistance parameters)
Insulator, band structure of, 14–17
  conduction in, 14, 15
Intrinsic conductivity, 18–21, 31
Intrinsic hole-electron densities, 29–32
Intrinsic region devices, 237–240
Intrinsic resistivity, 25
Intrinsic semiconductor, 18–20, 42–44
Inversion layer, 270–275
Ion, impurity, 22–25, 48, 297
Ionization energy, 102
Ionization and temperature, 22–29

Ionized acceptor, 22, 45, 49
Ionized donor, 23, 31, 44, 49

Junction, metal-metal, 256–263
  metal-semiconductor (see Contact, metal-semiconductor)
  (See also n-n^+ junction; p-n junction; p-p^+ junction)

Kirchhoff's law, 146–149, 163

Large-signal behavior, 126, 203–216
Leakage current (see Rectifier; Switch; Transistor)
Lewis and Randall free energy function, 37n.
Lifetime, minority carrier, 123, 209
Linear junction, 63, 69
Load line analysis, 126–128, 212–216
Loop currents, 148–153, 162–169
  (See also Equivalent circuit)

MADT, 198
Majority carrier, 21, 29, 53–57
  density of, 29–32
  injection of, 235, 249–254
Majority-carrier devices, 235, 243–254
Metal, 13
  band structure of, 13–17, 257–263
  conduction in, 13–15
  Fermi level in, 39, 42
  work function of, 260–263
Metal-metal contact, 256–263
Metal-semiconductor contact (see Contact, metal-semiconductor)
Minority carrier, 21, 29, 55–57
  collection of, 86, 112–114, 120
  density of, 19, 20, 29–32
  diffusion of, 52–57, 79–82, 105
  injection of, 56, 112–118
  lifetime of, 123, 209
  in transistor action, 21, 55
Mobility, of carriers, 18–21
  effect of temperature on, 20
  of electrons and holes, 19, 20
Multi-junction devices, 217–254
Multiplication, current, 73–75, 120–123, 206–212
Multivibrator, 212–215

n-type impurity, 21–27, 31, 92
Network, four-pole (-terminal), 140–143
n-i-p rectifier, 238–240

n-n^+ junction, 234–237
Nonlinear characteristic, 147, 288
Nonohmic, definition, 126n.
n-p-i-n transistor, 237
n-p-n transistor, 111–117
n-p-n-p (see p-n-p-n devices)

On-off resistance, 199–202, 229
Oscillator, 212–215
Output characteristic, 124–126, 203–208
Output conductance (see Hybrid parameters)
Output impedance, 286
  of grounded-base, 135–137, 153
  of grounded-collector, 135–137, 169
  of grounded-emitter, 135–137, 165
Output resistance (see Resistance parameters)

p-type impurity, 21–32, 92, 284
PADT, 198
Parameters, switch, summary, 225
  transistor (see Conductance parameters; Hybrid parameters; Resistance parameters)
Passivation, 294, 297
Passive circuit, 145
Pauli exclusion principle, 10, 14
Peak inverse voltage, 83, 185, 237
Phase reversal, 130, 137
Photocell, 107
Physical causes of failure, 296, 306–312
p-i-n rectifier, 238–240
Pinch-off current, voltage, 245–254
PIV, 83, 185, 237
p-n junction, 48–52
  abrupt, 63–68, 100
  alloy, 64, 94, 197
  barrier potential in, 57–68, 79, 81
  barrier width in, 62–65
  biased, 77–90
  breakdown in (see Breakdown in p-n junction)
  capacitance of, 64–70, 106
  carrier generation in (see Generation)
  as collector, 86, 112, 115
  conductivity modulation in, 90
  degeneracy in, 75, 102–106
  diffused, 64, 96, 197
  edge leakage in, 74
  electron energy in, 51
    (See also Electron energy)
  as emitter, 85–90, 104, 114
  epitaxial, 96, 197

p-n junction, equilibrium currents in, 52–57, 77–90
  fabrication of, 91–97
  Fermi level in, 51, 79, 103
  formation of, 48–52
  forward-biased, 81–89
  fused, 94
  graded, 96
  grown, 64, 92, 197
  impurity concentration in, 48, 69, 70
  injection in, 56, 85–90
  intrinsic region in, 50–52, 72
  I-V characteristic of, 82–85, 89
  linear, 63, 69
  melt-back, 95, 197
  micro-alloy, 197
  peak inverse voltage in, 83, 185, 237
  rate-grown, 93, 197
  recombination in, 54–57
  rectification in, 77, 80–85
  resistance in, 113
  reverse saturation current in, 82–90, 118–122
  reverse-biased, 67, 71, 79–90
  and space charge, 48–52
  step, 63–68, 100
  surface-barrier, 95
  symmetrical, 52–82
  and temperature, 70
  tunneling in, 75, 76
  unsymmetrical, 85–90
  width of, 62–70
p-n product, 29–32
p-n-i-p transistor, 237
p-n-p transistor, 111–117
p-n-p-n devices, 218–229
  diode, 229
  hook transistor, 218–220
  rectifier (controlled), 225–229
  tetrode switch, 220–225
Point-contact devices, 281–290
  capacitance of, 283
  collector in, 285, 289
  current gain (α) of, 281, 286–289
  emitter in, 285, 289
  emitter-collector spacing in, 285, 289
  forming contacts for, 284–286
  frequency response of, 281, 286
  as rectifier, 281–284
  resistance parameters, typical values of, 286
  reverse current in, 283, 284, 287
  spreading resistance in, 283
  theory of, 289, 290
  as transistor, 285–290
  Type A transistor, 285, 286
Poisson's equation, 60

Polycrystalline devices, 249–254, 281
Potential, 8, 57–62
  barrier, 57–68, 79–81
  contact, 261–267, 276
  dipole, 261
  energy, 12, 57–62
  Fermi, 40, 67
  Galvani, $261n$.
  gradient, 8, 53
  inner, 259, 270, 275
    bulk, 264–266
    surface, 266
  periodic, 58
  and space charge, 259–267
  surface, 259
  Volta, 261–267, 276
  (*See also* Work function)
Power, absolute maximum, 188, 192
Power dissipation, 186, 192, 203
Power gain, maximum, 137
  of grounded-base, 156–159
  operating, 126, 133–135
    and comparison of circuits, 133–137
    of grounded-base, 156–159, 286
    of grounded-collector, 170
    of grounded-emitter, 166, 167
p-p^+ junction, 230–235
Probability, distribution function, 33–46
  of occupation, 37–46

Quantum statistics, 14, 75

r parameters (*see* Resistance parameters)
Rating data, 188, 192
Recombination, in base region, 117–121, 204, 210
  of hole-electron pairs, 29–32, 49, 88
  in p-n junction, 54–57
  rate, 30, 31
Rectification, 2–5
  at contacts, 263–267
  efficiency, 84, 269, 281–283
  in p-n junction, 77, 80–85
  ratio, 284
Rectification series, 269
Rectifier (diode), 2, 77–90, 97–106
  alloy-junction, 98
  avalanche, 235–237
  body resistance of, 83, 89
  bonded (gold-bonded), 284
  breakdown of (*see* Breakdown in p-n junction)
  characteristic of, 83, 84, 89
    high current, 89
    negative resistance, 103–106

Rectifier (diode), and conductivity modulation, 90, 238, 239
  controlled, 225–229
  copper oxide, 274
  current limiter (varistor), 1–3, 244, 245
  diffused-junction, 98
  edge leakage in, 99
  epitaxial-junction, 99
  Esaki, 103–106
  fabrication of, 97–99
  forward current in, 54–57, 81–89
  forward-limiting resistance of, 82–85
  frequency cutoff in, 98
  fused-junction, 98
  grown-junction, 97, 98
  high-frequency, 99, 106
  injection in, 98
  large-area, 98, 225
  leakage current in, 227, 295
  metal-semiconductor, 263–267
  microwave diode, 282
  n-i-p (p-i-n), 238–240
  PIV, 83, 185, 237
  p-n-p-n controlled, 225–229
  power, 89, 225, 238
  reverse saturation current in, 82–90, 226
  selenium, 269
  small-signal behavior of, 85–89
  voltage-limiter (Zener), 2, 75, 100–103
Rectifier equation, 82–85, $267n$., 277
Reliability of devices, 292–313
  and aging, 288, 292, 294
  and burn-in, 305–307
  and failure rate, 296–298, 304–306
  and moisture, 294–297
  and process control, 294, 296
  and sampling, 295–297, 305
  and screening tests, 305–307
  and temperature, 294–299, 305
  and testing, 293–297
  and workmanship flaws, 294, 297, 305
Reliability of systems, 307–313
  and connections, failure of, 307–312
  and device-system interaction, 293, 307–310
Resistance, base, 138, 206, 242
  blocking, 222–229
  body, 83, 89
  bulk, 83, 90
  collector, 137
  emitter, 138
  forward-limiting, 82–85
  input (*see* Resistance parameters)
  load, 114, 138, 186
    and current gain, 168, 172–175
    and input impedance, 175–177

Resistance, mutual, 138, 145–150
  negative, 103–106, 208–216, 240–242
  output (*see* Resistance parameters)
  in *p-n* junction, 113
  saturation, 186, 205, 208
  temperature coefficient of, 3, 29
  of voltage source, 130, 138
Resistance parameters, 141–146
  conversion, to conductance
    parameters, 180
    to hybrid parameters, 181
  forward transfer resistance, 143–146
  of grounded-base, 143, 146, 177–179
  of grounded-collector, 167, 177–179
  of grounded-emitter, 161, 177–179
  input resistance, 143
  open circuit requirement for, 141–143, 154, 179
  output resistance, 143
  reverse transfer resistance, 143–146
  typical values of, 143, 177–179
Resistivity, 4, 9
  base, 115, 131, 206
  collector, 116
  effect of impurities on, 4, 21
  emitter, 115
  intrinsic, 25
  of semiconductors, 3, 4
  and temperature, 25–29
Reverse saturation current, 82–90, 118–122, 226
Reverse transfer resistance (*see* Resistance parameters)
Reverse voltage transfer ratio (*see* Hybrid parameters)
Rise time, 186, 192

Saturation, 205, 208, 213
Screening tests, 305–307
Segregation coefficient, 92–95
Semiconductor, band energy of, 15–17, 263–279
  classification of, 4
  compensated, 24, 25
  conduction in, 15–32
  definition, 3
  degeneracy in, 58, 102–106
  Fermi level in, 40–46, 279
  history of, 2, 9
  impurity, 21–32
  intrinsic, 18–20, 42–44
  n-type, 23, 24
  p-type, 22, 23
  resistivity of, 3
  surface of, 270, 279
  work function of, 264–276

Semiconductor-metal contact (*see* Contact, metal-semiconductor)
Silicon, electrical properties of, 20
Single-junction devices, 91–108
Small-signal behavior (*see* Rectifier; Switch; Transistor)
Solar battery, 107, 108
Space charge, in accumulation layer, 268
  barrier, 53, 54
  and charge density, 57–62
  and contact potential, 259–267
  and field, 53, 57–62
  and potential (internal), 259
  and saturation, 205
  in surface layer, 266–269
  and Volta potential, 261–267
Space-charge depletion layer, 62–65
  and capacitance, 65–70
  formation in *p-n* junction, 49–52
  impurity ions and, 49–53
  width of, 62–64
    effect of bias on, 64, 67, 72, 102
    impurities and, 62–70
Specification sheet, 2N332 transistor, 187–191
  2N2193 transistor, 192–196
Specifications, 184–198
Step junction, 63–68, 100
Storage time, 186, 192
Surface, clean, 259–270
  conductance, 271
  contamination, 270–274, 278
  dipole layer, 258–263
  double layer, 271–278
  impurities, 259–279
  inversion layer, 270–275
  leakage, 295
  potential, 259
  protection, 294, 297
  of semiconductor, 270, 279
  space charge, 266–269
  states, 270–279
    and band energy, 263
  traps, 264, 270–273
Surface barrier, 272–279
Switch, 199–203
  alpha cutoff in, 185–187, 287
  alpha greater than unity in, 207–212, 217–233
  astable, 212–215
  avalanche, 206–216, 234–237
  bistable, 212–216
  breakover voltage in, 207, 228, 229
  characteristic of, 200, 207, 212
    negative resistance, 208–216
  circuit, for grounded-base, 203–207
    for grounded-emitter, 200

Subject Index 327

Switch, and conductivity modulation, 210
  cutoff region of, 213
  double-base diode, 240–242
  fabrication of, 209
  fall time in, 186, 192
  and h parameters (*see* Hybrid parameters)
  injection in, 208–210, 234–236
  instability of, 212–215
  large-signal behavior of, 203–216
  leakage current in, 200, 202, 295
  monostable, 212–215
  as multivibrator, 212–215
  off state of, 200–202
  on-off resistance of, 200–202, 208, 229
  on state of, 200–202
  operating modes of, 212–216
  as oscillator, 212–215
  parameters, summary of, 225
  p-n-p-n diode as, 229
  p-n-p-n tetrode as, 220–225
  and r parameters (*see* Resistance parameters)
  rise time in, 186, 192
  saturation of, 205, 208, 213
  saturation resistance in, 186, 205, 208
  saturation voltage in, 185, 205
  small-signal behavior of, 200–202
  stability of, 200–202, 212–216
  sustaining voltage in, 208, 212, 229–233
Switching action of transistors, 199–216
Switching speed, 205, 229, 234–237
Symbols for transistors, 133, 134
Symmetrical p-n junction, 52–82

T network, 144
Temperature, commercial symbols for, 186
  effect of, on conductivity, 3, 25–29, 70
  on Fermi level, 40–44
  on p-n junction, 70
  maximum, 187, 192, 193
  storage, 186, 192
Temperature coefficient of resistance, 3, 29
Tetrode transistor, 242
Three-junction devices, 218–233, 237–240
Thyristor, 230–233
Transistor, alloy-diffused, 198
  alloy-junction, 198, 200, 206
  alpha for (*see* Alpha)
  alpha cutoff for, 185–187, 287
  as amplifier, 114, 129–138, 218
  avalanche, 206–216, 234–237
  base of (*see* Base)

Transistor, base transport efficiency in, 120, 123, 206
  and bias, 111–117, 133
  black box equivalent for (*see* Black box)
  breakdown voltage in, 120–123
  characteristic of, input, 210–216
    negative resistance, 208–216, 240–242
    output, 124–126, 203–208
  collector of (*see* Collector)
  collector efficiency in, 86, 87, 120
  collector multiplication in, 73–75, 120–123, 206–212
  commercial symbols for, 133
  and conductance parameters, 180, 181
  and conductivity modulation, 122
  conjugate, 117, 121
  connections for, comparison of, 129–138
  current flow in, 117–124
  current gain for (*see* Current gain)
  current multiplication in, 73–75, 120–123, 206–212
  current ratios in, 119–124, 172–175
  design requirements for, 115–117
  diffused-base, 197, 198
  diffused-junction, 198
  double-diffused, 197, 198
  edge leakage in, 198
  emitter of (*see* Emitter)
  emitter efficiency in, 120–122, 206–210
  epitaxial-junction, 193, 198
  equivalent circuit for (*see* Equivalent circuit)
  fabrication of, 197, 209
  failure of (*see* Failure of devices)
  field effect (unipolar), 246–249
  four-pole equivalent for, 140–143
  frequency cutoff in, 185–188
  frequency response of, 177–179
  and g parameters, 180, 181
  graded-junction, 96, 198
  grown-diffused, 242
  grown-junction, 198, 242
  and h parameters (*see* Hybrid parameters)
  high-frequency, 193–196, 238
  history of, 2, 3
  hook, 218–220
  and hybrid parameters (*see* Hybrid parameters)
  injection in, 111–115, 250–254
  and instability, 153, 155
  junction tetrode, 242
  large-signal behavior of, 126, 203–216
  leakage current in, 192, 295

Transistor, mesa, 197, 198
  micro-alloy-diffused (MADT), 198
  n-p-n, 111–117
  parameters, summary of, 137, 177
    typical values of, 138, 178
  and phase reversal, 130, 137
  planar, 193, 197, 198
  planar-passivated, 197, 198
  p-n hook, 218–220
  p-n-i-p (n-p-i-n), 237
  p-n-p, 111–117
  p-n-p-n diode (see Switch)
  p-n-p-n hook, 218–220
  p-n-p-n tetrode, 220–225
  post-alloy-diffused (PADT), 198
  power, 124–126
  and power gain (see Power gain)
  and r parameters (see Resistance parameters)
  rating data for, 188, 192
  reliability of, 292–313
  saturation of, 205, 208, 213
  saturation resistance in, 186, 205, 208
  saturation voltage in, 185, 205
  small-signal behavior of, 124, 125, 171–179, 200–202
  storage time in, 186, 192
  structure terminology for, 196–198
  surface-barrier, 198
  as switch, 199–216
  thin-film, 251–254
  as two-junction device, 111–128
  types, A, 285–287
    2N332, 187–191
    2N2193, 192–196
  unijunction (double-base) diode, 240–242
  unipolar, 246–249
  voltage gain for (see Voltage gain)
Transistor action, minority carriers in, 21, 55
  theory of, 111–138, 217–229
Traps, 264, 270–273
Tunnel diode, 103–106
Tunneling, 100–106, 249–254

Unipolar transistor, 246–249
Unsymmetrical junction, 85–90

Vacancy (see Hole)
Valence band, 13–17
Varistor, 1–3, 244, 245
Volta potential, 261–267, 276
Voltage, avalanche, 71–76, 120–123
  breakover, 207, 228, 229
  commercial symbols for, 185
  Fermi, 40, 67
  maximum, 186, 192
  peak inverse, 83, 185, 237
  pinch-off, 245–254
  saturation, 185, 205
  of signal generator, 130, 138
  sustaining, 208, 212, 229–233
  Zener, 72–76, 100–103
Voltage gain, 133–135
  and comparison of circuits, 133–138
  of grounded-base, 155, 159
  of grounded-collector, 169
  of grounded-emitter, 166
  maximum, 135, 155
Voltage-limiter (Zener) diode, 2, 75, 100–103

Work function, 257–279
  of metals, 260–263
  and rectification series, 269
  of semiconductors, bulk, 264–276
    surface, 266–276
  values of, 260
  and Volta potential, 261–266, 273–276

Y parameters, 180n.

Z parameters, 180n.
Zener breakdown (see Breakdown in p-n junction)
Zener diode, 2, 75, 100–103